I0896783

Contributions of Africa's Indigenous Knowledge to the Wave of Digital Technology:

Decolonial Perspectives

Eric Blanco Niyitunga
University of Johannesburg, South Africa

A volume in the Advances in Knowledge Acquisition, Transfer, and Management (AKATM) Book Series

IGI Global
PUBLISHER of TIMELY KNOWLEDGE

Published in the United States of America by
IGI Global
Information Science Reference (an imprint of IGI Global)
701 E. Chocolate Avenue
Hershey PA, USA 17033
Tel: 717-533-8845
Fax: 717-533-8661
E-mail: cust@igi-global.com
Web site: http://www.igi-global.com

Copyright © 2024 by IGI Global. All rights reserved. No part of this publication may be reproduced, stored or distributed in any form or by any means, electronic or mechanical, including photocopying, without written permission from the publisher.
Product or company names used in this set are for identification purposes only. Inclusion of the names of the products or companies does not indicate a claim of ownership by IGI Global of the trademark or registered trademark.

Library of Congress Cataloging-in-Publication Data

CIP PENDING

Contributions of Africa's Indigenous Knowledge to the Wave of Digital Technology: Decolonial Perspectives
Eric Blanco Niyitunga
2024 Information Science Reference

ISBN: 9781668478516 (h/c)
ISBN: 9781668478554 (s/c)
eISBN: 9781668478523

This book is published in the IGI Global book series Advances in Knowledge Acquisition, Transfer, and Management (AKATM) (ISSN: 2326-7607; eISSN: 2326-7615)

British Cataloguing in Publication Data
A Cataloguing in Publication record for this book is available from the British Library.

All work contributed to this book is new, previously-unpublished material.
The views expressed in this book are those of the authors, but not necessarily of the publisher.

For electronic access to this publication, please contact: eresources@igi-global.com.

Advances in Knowledge Acquisition, Transfer, and Management (AKATM) Book Series

ISSN:2326-7607
EISSN:2326-7615

Editor-in-Chief: Murray E. Jennex, San Diego State University, USA

MISSION

Organizations and businesses continue to utilize knowledge management practices in order to streamline processes and procedures. The emergence of web technologies has provided new methods of information usage and knowledge sharing.

The **Advances in Knowledge Acquisition, Transfer, and Management (AKATM) Book Series** brings together research on emerging technologies and their effect on information systems as well as the knowledge society. **AKATM** will provide researchers, students, practitioners, and industry leaders with research highlights surrounding the knowledge management discipline, including technology support issues and knowledge representation.

COVERAGE

- Cognitive Theories
- Cultural Impacts
- Information and Communication Systems
- Knowledge Acquisition and Transfer Processes
- Knowledge Management Strategy
- Knowledge Sharing
- Organizational Learning
- Organizational Memory
- Small and Medium Enterprises
- Virtual Communities

IGI Global is currently accepting manuscripts for publication within this series. To submit a proposal for a volume in this series, please contact our Acquisition Editors at Acquisitions@igi-global.com or visit: http://www.igi-global.com/publish/.

The Advances in Knowledge Acquisition, Transfer, and Management (AKATM) Book Series (ISSN 2326-7607) is published by IGI Global, 701 E. Chocolate Avenue, Hershey, PA 17033-1240, USA, www.igi-global.com. This series is composed of titles available for purchase individually; each title is edited to be contextually exclusive from any other title within the series. For pricing and ordering information please visit http://www.igi-global.com/book-series/advances-knowledge-acquisition-transfer-management/37159. Postmaster: Send all address changes to above address. Copyright © 2024 IGI Global. All rights, including translation in other languages reserved by the publisher. No part of this series may be reproduced or used in any form or by any means – graphics, electronic, or mechanical, including photocopying, recording, taping, or information and retrieval systems – without written permission from the publisher, except for non commercial, educational use, including classroom teaching purposes. The views expressed in this series are those of the authors, but not necessarily of IGI Global.

Titles in this Series

For a list of additional titles in this series, please visit:
http://www.igi-global.com/book-series/advances-knowledge-acquisition-transfer-management/37159

Social Research Methodology and Publishing Results A Guide to Non-Native English Speakers
Candauda Arachchige Saliya (Sri Lanka Institute of Information Technology, Sri Lanka)
Information Science Reference • copyright 2023 • 326pp • H/C (ISBN: 9781668468593) • US $205.00 (our price)

Knowledge Management and Research Innovation in Global Higher Education Institutions
Lawrence J. Jones-Esan (York St. John University, UK) Vipin Nadda (University of Sunderland, UK) and Kendra S. Albright (Kent State University, USA)
Information Science Reference • copyright 2023 • 258pp • H/C (ISBN: 9781668436523) • US $210.00 (our price)

Innovative Technologies for Enhancing Knowledge Access in Academic Libraries
Tlou Maggie Masenya (Durban University of Technology, South Africa)
Information Science Reference • copyright 2022 • 317pp • H/C (ISBN: 9781668433645) • US $215.00 (our price)

Understanding, Implementing, and Evaluating Knowledge Management in Business Settings
Tereza Raquel Merlo (University of North Texas, USA & Merlo Management Consulting, USA)
Business Science Reference • copyright 2022 • 307pp • H/C (ISBN: 9781668444313) • US $215.00 (our price)

Cases on Developing Effective Research Plans for Communications and Information Science
María-Victoria Carrillo-Durán (University of Extremadura, Spain) and Margarita Pérez Pulido (University of Extremadura, Spain)
Information Science Reference • copyright 2022 • 466pp • H/C (ISBN: 9781668445235) • US $215.00 (our price)

Handbook of Research on the Global View of Open Access and Scholarly Communications
Daniel Gelaw Alemneh (University of North Texas, USA)

For an entire list of titles in this series, please visit:
http://www.igi-global.com/book-series/advances-knowledge-acquisition-transfer-management/37159

IGI Global
PUBLISHER of TIMELY KNOWLEDGE

701 East Chocolate Avenue, Hershey, PA 17033, USA
Tel: 717-533-8845 x100 • Fax: 717-533-8661
E-Mail: cust@igi-global.com • www.igi-global.com

This book is dedicated to Funmi Olonisakin, the Vice-President International, Engagement and Service (IES) at King's College London and the mother of leadership, development and security in Africa, and founder of African Leadership Centre.

Table of Contents

Preface xiii

Chapter 1
Africa's Indigenous Knowledge: Essential Ingredient for Digital Technology 1
Eric Blanco Niyitunga, University of Johannesburg, South Africa

Chapter 2
Africa at the Center of Digital Technology 47
Cleophas Gwakwara, University of Johannesburg, South Africa

Chapter 3
Effects of the Trans-Atlantic Slave Trade and Colonialism on Indigenous Knowledge for Digital Technology in Africa 70
Eric Blanco Niyitunga, University of Johannesburg, South Africa
Logic L. Modibedi, Johannesburg Business School, South Africa

Chapter 4
Revisiting the Effects of Colonialism on Africa's Indigenous Knowledge........ 103
Thivhavhudzi Muriel Badugela, University of Venda, South Africa

Chapter 5
Coloniality Forces Within the Cold War and Neocolonialism: A Hindrance to Africa's Contributions to Digital Technology 125
Eric Blanco Niyitunga, University of Johannesburg, South Africa

Chapter 6
Green Energy Solutions in Africa: Rediscovering Indigenous Knowledge for the Fourth Industrial Revolution 159
Joshua Mawere, University of Venda, South Africa

Chapter 7
Technology for Digitalization in Africa: The Contribution of Africa's Raw Minerals 178
Sandile Moloi, University of Johannesburg, South Africa
Dimakatso John Manthosi, University of Pretoria, South Africa

Chapter 8
Digital Technology and Sustainable Development in Africa 193
Mkululi Gacula, University of Johannesburg, South Africa

Chapter 9
Harnessing Digital Technology for Economic Development in Africa 232
Victor H. Mlambo, University of Johannesburg, South Africa
Xolani Thusi, University of Limpopo, South Africa
Muzi Shoba, Nelson Mandela University, South Africa
Hv Mlambo, University of Zululand, South Africa
Hlanganani Mnguni, Hlanganani Mnguni University of Zululand, South Africa
Lethiwe Mbongwa, Daleview Secondary School, South Africa

Chapter 10
Africa's Contributions to Digital Technology: A Case of M-Pesa Technology in Kenya 260
Patrick Makau Musembi, Daystar University, Kenya

Conclusion 290

Compilation of References 302

About the Contributors 356

Index 360

Detailed Table of Contents

Preface...xiii

Chapter 1
Africa's Indigenous Knowledge: Essential Ingredient for Digital Technology 1
Eric Blanco Niyitunga, University of Johannesburg, South Africa

Indigenous knowledge refers to knowledge systems unique to a given culture. Tharakan argued that indigenous knowledge is differentiated from the modern and international scientific knowledge systems. The current universities and institutions of learning are built on indigenous knowledge. The roots of today's social sciences, science, and technology rest on indigenous knowledge as it informs scientific research conducted in different institutions of higher learning, such as universities and research institutions. Therefore, indigenous knowledge can be seen as a component of society, part of humanity's scientific and technological advancements, which generates knowledge that cannot be achieved orally but gained through a rigorous academic investigation.

Chapter 2
Africa at the Center of Digital Technology 47
Cleophas Gwakwara, University of Johannesburg, South Africa

Digital technology is defined as electronic technology that produces, stores, and processes data. Digital technology has evolved over time with several global players, including those in Africa, making varied but significant contributions within their capacities and digital spaces. A critical question that has been asked is to locate and clarify the myths and misconceptions surrounding the contribution of African Indigenous knowledge (AIK) to modern digital technology. Voronkova et al. assert that the development of digital technology was dominated by the United States (US) and Germany during the second industrial revolution following the widespread use of electricity from the second half of the 19th century to the early 20th century in that order.

Chapter 3
Effects of the Trans-Atlantic Slave Trade and Colonialism on Indigenous Knowledge for Digital Technology in Africa 70
Eric Blanco Niyitunga, University of Johannesburg, South Africa
Logic L. Modibedi, Johannesburg Business School, South Africa

Over time, Indigenous knowledge has preserved distinctive understandings rooted in cultural experience and creativities that guide relations among human, non-human, and other-than-human beings in specific ecosystems. These understandings and relations constitute a system broadly identified as Indigenous knowledge, also called traditional or Aboriginal knowledge. Archaeologists conducting excavations in Indigenous locales may uncover physical evidence of Indigenous knowledge (e.g., artifacts, landscape modifications, ritual markers, stone carvings, faunal remains).

Chapter 4
Revisiting the Effects of Colonialism on Africa's Indigenous Knowledge........103
Thivhavhudzi Muriel Badugela, University of Venda, South Africa

In Africa's history, colonialism deeply impacted indigenous knowledge, influencing cultural identity, knowledge transmission, and societal dynamics. Examining disruptions in education, governance, and environmental practices, the review synthesises insights from scholarly works to elucidate the marginalisation and erosion of traditional systems. It places particular emphasis on cultural identity and knowledge transmission. Simultaneously, the review explores the transformative potential of the fourth industrial revolution (4IR) in addressing global challenges, utilising a qualitative lens to extract nuanced perspectives. Advocating for a holistic approach, the review underscores the qualitative integration of indigenous knowledge into 4IR development, emphasising technology's ethical and culturally sensitive application. This qualitative research perspective aims to contribute valuable depth to our understanding of the intricate interplay between historical legacies, indigenous knowledge, and contemporary technological advancements for a more inclusive and sustainable global future.

Chapter 5
Coloniality Forces Within the Cold War and Neocolonialism: A Hindrance to Africa's Contributions to Digital Technology 125
Eric Blanco Niyitunga, University of Johannesburg, South Africa

The troubling question that one needs to answer is whether the African continent's supply of raw materials, which depends on the functionality and advancement of digital technology, is not a significant contribution to the wave of digital technology the global world faces today. If that is reality, how is it that Africa's contributions to digital technology are not recognized, and who will make the world recognize them? This is a troubling question that needs to be answered. The aim of this chapter is to examine coloniality forces within the Cold War and Neocolonialism events and expose factors emerging from them that have hindered Africa's contributions to the digital technology.

Chapter 6
Green Energy Solutions in Africa: Rediscovering Indigenous Knowledge for the Fourth Industrial Revolution 159
Joshua Mawere, University of Venda, South Africa

This study robustly explores green energy solutions in Africa, underscoring the imperative rediscovery of indigenous knowledge amid the fourth industrial revolution (4IR). It aims to underscore the manifold advantages of incorporating indigenous knowledge for developing sustainable energy solutions across the continent. Addressing a central problem—the energy deficit exacerbated by Eurocentric systems, overlooking Africa's unique needs—the study uses a secondary data analysis methodology, drawing insights from diverse sources including academic articles, policy documents, and reports from international organisations. Rooted in Afrocentricity theory, the research adopts a firm decolonial perspective, unveiling cultural and historical dimensions crucial for seamlessly integrating indigenous knowledge into green energy practices. The investigation explores the transformative potential of minerals and digital technology in shaping innovative, contextually relevant solutions.

Chapter 7
Technology for Digitalization in Africa: The Contribution of Africa's Raw Minerals 178
Sandile Moloi, University of Johannesburg, South Africa
Dimakatso John Manthosi, University of Pretoria, South Africa

Africa has been on the development trajectory since its liberation from colonialism. Scholars, policy makers, and the citizens of Africa unanimously agree that freedom means nothing without development, and development means nothing without freedom. Over the years, Africans have been in search of development modals and strategies that suits the conditions of the peoples of Africa, which will remedy the historical injustices. As the fourth industrialization (4IR) unravels, many questions are raised: (1) Is 4IR compatible to Africa's developmental agenda? (2) Can 4IR help Africa tackle poverty and inequality, enhance inclusion, and foster sustainable development?

Chapter 8
Digital Technology and Sustainable Development in Africa 193
Mkululi Gacula, University of Johannesburg, South Africa

The role of digital technology in promoting sustainable development and economic growth in Africa has remained a mystery in Africa. Africa is endowed with numerous raw materials and natural resources that should enhance sustainable development in Africa. However, because of coloniality within science and technology, Africa has not benefitted from its indigenous knowledge and technology, and Africa's contributions to the wave of digital technology has not yet been understood nor recognized. The current literature presents that Africa is still behind in technology and science and yet technological innovation started in Africa.

Chapter 9
Harnessing Digital Technology for Economic Development in Africa............. 232
Victor H. Mlambo, University of Johannesburg, South Africa
Xolani Thusi, University of Limpopo, South Africa
Muzi Shoba, Nelson Mandela University, South Africa
Hv Mlambo, University of Zululand, South Africa
Hlanganani Mnguni, Hlanganani Mnguni University of Zululand, South Africa
Lethiwe Mbongwa, Daleview Secondary School, South Africa

According to Ambassador Albert Muchanga, the African Union's Commissioner for Economic Development, Trade, Tourism, Industry, and Minerals, digital technologies could present Africa with a significant opportunity to open new avenues for swift, equitable economic growth and the creation of jobs. For Africa, digital transformation has the potential to be revolutionary. It is a chance to advance industrialization and economic growth while reducing poverty and enhancing people's quality of life. Utilizing digital services and technology will support Agenda 2063 of the African Union. Digital technology can spur innovation, economic growth, and employment creation in many critical economic sectors. It also makes African markets more interconnected with the rest of the world and with each other.

Chapter 10
Africa's Contributions to Digital Technology: A Case of M-Pesa Technology in Kenya 260
Patrick Makau Musembi, Daystar University, Kenya

Historical literature shows that Africa is the cradle of mankind and has the world's oldest record of science and human technological achievements. The oldest stone tools in the world have been recorded in Eastern Africa, and later evidence for tool of technological production by humans' hominin ancestors was recorded in West, Central, Eastern, and Southern Africa. For example, some of the earliest inventions and African technology include hand axes, fishing hooks, bows and arrows, and even boats. Many of the African inventions in history pre-date the use of the written word by tens of thousands of years. These earliest technological advancements became the basis and foundations of the digital technologies the world experiences today.

Conclusion 290

Compilation of References 302

About the Contributors 356

Index 360

Preface

The idea of this book was conceptualized by my high school history teacher, who taught us that Africa was primitive and dark before the coming of colonization. The idea was brought back to life during my teaching of African Studies at the University of Johannesburg, in the School of Leadership under the tutorship of Dr Sydney F. Mufamadi, where I read historical books that transcend the trans-Atlantic slave trade. To my surprise, I found that Africa was neither dark nor primitive because the continent is the mother of all civilizations, and contemporary Western Philosophy, which spread from Greek philosophy, is a stolen legacy of Egyptian philosophy (James, 2009). Education started in Africa before it moved to the Western world, and Christianity first visited Africa before it spread to the Western World. Christianity, which is associated with enlightenment, entered Africa during the first century Christianity. The biblical evidence shows that Christianity started to spread from its origins in Palestine all the way to take root in North-East Africa not long after Pentecost (Hildebrandt 1981:5). Groves (1948:58) argued that the flight into Egypt by Joseph and Mary with the infant Jesus shows the accessibility of Africa from Palestine. The Biblical evidence in the book of Acts 8:26-40 shows that an African by the name of Eunuch, on his way from Palestine to his home Abyssinia (Ethiopia) from a region overlapping with present-day Egypt-Sudan, got converted to Christianity, be thus the first person to bring Christianity to Africa in the first century.

This interest was also intensified by my research interests in Africa's international relations and diplomacy to understand the role of Africans in the search for Pax-Africana. This is a peace that is defined and run by Africans themselves. This investigation required to move beyond the trans-Atlantic slave trade to understand how Africa was and the impact of the trans-Atlantic slave trade on African indigenous knowledge and technology. Emenyonu (1987:10) found that "Literature had existed in Africa before the coming of the Europeans to the Continent. History shows that the name Africa is a colonial name that emanated from outside Africa. The original name of Africa was *Alkebulan*, which meant Africa, the mother of mankind, or *the Garden of Eden*. Moving forward, Africa, the mother of mankind, is the origin of technology. It means that technology grew out of Africa and as civilizations moved to other parts of the world from Africa. It also means that Africa was not dark or primitive because all the other continents received civilizations that had their origins in Africa. The history of ancient Africa that existed before the trans-Atlantic slave trade and colonization demonstrated the complexity of Africa's government and civilization throughout its various empires (Muhammad Esq (2003:888).

The slogan, rather than the myth that engulfs Western education in Africa that Africa was dark and primitive before the trans-Atlantic slave trade and colonialism, has worked well to justify the Western world for the atrocities they committed on the Africans and their indigenous knowledge and technology. The slogan has eased the Western world's psyche on those evils done to the Africans and has equally led Africa to accept their misrepresentations and hunt to death anyone who challenges the myth. Muhammad Esq (2003) argued that the trans-Atlantic slave trade is a forgotten crime against humanity. It has been found that many of the European invaders in Africa were appalled to witness that Africans ruled themselves and possessed their own functional system of government.' However, these truths and evidence were burned to death, and a Western education was imposed on Africans to propagate and continue colonialism claims that Africans were barbaric and uncivilized. This misconception was a concerted effort aimed at a superimposition of the European psyche over that of the Africans, often strategically orchestrated through the colonially established or post-colonially controlled education systems (Heshmati, 2019). African indigenous knowledge and technology, which is the basis of digital technology advancement and upon which the formal education system of any society is constructed, have been consistently and intentionally ridiculed to a primitive and backward position.

African indigenous knowledge (IK) refers to the ancient African knowledge systems developed and passed down through generations within African communities (Niang, 2017). The knowledge invented science and technology, and the building of the sophisticated pyramids resulted from that knowledge. Through indigenous African knowledge, mathematics controlling all digital technology was discovered and invented. Coloniality through Western education has made most Africans fail to understand that many modern high-school-level mathematics concepts were first invented in Africa. This means the first model and method of mathematic counting was first developed in Africa before it spread to other parts of the world. It was found that Ancient Egyptians (a black-populated society) scripted textbooks about mathematics that included division and multiplication of fractions and geometric formulas in calculating the area and volume of shapes (Woods, 1988). Distances and angles were calculated, algebraic equations were solved, and mathematically based predictions were made of the size of floods of the Nile (Woods, 1988).

The ancient Egyptians considered a circle to have 360 degrees and estimated Π at 3.16 (Woods, 1988). Later, in pre-colonial Africa, people from the present-day Democratic Republic of Congo (DRC) and Yoruba people in present-day Nigeria developed their own numeration system. The Yoruba system was based on units of 20 (instead of 10) and required an impressive amount of subtraction to identify different numbers. Scholars have lauded this system, as it required much abstract reasoning (Zaslavsky, 1983).

African indigenous knowledge made ancient African astronomers of the highest caliber. Several ancient African cultures birthed discoveries in astronomy. Many of these foundations and inventions are still holding the modern world. This means that the current digital technologies rely on them. Though some of these indigenous technologies were highly advanced, their mode of discovery still cannot be understood or fathomed by today's technological inventions (Zaslavsky, 1983). Ancient Egyptians charted the movement of the sun and constellations and the moon's cycles. They divided the year into 12 parts and developed a yearlong calendar system containing 365 ¼ days (Woods, 1988). Clocks were made with moving water, and sundial-like clocks were used (Woods, 1988). It has been argued that from inventing our modern calendar to inventing writing and creating surgical instruments similar to those used today, the ancient Egyptians were truly masters of invention (Wendorf, 2023). A structure known as the African Stonehenge in present-day Kenya (constructed around 300 B.C.) was a remarkably accurate calendar (Lynch& Robbins, 1978).

The Dogon people of Mali amassed a wealth of detailed astronomical observations (Lynch& Robbins, 1978). Many of their discoveries were so advanced that some modern scholars made them the foundation of digital technologies the world experiences today. However, because of the trans-Atlantic slave trade, colonialism, and coloniality, ancient Africa's contributions to digital technology are accredited to the Western scholars who never championed them. The ancient African's inventions and discoveries instead of space aliens or unknown European travelers, even though the Dogon culture is steeped in ceremonial tradition centered on several space events. The Dogon knew of Saturn's rings, Jupiter's moons, the spiral structure of the Milky Way, and the orbit of the Sirius star system. Hundreds of years ago, they plotted orbits in this system accurately through 1990 (Adams, 1983). They knew this system contained a primary and secondary star (now called Sirius B) of immense density and not visible to the naked eye (Adams, 1983).

The African indigenous knowledge and technology championed the inventions and practicality of metallurgy and tools that make contemporary science and technology. Many advances in metallurgy and toolmaking were made across ancient Africa. These include steam engines, metal chisels and saws, copper and iron tools and weapons, nails, glue, carbon steel, bronze, and art (Van Sertima,

1983; Brooks, 1971). For example, the advances in Tanzania, Rwanda, and Uganda between 1,500 and 2,000 years ago surpassed those of Europeans then and were astonishing to Europeans when they learned of them (Wendorf, 2023). Ancient Tanzanian furnaces could reach 1,800°C- 200 to 400°C warmer than the Romans (Shore,1983). From inventing our modern calendar to inventing writing and creating surgical instruments like those used today, the ancient Egyptians were truly masters of invention (Wendorf, 2023).

African indigenous knowledge and technology championed architecture and engineering. Various past African societies created sophisticated built environments. History reveals evidence of the engineering feats of the Egyptians. The bafflingly raised obelisks and the more than 80 pyramids. The largest of the pyramids covers 13 acres and is made of 2.25 million blocks of stone (Woods, 1988). Later, in the 12th century, before the trans-Atlantic slave trade and much farther south, there were hundreds of great cities in Zimbabwe and Mozambique. Massive stone complexes were the hubs of cities there. One included a 250-meter-long, 15,000-ton curved granite wall (Asante et al. 1983). The cities featured huge castle-like compounds with numerous rooms for specific tasks, such as iron-smithing (Asante et al. 1983). In the 13th century, just a century before the trans-Atlantic slave trade, the empire of Mali boasted impressive cities, including Timbuktu, with grand palaces, mosques, and universities (Van Sertima, 1983).

African indigenous knowledge and technology championed medical science technology. Several ancient peoples employed many technological medical treatments used today throughout Africa (Brier & Hobbs, 2013). Before the European invasion of Africa, medicine in what is now Egypt, Nigeria, and South Africa, to name just a few places, was more advanced than medicine in Europe (Brier & Hobbs, 2013). Some of these practices were the use of plants with salicylic acid for pain (as in aspirin), kaolin for diarrhea (as in Kaopectate), and extracts that were confirmed in the 20th century to kill Gram-positive bacteria (Van Sertima, 1983). Other plants used had anticancer properties, caused abortion, and treated malaria, and these are as effective as many modern-day Western treatments (Van Sertima, 1983). Furthermore, African indigenous knowledge led to the discovery of ouabain, capsicum, physostigmine, and reserpine. Medical procedures performed in ancient Africa before they were performed in Europe include vaccination, autopsy, limb traction and broken bone setting, bullet removal, brain surgery, skin grafting, filling of dental cavities, installation of false teeth, what is now known as Caesarean section, anesthesia and tissue cauterization (Woods, 1988; David, 2007). In addition, African cultures performed surgeries under antiseptic conditions universally when this concept was only emerging in Europe (Van Sertima, 1983).

African indigenous knowledge and technology led to numerous navigation technologies. The colonized Western universities have taught us that Europeans were the first to sail to the Americas. However, evidence suggests that ancient Africans sailed to South America and Asia centuries before Europeans (Van Sertima, 1983). This evidence has been buried to death, and the African continent produced genius that set the foundation of the modern world. digital technology keeps being referred to as a dark and primitive continent. Thousands of miles of waterways across Africa were trade routes. Many ancient African societies-built boats, including small reed-based vessels, sailboats, and grander structures with many cabins and cooking facilities (Bunson, 1991). The Mali and Songhai built boats that were 100 feet long and 13 feet wide and could carry up to 80 tons (Van Sertima, 1983). Currents in the Atlantic Ocean flow from this part of West Africa to South America (David, 2007).

Genetic evidence from plants and descriptions and art from societies inhabiting South America at the time suggest small numbers of West Africans sailed to the east coast of South America and remained there (Van Sertima, 1983). Contemporary scientists have reconstructed these ancient vessels and their fishing gear and completed the transatlantic voyage. Around the same time they were sailing to South America, the 13th century, these ancient peoples also sailed to China and back, carrying elephants as cargo (Van Sertima, 1983). People of African descent come from ancient, rich, elaborate cultures that created a wealth of technologies in many areas. However, trans-Atlantic slave trade, colonialism, and coloniality events have sidelined African indigenous knowledge and indigenous technology on science and technology and have hindered Africa's contributions to the wave of digital technology from the first, second, third, fourth, and fifth industrial revolutions. The earliest human inventions originated in Africa, such as the hand-axe and manipulating fire. People in Africa developed inventions used up to the modern day, including fishhooks, bows and arrows, and boats. The most important center of learning in the ancient world was the Library of Alexandria in Egypt. Many of the most important Greek contributions to math, science, medicine, astronomy, and more were made in Alexandria. The first evidence of both mathematics and astronomy is in Africa. Using historical records and evidence from Timbuktu and even traditional games such as Mancala as evidence, it is undeniable that Africa had a rich and complex indigenous tradition of mathematics long before the arrival of the Arabs or later European colonizers.

African indigenous knowledge and technology encompass various domains, including agriculture, natural resource management, healthcare, and cultural practices, and play a crucial role in promoting sustainable development across the continent (Altieri, 2004). Despite all the above great achievements by Ancient Africans through their indigenous knowledge that have become influential in the current digital technology advancements, indigenous knowledge has been ignored and dismissed from science and technology. It has been seen and viewed as the odds and limits of scientific knowledge. Some proponents full of Western education have non-ashamedly argued that African indigenous knowledge that indigenous knowledge should be prevented (Aikenhead & Ogawa, 2007). Other scholars have argued that the most efficient way of strengthening indigenous knowledge is integrating it into Western education in Africa (Regmi & Fleming, 2012; Zinyeka, 2013; Zinyeka, Onwu, & Braun, 2016). Other scholars have argued that merging Western education and African indigenous knowledge is challenging but prefer to be happy with Western education (Aikenhead, 2006; Meyer & Crawford, 2011). They argue that African indigenous knowledge is more primitive and backward as it emanates from the so-called primitive and dark continent, which advances the idea that Western education is the best for Africa (Aikenhead, 2006; Meyer & Crawford, 2011). Regmi and Fleming (2012) argued that teachers in Africa must recognize that indigenous knowledge and classroom scientific knowledge can synchronize and support each other (Regmi & Fleming, 2012).

Since the imposition of Western education in African territory to continue its colonialism mission that uprooted Africa's educational system that was based on indigenous knowledge, Africa has been referred to as a dark continent inhabited by primitive people. From Western literature in secondary schools to university, Africa is portrayed as a dark continent. This continent is behind civilization, yet it is the mother of all civilization. Africa was long known as the "dark continent," but in the words of historians Paul Bohannan and Philip Curtin (1995), the darkness was in the ignorance of the outside world, not in Africa. This ignorance is the colonial tool used to deny the boast of her contributions to the wave of digital technology. Charles Darwin is among the first thinkers to label Africa as the likely cradle of humanity, and Diop and Ben-Jochannan, African scholars in Diaspora, were the second to find that Africa is the mother of Western civilizations (Diop, 1974; Ben-Jochannan, 1988). Jochannan argued that Africa originated the major Western religions (Ben-Jochannan, 1991). James (2009) presented that Africa is the birthplace of Greek philosophy, which later became Western philosophy and formal education. Africa is the cradle of mankind (Semaw et al., 1997; Delagnes and Roche, 2005), meaning that the continent is the birthplace of modern human beings. thus mainly the origins of civilizations and technology. Technological advancements cannot be left out or separated from civilizations.

Since the level of technology measures civilization and cannot be detached from people, Africa thus becomes the origin of technology. This enables one to understand that Africa remains the origin of technology. This means that African indigenous knowledge led to indigenous technology and thus became the starting point of today's technological advancement. Moreover, civilization and technology cannot be detached from the people; this makes Africa home to the world's greatest technologies that influenced civilizations, such as the Egyptian and Nubian societies that shaped the modern world, emerged (Bell, 1976; Ben-Jochannan, 1989). History proves that education that brought about technology that civilized the world started in Africa (Rodney, 1972). The first university to offer degree certificates was in Africa and was established in 898 AD in Morocco. Historical studies also show that mining and metallurgy technology evolved to civilize the world, which became today's digital technology started in Africa (Holl, 2020). Metallurgical processes in Africa were achieved and explained by drawing upon indigenous natural and social order theories. Metallurgical explanations had a marked influence on the social organization of production. Today's technology and engineering are built upon African indigenous technology that brought about metallurgical skills that are still in use today. Van Sertima (1983) argued that "the nerve of the world has been deadened for centuries to the vibrations of African genius."

Pre-colonial African societies were not homogenous. There were marked contrasts between the Ethiopian empire and the hunting groups of the Mbuti in the Congo or between empires of the Western Sudan and the Khoisan hunter-gatherers of the Kalahari Desert. Moreover, there were also striking contrasts within any given geographical area. For example, the Ethiopian empire embraced literate feudal Amharic noblemen, Kaffa cultivators, and Galla pastoralists. The empires of Western Sudan had sophisticated, educated Mandinga townspeople, small communities of Bozo fishers, and nomadic Fulani herders. There were considerable differences even among clans and lineages that appear roughly similar. These communities had their own form of civilization and designed technology that was appropriate to their economic activities. This technology constituted civilization and was embedded in African indigenous knowledge, technology, culture, and politics.

Africa created its indigenous technology using its scientific knowledge. Each African society, from the East to the West and from the North to the South, had its scientific knowledge, which was put into practice and worked for the good of its people. The tools with which societies worked and how they organized their labor are important indices of social and technological development. This means that pre-colonial Africa had scientific and technological tools designed and used to enhance the quality of lives of indigenous people in African communities. As Africa is the mother of all civilizations and the birth of mankind, this indigenous scientific and technology became the foundation of the digital technology we see today. Some traditional skills and techniques were used in producing arts and crafts, blacksmithing, iron smelting, carding and weaving, and brewing, among others, that summed up indigenous African indigenous technology. These technologies are still adopted in the contemporary world of digital technology, and while they are Africa's contribution, they are still hidden, and Africa is still viewed as primitive.

Gennaioli and Rainer (2005) argued that African pre-colonial institutions shaped modernization due to their continuity in the periphery, especially in rural areas, and that such continuity dates to the end of the 19th century when the massive European colonization of Africa began. Gennaioli and Rainer (2005) argued that pre-colonial Africa was disrupted by its course of technological development and advancements. However, the disruptions and the darkness the trans-Atlantic slave trade and colonialism plunged into Africa, the remains of African indigenous knowledge and technology have not yet defended Africa's contributions to digital technology. This is because those events introduced corruption, mental slavery, and mental colonialism, which has persisted through coloniality forces, to the fact that many Africans admire becoming whites and behaving like whites. Pre-colonial African indigenous technology was relevant in organizing labor and producing goods and services that responded to the needs of each African society.

UGLY EFFECTS OF TRANS-ATLANTIC SLAVE TRADE AND COLONIALISM ON AFRICAN INDIGENOUS KNOWLEDGE

Despite suffering through the horrific system of slavery sharecropping, early African Americans made countless contributions to science and technology (Kresge,2011). Similarly, Azikiwe (1968:7) argued that the trans-Atlantic slave trade played a "shameful role in depopulating Africa; capitalism denuded it of its wealth; colonialism deprived it of its birthright, and imperialism emasculated its will to live as a human being and to enjoy its fair share of the bounties of the good earth." It has been indeed argued that the trans-Atlantic slave trade and colonialism are "economies of slavery that harmed Africa and enriched the Europeans (Senghor, 1958:40). Soyinka (2012) argued in a media interview that Africa's underdevelopment is the outcome of "the twin evils of slavery and colonialism inflicted by the Western world." Adeniji (2012) argued that slavery and colonialism are two monumental historical events that dislocated Africa and aborted the continent's social, political, economic, and technological advancement.

Ojo (2015:109) argued that during the 2012 World Summit on Sustainable Development held in Johannesburg, African leaders took a unanimous stand against the trans-Atlantic slave trade and colonialism that harmed Africa and retarded technological advancement, which in turn promoted and sustained Africa's development tragedy. Ojo (2015: 110) further argued that the trans-Atlantic slave trade "is responsible for Africa's socio-economic inertial, political ineptitude, technological backwardness, and sundry dislocations." Nunn (2008) showed the link between the slave trade, the legacy of extractive colonial institutions, and Africa's underdevelopment that affected digital technology. Nunn (2007:145) further "linked Africa's current under-development to colonial rule and the trans-Atlantic slave." Nunn (2008) and Nunn (2011) argued that the trans-Atlantic slave trade pillaged Africa and prospered in Europe and the Americas so that to date. Through coloniality, which takes forces from the legacies of both the trans-Atlantic slave trade and colonialism, Africa's underdevelopment and poverty have persisted. This has hindered the ability of the continent to echo its contributions to the world of digital technologies. Yet, Africa is the greater contributor to the raw materials and natural resources needed to advance digital technology.

Lovejoy (1974:65) pointed out that "many African communities relocated as far from the slavers' route as possible, and in the process, the growth of their indigenous technology was hindered. Thus, their technological and economic development was hindered as they devoted their energy to hiding and defending themselves (Lovejoy, 1974:65). In removing millions of able-bodied men and women from Africa, the Atlantic slave trade knocked off the premise of Africa's indigenous knowledge and technology thus headed to the poverty trajectory. Alcott (2015) pointed out that:

The transatlantic slave trade caused the forced removal from Africa by millions of Africans. This number included a large percentage of skilled tradesmen and women from various occupations and professions who were making their contributions to African societies...Slave trading undermined the Gold Coast economy and destroyed the gold trade. Slave raiding and kidnapping made it unsafe to mine the gold or to travel with gold. The Europeans' demand for slaves made raiding for captives more profitable than gold mining. The transatlantic slave trade encouraged Africans to wage war against one another and conduct raids instead of building more peaceful links.

Contemporary African societies bear the imprint of the legacy of the trans-Atlantic slave trade, colonialism, and coloniality. Still, they are also marked both by their pre-colonial heritage and their different postcolonial experiences. African indigenous knowledge and technology are almost nowhere to be found, and the current education in Africa produces Western knowledge that reflects the political, economic, and historical intrusions that resulted from colonial contact with Europeans. Today, Africa's science and technology are judged from the perspectives of its former colonizers. There has been a tendency to look at Africa as if it had never possessed any science and technology before Europeans set foot on the continent. This is because, in almost all Western educational systems, Africa is portrayed as a primitive and dark continent, preventing scholars and pupils from understanding beyond the trans-Atlantic slave trade and colonialism. Colonial thought on indigenous African communities and societies was eminently used to reconstruct indigenous peoples as homogenous and unchanging (Brownlie, 2005).

Unfortunately, European colonization interfered with and destroyed the social relationships of production and the development of indigenous technological advancements that were in place. Basil Davidson (1992) argued that colonialism prevented the potential for the natural maturing of pre-colonial African institutions. These postcolonial sentiments are also expressed by Onipede (2010), who stated that colonialism brought about colonial rule in Africa and, by extension, imperialism, which prevented the growth of industrialization in Africa. Onipede (2010) further argued that the nature of imperialism is fortuitous, transferring the wealth of underdeveloped nations to the Western world and undermining them through capital and human exploitation. The current lack of technology, which has underdeveloped Africa, could not be explained without reference to colonialism and contemporary neocolonialism through coloniality. Post-colonial Africa still suffers from deskilling and disturbances during development, heinously affected by colonial regimes.

Rodney (1972) noted that colonialism and imperialism corrupted indigenous knowledge and technologies in Africa. This led to the recolonization of indigenous knowledge and technology in Africa, which has hindered its contributions to the wave of digital technology. As Ngugi wa Thiong'o (1981) argued, colonial education came with the idea of Europeanizing Africans into the colonial civil service. Mazonde (1995) noted that the colonial educational emphasis was on liberal arts, as there was little instruction in the technical, vocational, and professional fields. So, for most of Africa, instructional technology is not a feature of their educational offerings during the pre-trans-Atlantic slave trade and colonial period.

The book's central thesis is based on the contributions of African indigenous knowledge to the wave of digital technology, focusing on the decoloniality perspective. Indigenous knowledge can be differentiated from the modern scientific knowledge system (MSKS) and international knowledge systems. While indigenous knowledge is based on the people's cultures and experiences and is practical, development-oriented, and futuristic; international knowledge is colonial in nature, produces colonized minds, and promotes subordination. International knowledge is not development but rather promotes Western culture, views, and experiences in Africa, which are enemies of technological advances. Hence, international knowledge humpers Africa's contributions to the wave of digital technology and hinders Africa's efforts to revert to African indigenous technology to use the current digital technology to promote sustainable development.

This book is, therefore, a timely investigation of African indigenous knowledge and technology. The book aims to disprove the colonial statement trending in Western literature that Africa was a primitive and dark continent to provide a more comprehensive investigation of African history and its contributions to the wave of the Fourth Industrial Revolution and the wave of digital technology. This investigation enables us to comprehend Africa from the realm of technology advancement and note that Africa is the origin of digital technology shaping the modern world. This book questions whether Africa has contributed to the wave of digital technologies from the first, second, third, and fourth Industrial Revolutions, which have influenced the use of digital technologies in producing goods and services and distributing them to the citizens. This means that the book questions the existing literature, starting with history, to understand whether the African indigenous knowledge and technologies that influenced the ancient empires have impacted the wave of today's digital technology.

Thus, in line with what is now called the Fourth Industrial Revolution, this book provides a framework to explore the role of African indigenous knowledge and technology in digital technology. The book also explores the effects of the trans-Atlantic slave trade, colonialism, and coloniality on the materialization of African indigenous knowledge to lead to technological advancement. The book also explores the contrast between the pre-trans-Atlantic slave trade, pre-colonial African education, and Western or colonial education on the wave of digital technology. While the former type of education promoted science and technology, which influenced the current digital technologies, the latter created mental slavery and mentally colonized epistemologists that hindered Africa's contributions to the wave of digital technology. Questions that the above have generated include:

- What is the nature of indigenous African knowledge, and why is it an essential ingredient and element for the growth of digital technology?
- In what ways are raw materials and natural resources in Africa, make the continent to be at the center of technology such that one cannot talk to the wave of digital technology without Africa?
- What are the effects of the trans-Atlantic slave trade, colonialism and coloniality on indigenous knowledge of digital technology in Africa? In order words, to what extent are trans-Atlantic slave trade and colonialism, as well as modern coloniality hampering Africa's contributions to the wave of digital technology?
- In other words, in what way will the understanding of Africa's indigenous knowledge systems enable us to improve upon its paradigms and development models through the existing education system?

- What makes decolonizing digital technology in Africa significant and essential in contemporary society?
- What are the key cases of the creation of African home-grown digital technologies, and how have these technologies influenced economic growth and sustainable development, and what lessons can be learned from them?

These questions were investigated and answered in various chapters that make up this book.

It must be stated that the current study is not primarily concerned with the African government's budget allocation to education. Neither is it concerned with school enrolment, gender and ethnic issues in school attendance, and the length of time students stay in school, among other things. Its primary objective is to provide conceptual resources and to direct attention to the continuing disconnect between Africa's own knowledge systems and those imported and superimposed on the region through the education curricula, which the study considers a major source of the region's continuing development predicament. The discussion of the above truths is carried out in eleven chapters.

In Chapter 1, Eric Niyitunga investigates issues related to Africa's indigenous knowledge and its significance in the wave of digital technology. Indigenous science plays a crucial role in the development of science and technology. It provides rich contexts for understanding the relationship between sociocultural life and environmental ethics. Indigenous science is based on the original knowledge of local communities and is aligned with their culture and location. It encompasses scientific attitudes, products, and processes, making it a valuable resource for science learning. Modern science and technology take advantage of indigenous knowledge that advances traditional medicines and convert them into powerful drugs using the methods of modern science, computer heuristics, combinatorial chemistry, rational drug design based on pharmacophore model generation, computer-aided compound selection, and state-of-the-art analytical tools (Arunachallam, 2007). It is important to note that the development of some of the complicated technologies we see today copied the original ideas from indigenous knowledge in Africa. This indicates that today's digital technology advancement owes Africa's indigenous knowledge.

Cleophas Gwakwara in Chapter 2 assessed the factors that make African be at the center of modern digital technology and yet her contributions are neither recognized nor considered in the wave of the digital technology. Cleophas Gwakwara in this chapter argues that many modern high-school-and University level concepts in mathematics were first developed in Africa, developed and typified as it was the first method of counting. This means that Africa was and has been the origins of the science and technology on which digital technology is built. The Dogon people of

Mali amassed and digested a wealth of detailed astronomical observations (Adams, 1983). Precolonial Africa was a major hub for trade and manuscripts and Africans had advanced knowledge and technology in areas such as astronomy, mathematics, agriculture, navigation and sailing and medicine.

Cleophas Gwakwara finds that Africa created its own indigenous technology using its own scientific knowledge which became the foundation for the current digital technology. Pre-colonial Africans were technologists, to build structures, possessed skills and expertise in architecture and construction of boats and sailing. They also carried out mathematical and numerical computations (Sydella, 2013). In this regard, Africa's contribution to digital technology has not received due consideration. The researchers argue that despite the myths and misconceptions around Africa's role in the advancement of digital technology, Africa remains at the centre of digital technology. There should be a deliberate effort to decolonize and demystify modern digital technology to recognize Africa's contributions. Using an exploratory, narrative and descriptive approach, it is evident that Africa is at the center of the Fourth Industrial Revolution (4IR). Cleophas Gwakwara concludes by saying that there is need to harness indigenous knowledge and address discrimination and racism withing the digital technology, particularly 4IR. By recognizing and valuing Africa's role in shaping the digital age, the continent would continue to actively contribute to global problem-solving, innovation, skills and technology to realise sustainable development.

In Chapter 3, Eric Niyitunga provides a more comprehensive understanding of the effects of the Atlantic slave trade and colonialism on Africa's indigenous knowledge. He explained factors that make trans-Atlantic to repress all the technological advancements the Africa had achieved, putting Africa on the periphery, and calling it a dark and primitive continent. investigates the effects of colonial forces within the Cold War and neocolonialism. He discusses factors emerging from those forces and presents the manner in which those factors hinder Africa's contributions to digital technology. Eric Niyitunga's findings show that the trans-Atlantic slave trade demolished Africa's legal systems and that the existing uneven Africa's political structures and socio-economic that hinder Africa's contributions to the wave of digital technology emerge from the legacy of trans-Atlantic slave trade.

Eric Niyitunga and Logic L Modibedi discussed in the chapter that the trans-Atlantic slave trade affected the growth of African indigenous technology and led to the stagnation of Africa's technological development. The trans-Atlantic slave trade affected Africa's economy and brought about a loss of industry, skills, technological invention, and production of Africans (Rodney, 1972). They presented that the trans-Atlantic slave trade brought about a total loss of development opportunity and made Africans lose the ability to achieve self-sustaining growth and the growth of technology. The trans-Atlantic slave trade enslaved young, energetic Africans who would have secured economic and digital technology opportunities for the continent. He found that colonialism affected the growth of digital technology by imposing direct control of the African economy and administration to protect the capital they had transported from Europe to be invested in Africa. They explained the need to decolonize education and educational systems in Africa to Advance digitalization in Africa. This is because education is at the center of digitalization. Yet, Africa's inherited colonial education systems demolished African indigenous knowledge and hinders Africa's contributions to the wave of digital technology.

In Chapter 4, Thivhavhudzi Muriel Badugela revisits the effects of colonialism on African indigenous knowledge. He found that Africa's history, colonialism deeply impacted indigenous knowledge, influencing cultural identity, knowledge transmission, and societal dynamics." Examining disruptions in education, governance, and environmental practices, the review synthesises insights from scholarly works to elucidate the marginalisation and erosion of traditional systems. It places particular emphasis on cultural identity and knowledge transmission. Simultaneously, the review explores the transformative potential of the Fourth Industrial Revolution (4IR) in addressing global challenges, utilising a qualitative lens to extract nuanced perspectives. Advocating for a holistic approach, the review underscores the qualitative integration of indigenous knowledge into 4IR development, emphasising technology's ethical and culturally sensitive application. This qualitative research perspective aims to contribute valuable depth to our understanding of the intricate interplay between historical legacies, indigenous knowledge, and contemporary technological advancements for a more inclusive and sustainable global future.

In Chapter 5, Eric Niyitunga assesses coloniality within digital technologies discriminating other technological contributions that are non-western to flourish. Digital technology carries within themselves a colonial environment and structures that make them to be colonial in today's societies (Akrich, 1997). This analysis needed a careful assessment, as the failure to understand that digital technologies promote and enable colonial practices. Mbembe (2017) argued current technologization of the world is a means of spreading Western culture and coloniality to the rest of the world. Mbembe (2019) further described that a computable speed regime is colonial power on its own because it is organised around the forces and energy derived from computability.

This self-perpetuating regime thrives on normative descriptions in a Western mathematical language produced by the international knowledge resulting from western education, which in turn, transposed into algorithms for derivation. Mbembe (2019) focused on quantifiable information and governance through calculations to how technology is used to set up mobile, portable, omnipresent, and ubiquitous delineations and borders that are controlled by the colonialists. Digital technology functions as the means by which the West manages and colonise the world (Adamu, 2021; Mamdani, 2012). Zenon Ndayisenga showed that digital technology is colonised and thus needs to de decolonized so that other non-western contributions are acknowledged and brought on board. His argument are in agreement with Dourish and Mainwaring (2012) who argued that the underpinning ubiquitous digital technology through computing are inherently colonial. Technologies carry expressed values, from the site of their conceptualisation to the place of deployment (Feenberg, 2005). For example, this colonial mastery in digital technology is easily seen when one assesses drones or killer robots, but less obvious when implementing digital apparatus like mobile phones and Wi-Fi routers or interacting with applications that exist by virtue of the dashboard.

In Chapter 6, MMX investigates the green energy solutions in Africa in the effort to rediscover and discuss the role of indigenous knowledge for 4IR. Taking a decolonial perspective, MMX delves into the latent capabilities of AIK in propelling the development of green energy solutions. The integration of traditional knowledge systems into sustainable development has garnered increasing recognition, with AIK systems emerging as reservoirs of invaluable wisdom that can harmonise with modern science and technology. An essential acknowledgment is that Africa possesses indigenous knowledge (IK) capable of contributing significantly to the 4IR, independently of Eurocentric ideologies. MMX showed in this chapter that the integration of indigenous knowledge systems (IKS) and practices has the potential to yield innovative solutions that resonate with the diverse cultural identities of Africa's populations. MMX shows that the utilization of IK in sustainable development is inherently transformative, ethically grounded, and conducive to decolonisation.

Amidst a global shift towards sustainable energy to combat climate change, Africa finds itself at the forefront of an opportunity to lead in green energy solutions, leveraging its vast renewable energy sources. MMX argued that unfortunately understanding and recognizing the full exploitation of the potential of IK in Africa and implementing green energy solutions on the continent has remained a mystery. The revitalization of IK systems can serve as a guiding force for developing locally tailored and sustainable energy systems. By acknowledging the value of traditional wisdom, Africa has the potential to empower local communities to actively participate in shaping their energy landscape. This would ensure that local communities gain reliable access to energy, thereby fostering sustainable development at high speed.

In Chapter 7, Sandile Moloi and Dimakatso Manthosi assessed the role of Africa's raw materials and natural resources to the wave of digital technology.

They assess and investigate the role of Africa's minerals and raw materials in the manufacturing of computers, smartphones, laptops and concluded that African has a lot of contributions to the growth of digital technology as well as their functionalities. They showed in this chapter that 4IR represents the emerging technologies, which are blurring the traditional boundaries between the physical, digital, and biological worlds. They argue that this fusion is driven by artificial intelligence (AI), robotics, the Internet of Things (IoT), genetic engineering, quantum computing, and more. They also explain that digital technologies are key components of 4IR and have the potential of unlocking sustainable industrial innovation in Africa. These digital technologies were shown to facilitate inter-linkages in the enhanced production process. Technological advancement is frequently linked to the development of novel and useful products through ingenuity and invention. They found that beneath the Africa soil lies a wealth of mineral resources of enormous value. Africa is home to about 30 percent of the world's mineral reserves, 12 percent of the world's oil and 8 percent of the world's natural gas reserves according to the United Nations. In 2019, the continent produced almost 1 billion tonnes of minerals worth $406bn. These African mineral resources often end up in industrialised economies where they are engineered into various technologies which includes digital technologies.

In Chapter 8, Mkululi Gacula assesses the role of digital technology in promoting sustainable development in Africa. He argues that Africa, being at the centre of raw minerals needed for digital technology, stands a high change to lead digital would benefit from these minerals to achieve sustainable development. Mkululi chapter shows that digital technologies are new engine of sustainable development because they lead to social change and transformation of enterprises. They enable the achievement of quality and efficiency of the economy, hence providing an ample opportunity for countries to achieve sustainable development. The application of digital technologies applied to daily businesses activities promote innovation thus leads to sustainable development. Mkululi chapter shows that digital technology in Africa is colonised and thus needs to be decolonized to enable the realization of digital technology application in Africa to achieve economic growth that leads to sustainable development. Mkululi explain extensively the effects of trans-Atlantic slave trade and colonialism on digital technology adoption (DTA) in Africa. He shows that the two events are to be blamed for exploiting Africa, and introducing mental slavery which still exists in Africa. Due to mental slavery and colonized mind and attitude, digital technology in Africa has not been flourish and lead to sustainable development. Mkululi concludes in the chapter that digital technologies is an ingredient for sustainable development. However, as Africa is now, the need to address the legacies of trans-Atlantic and colonialism must be taken into consideration.

In Chapter 9, Victor H Mlambo, Xolani Thusi, Muzi Shoba, Mandisi Matyana, Hlanganani Mnguni, and Lethiwe Mbongwa explore the need to harness digital technology for economic development in Africa. They noted that digital technologies, from computers to apps to machine learning, offer new opportunities for people, businesses, and jobs. Evidence presented in this report demonstrates that internet availability increases jobs and reduced poverty in African countries. They examine how harnessing digital technology can be a catalyst for economic development in Africa. They relied on the on secondary data and employed a narrative literature review as a research methodological approach. Their findings in this chapter revealed that the emergence of digital technologies presents a chance to open up fresh avenues for swift economic expansion and encourage upward economic mobility. Digital technologies have the potential to improve quality of life by increasing access to basic services, expanding global markets, establishing network economies, and reducing travel costs. They also result in significant productivity gains.

Chapter 10 is written by Patrick Musembi explores the role of M-PESA – digital mobile banking technology that originated in Africa during the current era of post-modernity in promoting sustainable development. Patrick Musembi's chapter examines the ways in which MPESA technology has shaped the modern world of digital technology. Patrick Musembi shows that there is overwhelming evidence that Kenyans are using M-PESA in a variety of ways to enhance their quality of life including financial inclusion, improving access to healthcare and education. M-PESA technology is also supporting the growth of small businesses, boosting women's economic opportunities, reducing poverty. M-PESA technology in Kenya has promoted and enhanced women empowerment which in turn leads to the achievement of sustainable economic growth that leads to sustainable development. Patrik Musembi found that since the inception of M-PESA technology, there has been promotion of socio-economic and political economic growth in Kenya. M-PESA has empowered more Kenyans to be in control over their lives including health, economic and political situations, and self-respect, dignity, and community engagement. Patrick Musembi also shows in the chapter that despite its documented success, MPESA technology has brought potential negative effects such as reinforcing patriarchy and gender inequalities, weakened social ties between migrants and their home communities and increased women's workloads. M-PESA technology services has increased the risk of domestic violence and abuse, limiting women's access to formal financial services, and has promoted disintegration of family and the clan system.

Finally, by conducting a deeper exploration of African indigenous and digital technology, it became vividly that neither was Africa a dark nor primitive continent. It was found that Africa is the mother of all civilizations which means that Africa civilized the world and making it a home for the earliest technologies. African indigenous knowledge influenced technology which later influenced digitalization and innovation. Innovation and digitalisation sits at the intersection of the natural sciences, raw material and natural resources which are mostly found in Africa. The perception of universalizing digital technology needs to be interrogated, challenged, and decolonized because it is serving as colonial tool used to subordinate other states. In its decolonized status, it is a threat to digital technology and the failure to recognize the Africa's contributions to digital technology must be recognized as colonial practices. The imposition of the western education on and the universalization of science and technology suffocates the contributions African indigenous knowledge to digitalization. The need to acknowledge African indigenous knowledge and decolonize it is requirement for the survival of the Africa's contributions to the wave of digital technology.

Eric Blanco Niyitunga
University of Johannesburg, South Africa

REFERENCES

Adams, H. (1983). African Observers of the Universe: The Sirius Question. *Blacks in Science: Ancient and Modern*, 27–46.

Adamu, M. S. (2021). Rethinking technology design and deployment in Africa: Lessons from an African standpoint. AfriCHI 2021.

Adeniji G. (2012, Dec. 19). Slavery, colonialism haunting Africa – Soyinka. *Punch.*

Aikenhead, G. S. (2006). *Science education for everyday life: Evidence-based practice*. Teachers College Press. Retrieved from https://www.tcpress.com/scienceeducation-for-everyday-life-9780807746349

Akrich, M. (1997). The de-scription of technical objects. In W. E. Bijker & J. Law (Eds.), *Shaping Technology/Building Society* (pp. 205–224). The MIT Press.

Alcott, W. (2015). *The Underdevelopment of Africa by Europe*. Revealing Histories. Retrieved at http://www.revealinghistories.org.uk/africa

Asante, M., & Asante K. (1983). Great Zimbabwe: An Ancient African City-State. *Blacks in Science: Ancient and Modern,* 84-91.

Azikiwe, N. (1968). The Future of Pan-Africanism. *Presence Africaine*, *12*, 7.

Bidwell, N. J. (2020). Wireless in the weather-world and community networks made to last. In PDC'20: Vol.1, June 15-20, Manizales, Colombia (pp. 126–136). doi:10.1145/3385010.3385014

Brier, B., & Hobbs, H. (2013). *Ancient Egypt: Everyday Life in the Land of the Nile*. Sterling.

Brooks, L. (1971). *African Achievements: Leaders*. Civilizations and Cultures of Ancient Africa.

Bunson, M. (1991). *The Encyclopedia of Ancient Egypt*. Gramercy Books.

David, R. (2007). *Handbook to Life in Ancient Egypt Revised*. Oxford University Press.

Dourish, P., & Mainwaring, S. D. (2012). Ubicomp's colonial impulse. UbiComp'12. doi:10.1145/2370216.2370238

Feenberg, A. (2005). Critical Theory of Technology: An Overview. *Tailoring Biotechnologies*, *1*(1), 47–64.

GennaioliN.RainerI. (2005). The Modern Impact of Precolonial Centralization in Africa. doi:10.2139/ssrn.848164

Groves, C. P. (1948). *The Planting of Christianity in Africa* (Vol. 1). Lutterworth Press.

Harding, S. (1998). *Is science multicultural? Postcolonialisms, feminisms, and epistemologies*. Indiana University Press.

Hildebrandt, J. (1981). *History of the Church in Africa. Achimota*. Christian Press.

James GM. (2009). Stolen Legacy: Greek Philosophy is Stolen Egyptian Philosophy. *The Journal of Pan African Studies*.

Jeater, D. (2018). Academic standards or academic imperialism? Zimbabwean perceptions of a hegemonic power in the global construction of knowledge. *African Studies Review*, *61*(2), 8–27. doi:10.1017/asr.2017.132

Kresge, N. (2011). *A history of black scientists*. ASBMB Today.

Lovejoy, P. E. (1974). *International Slave Trade: Causes and Consequences*. York University Press.

Lynch, B. M., & Robbins, L. H. (1978). Namoratunga: The First Archeoastronomical Evidence in Sub-Saharan Africa. *Science*, *4343*(4343), 766–768. doi:10.1126/science.200.4343.766 PMID:17743241

Mamdani, M. (2012). *Define and rule: Native as political identity (Kindle)*. Harvard University Press. doi:10.4159/harvard.9780674067356

Mbembe, A. (2017). Strategic Narratives of Technologies and Africa. Academic Press.

Mbembe, A. (2019). Bodies as borders. *From the European South, 4*, 5–18.

Meyer, X., & Crawford, B. A. (2011). Teaching science as a cultural way of knowing: Merging authentic inquiry, nature of science, and multicultural strategies. *Cultural Studies of Science Education*, *6*(3), 525–547. Advance online publication. doi:10.1007/s11422-011-9318-6

Nathan, N. (2011). The Slave Trade and Origins of Mistrust in Africa. *The American Economic Review*, *101*(7), 3221–3252. doi:10.1257/aer.101.7.3221

Nunn, N. (2007). Historical Legacies: A Model caineLinking Africa's Past to its Current Underdevelopment. *Journal of Development Economics*, *83*(1), 145. doi:10.1016/j.jdeveco.2005.12.003

Nunn, N. (2008). Shackled to the Past: The Causes and Consequences of Africa's Slave Trades. In Natural Experiments of History (pp. 142–184). Harvard University Press.

Ojo, E. O. (2015). The Atlantic Slave Trade and Colonialism: Reasons for Africa's Underdevelopment? European Scientific Journal, 11(17).

Regmi, J., & Fleming, M. (2012). Indigenous knowledge and science in a globalized age. *Cultural Studies of Science Education, 7*(2), 479-484. doi:10.1007/s11422-012-9389-z

Senghor, L. S. (1958). A Community of Free and Equal Peoples with the Mother Country. *Western World*, (18), 40.

Shore, D. (1983). Steel-Making in Ancient Africa. *Blacks in Science: Ancient and Modern,* 157-162.

Soyinka, W. (2012). *Slavery, Colonialism – Responsible for Africa's Underdevelopment*. ThisDay Journal.

Sydella, B. (2013). *Great achievements in science and technology in ancient Africa*. ASBMB.

Van Sertima, I. (1983). The Lost Sciences of Africa: An Overview. *Blacks in Science: Ancient and Modern,* 7-26.

van Stam, G. (2020). Power inequities. Observations on the development of information and communication technologies, from an African place. In *WebSci'20 Companion: 12th ACM Conference on Web Science Proceedings* (pp. 54–59). 10.1145/3394332.3402830

Wendorf, M. (2013). *Ancient Egyptian Technology and Inventions*. https://interestingengineering.com/lists/ancient-egyptian-technology-and-inventions

Woods, G. (1988). *Science in Ancient Egypt*. Franklin Watts.

Zaslavsky, C. (1983). The Yoruba Number System. *Blacks in Science: Ancient and Modern,* 110-127.

Zinyeka, G. (2013). Onwu and Mosimege on Indigenous knowledge systems and science and technology education: A dialogue in some remaining issues. *Greener Journal of Educational Research, 3*(9), 432–437.

Zinyeka, G., Onwu, G. O. M., & Braun, M. (2016). A truth-based epistemological framework for supporting teachers in integrating indigenous knowledge into science teaching. *African Journal of Research in Mathematics, Science and Technology Education, 20*(3), 256-266. doi:10.1080/18117295.2016.1239963

Chapter 1
Africa's Indigenous Knowledge:
Essential Ingredient for Digital Technology

Eric Blanco Niyitunga
University of Johannesburg, South Africa

ABSTRACT

Indigenous knowledge refers to knowledge systems unique to a given culture. Tharakan argued that indigenous knowledge is differentiated from the modern and international scientific knowledge systems. The current universities and institutions of learning are built on indigenous knowledge. The roots of today's social sciences, science, and technology rest on indigenous knowledge as it informs scientific research conducted in different institutions of higher learning, such as universities and research institutions. Therefore, indigenous knowledge can be seen as a component of society, part of humanity's scientific and technological advancements, which generates knowledge that cannot be achieved orally but gained through a rigorous academic investigation.

INTRODUCTION

Indigenous knowledge and technology are integral parts of the culture and history of a local community. Learning from local communities about indigenous science and technology is crucial because it enriches a society's development process and progress. Understanding how people and societies acquire and use knowledge is

DOI: 10.4018/978-1-6684-7851-6.ch001

Copyright © 2024, IGI Global. Copying or distributing in print or electronic forms without written permission of IGI Global is prohibited.

essential for improving people's lives, especially those of the poorest (The World Bank 1998). Indigenous knowledge refers to knowledge systems unique to a given culture (Ellen & Harris, 1996). Tharakan (2017) argued that indigenous knowledge is differentiated from the modern and international scientific knowledge systems. The current universities and institutions of learning are built on indigenous knowledge. The roots of today's social sciences, science, and technology rest on indigenous knowledge as it informs scientific research conducted in different institutions of higher learning, such as universities and research institutions (Tharakan, 2017). Therefore, indigenous knowledge can be seen as a component of society and part of humanity's scientific and technological advancements, which generates knowledge that cannot be achieved orally but is gained through a rigorous academic investigation. This means that academic investigation, whether science, social science, or technology, is based on and inspired by Indigenous knowledge. Unfortunately, due to existing coloniality forces that brought about Western education based on Western indigenous knowledge, Africa's indigenous knowledge has been overlooked, despised, and dismissed from science and technology.

The current science curriculum represents Western worldviews and indigenous knowledge, which is unknown by Africans, who seldom forget social activity in the real-life environment and alienate scholars from their roots (De Beer & Whitlock, 2009). This abandonment of African indigenous knowledge and adoption of Western education based on Western indigenous knowledge impacts the imbalances in societies where cultural values and local wisdom are abandoned. It also propagates mentally enslaved scholars who are Western-minded, spreading the interest of the Western world instead of Africans. Aikenhead and Ogawa (2007) noted that any education on the continent not informed by Africa's indigenous knowledge would prevent the non-indigenous community (Aikenhead & Ogawa, 2007). This implies that Africa has left its primary roots of education, which would inspire the continent's contributions to education, and adopted a Western form that limits its possibility of producing any contributions to the world of science and technology.

This chapter aims to assess the role of Africa's indigenous knowledge as an ingredient for digital technology advancement and growth. The chapter is concerned with assessing Africa's indigenous knowledge and the factors and forces that inhibited the growth of Africa's indigenous knowledge to promote digital technology in Africa. The issue that this chapter is raising is the need to consider the indigenous perspective in digital technology and understand the factors that have prevented Africa's contributions to the wave of digital technology. The indigenous perspective on digital technology has been direly overlooked and under-researched hitherto. To achieve this aim, the chapter is divided into six main sections. The first section defines and conceptualizes indigenous knowledge and explains the types and characteristics of indigenous knowledge. The second section assesses the significance of indigenous knowledge and the process of knowledge acquisition and transfer. The third section explains the nexus between indigenous knowledge and technology and the place indigenous knowledge has in advancing appropriate technology for sustainable development. The fourth section investigates the effects of the trans-Atlantic slave trade, colonialism, and coloniality on indigenous knowledge. The fifth section investigates the factors that make the inherited colonial education in Africa during this era of coloniality serve as a destructive force of indigenous knowledge. This paves the way to understanding the need to decolonize indigenous knowledge to enable the world to see Africa's contributions to digital technology. The sixth section consists of the chapter's concluding remarks.

INDIGENOUS KNOWLEDGE DEFINED

Indigenous knowledge is the traditional, cultural, and community knowledge produced and owned by local people in their specific communities and passed on from one generation to the next generation through practices and oral channels (Govender et al., 2013). Indigenous knowledge refers to the understandings, skills and philosophies developed by societies with long histories of interaction with their natural surroundings (Bah & Scott, 2022). Local knowledge refers to the understanding and skills developed by individuals and populations specific to the places where they live (Bah & Scott, 2022). Indigenous and local knowledge are inherently valuable but have only recently begun to be appreciated in Western scientific assessment processes in their own right (Ford et al., 2016). In the past, these often-endangered ways of knowing have been suppressed or attacked (Mustonen, 2014; Mustonen et al., 2021).

Indigenous knowledge is made of two words: Indigenous and knowledge. Schumpeter (1912) sees knowledge as new combinations of existing means of production. Weitzman (1998, p. 334) introduced the 'germinal ideas' in knowledge and argued that knowledge is the mental ability and capability of an individual or region, whether codifiable or not. The word indigenous originates naturally in a particular place: native. It also refers to the people inhabiting or existing in a land from the earliest times or before trans-Atlantic slave traders and colonialists' arrival. Indigenous people are associated with science and technology. Indigenous science and technology are those that Indigenous people developed independently of Western science (Ninnes, 2000). This means that indigenous relate to people who have a long-standing and complex relationship with a local area, and 'science' means a systematic approach to acquiring knowledge of the natural world. Thus, indigenous science is the process by which Indigenous people build their empirical knowledge of their natural environment (Snively & Corsiglia, 2001).

Traditional indigenous knowledge can be defined as a network of knowledge, beliefs, and traditions intended to preserve, communicate, and contextualize Indigenous relationships with culture and landscape over time. One must distinguish "knowledge" as factual data, "belief" as religious concepts, and "tradition" as practice, but these terms are often used imprecisely and interchangeably to describe indigenous epistemologies (Bruchac, 2014). Epistemology is the philosophical study of knowledge (O'hara, 2001). An epistemologist studies the nature of knowledge, its characteristics, and functions (Pardi, 2011). Plato defined knowledge as "justified true belief" (Welbourne, 2001; O'hara, 2001; Pardi, 2011). Nozick (1981, p. 208) noted that "knowledge is not simply true belief." Justification of perceptions, ideas, beliefs, actions, and behavior is needed to turn a true belief into knowledge (David, 2001). Indigenous knowledge is conveyed formally and informally among kin groups and communities through social encounters, oral traditions, ritual practices, and other activities. They include oral narratives that recount human histories; cosmological observations and modes of reckoning time; symbolic and decorative modes of communication; techniques for planting and harvesting; hunting and gathering skills; specialized understandings of local ecosystems; and the manufacture of specialized tools and technologies (such as flint-knapping, hide tanning, pottery-making, and concocting medicinal remedies) (Bruchac, 2014).

The popular perspective about indigenous knowledge has been that uncodified knowledge is often perceived as incompatible with modern scientific methods and is, therefore, not yet incorporated into modern science (Warren, 1991). This perception is an effect of colonialism to deny Africa a voice in the world of science and technology because Africa is the mother of mankind, and modern science and technology are built on the indigenous science and technology that evolved from Africa and spread to other parts of the world. Warren (1991) further noted that the popular perspective about indigenous knowledge is local knowledge. That is a knowledge which is unique to a given culture or society (Warren, 1991). Indigenous knowledge is the basis of modern knowledge acquired from the universities' or colleges' systems generated through education and research centers. This knowledge is the basis for local-level decision-making in agriculture, health care, food preparation, education, natural resource management, and other activities in rural communities (Warren, 1991).

Indigenous knowledge is the basis for decision-making, and unfortunately, this knowledge was destroyed and demolished by the trans-Atlantic slave traders and Western colonialists. It is important to note that current universities, research institutions, and private firms harness indigenous knowledge systems and resources (Nwagwu, 2023, pp. 738-739). A typical example is the activities in traditional medicine and the local ways of manufacturing cloths and tools like machetes and hoes that still exist in many African communities. These local or traditional ways have retained and maintained indigenous knowledge roles in technology. Its content and methodology operate alongside many universities' current technology system counterparts (Nwagwu, 2023, p. 739). The remnant of indigenous knowledge has also led to the transformation of many traditional activities, such as processing and packaging of some health care and food products, into small-scale firms, with their products retaining local content while gaining significant international standards (Nwagwu, 2023). Arunachallam (2007) argued that the activities of multinational pharmaceutical companies send out scouts to the Amazon and the jungles of Africa in search of plants that have been used in traditional pharmacopeia and to learn about their use from indigenous medical practitioners.

For example, modern science and technology are informed by indigenous knowledge that advances indigenous knowledge systems and converts them into powerful digital technologies and methods for modern scientific and technology research inquiries. It is important to note that developing some of the complicated technologies we see today copied the original ideas from indigenous knowledge in Africa. This indicates that today's digital technology advancement owes Africa's indigenous knowledge. This puts Africa at the center of modern mainstream digital technology advancement. Nwagwu (2023) argued that Africa's indigenous knowledge systems are, by and large, visible within highly developed modern digital technology knowledge systems. That means that Africa's contributions to the wave of digital technology are enormous, yet current literature has not recognized such contributions.

Indigenous knowledge is "knowledge upheld by people who "shared the same Indigenous particulars" (Chiwanza et al., 2013, p. 54). Indigenous knowledge includes all forms of knowledge, whether conforming to modern scientific ethos or not, but which have constituted the evidence of the knowledge enterprise of a people. It is a subset of the knowledge space that represents the capability of an individual or region. In this regard, indigenous knowledge should include undeveloped and uncodified indigenous knowledge practices, modern knowledge developed using indigenous methodologies, indigenous knowledge developed using modern methodologies, and modern knowledge developed using modern methodologies (Nwogwu, 2007). The United Nations Development Programme (UNDP) and Third World Network of Scientific Organizations (TWNSO) (1999) have argued that indigenous knowledge played a key and vital role in the development of the digital development we see today. A holistic perspective of indigenous knowledge from a free mind would see the development of local knowledge systems and give one a picture of the role of indigenous knowledge scholarship in the current digital technology.

African indigenous knowledge systems are crucial to modern science and technology, as well as the development of digital technology. Ngulube (2002) noted that indigenous knowledge is mainly tacit and derived from local experiments, innovations, creativity, and experiences embedded in the minds and activities of communities with long histories of close interaction. This knowledge serves as the basis for digital technological advancement. It serves as problem-solving, communication, teaching, and decision-making in the indigenous communities where it is embedded (Gumbo & Msila, 2017). Indigenous knowledge has also been the basis for agriculture, education, health care, and many other activities that sustain a society and its environment in many parts of the world for centuries (Senanayake, 2006). Indigenist thinkers have advocated for recovering and promoting indigenous knowledge systems as necessary in decolonizing indigenous nations and their relationships with governments, whether those strategies are applied to political systems, governance, health and wellness, education, or the environment (Churchill, 1996). The contributions of this knowledge to digital technology need to be acknowledged.

INDIGENOUS KNOWLEDGE CONCEPTUALIZED

Greiner (1998:1) conceptualizes indigenous knowledge as "the unique, traditional, local knowledge existing within and developed around specific conditions of women and men indigenous to a particular geographic area." Warren et al. (1995, p. xv) noted that indigenous knowledge is "the local knowledge that is unique to a given culture or society; it contrasts with the international knowledge system, which is generated through the global network of universities and research institutes." Kiggundu (2007, p. 42) conceptualizes the term indigenous knowledge as "local knowledge that exists because of interactions with the environment by members of a community within a geographical area. Indigenous knowledge covers all fields of human endeavor including, but not limited to, agriculture, environment, pharmacology, health, trade and economic, political systems."

Indigenous knowledge is the traditional, cultural, and community knowledge produced and owned by local people in their specific communities and passed on from one generation to the next generation through practice and oral channels (Govender et al., 2016). More importantly, Ngulube (2002) describes indigenous knowledge as mainly tacit and derived from local experiments, innovations, creativity, and experiences embedded in the minds and activities of communities with long histories of close interaction. This knowledge serves as the basis for problem-solving, communication, teaching, and decision-making in the indigenous communities where it is embedded (Furutnani et al., 2018). Indigenous knowledge has also been the basis for agriculture, education, health care, and many other activities that sustain a society and its environment in many parts of the world for centuries (Senanayake, 2006). Indigenist thinkers have advocated for recovering and promoting indigenous knowledge systems as necessary in decolonizing indigenous nations and their relationships with governments, whether those strategies are applied to political systems, governance, health and wellness, education, or the environment (Churchill, 1996). This knowledge, therefore, needs to be always safeguarded and decolonized for the benefit of indigenous communities (Sithole, 2007).

Indigenous knowledge can best be described in terms of the following: people, place/context, language, knowledge, culture, practices, and dynamism (Shava, 2013). Indigenous people (the knowers) create knowledge from their experiences during interactions with the lived environment and among themselves. This knowledge is derived from and related to the local context in which the indigenous people live and is embedded in their culture and embodied in their practices (Shava, 2013). Indigenous knowledge is passed on from generation to generation orally (through stories, narratives, poetry, and songs), visually (through the arts, such as 'bushmen' paintings or rock art, craftware, traditional rituals, and dance), and practically (through such activities as crop farming, keeping livestock, hunting, traditional healing, making crafts and artifacts). Indigenous language is the primary medium for representing and transmitting indigenous knowledge (Gumbo & Msila, 2017). Indigenous knowledge is created over time about changes in the local environment that bring about changes in the needs of local communities as they adapt to the local context (Gumbo & Msila, 2017). This, therefore, makes indigenous knowledge dynamic in nature.

Indigenous knowledge is the knowledge used by local people to make a living in a particular environment (Warren, 1991). It is linked to their survival in society as it promotes the way local people earn a living. One can say today that indigenous knowledge is vital in the disciplines of sustainable development because it includes indigenous technical knowledge, traditional environmental knowledge, rural knowledge, local knowledge, and farmer's or pastoralist's knowledge (Zinyeka, Onwu, & Braun, 2016; Handayani et al., 2018). In the context of education, the most efficient way of strengthening indigenous knowledge is integrating the knowledge into school science (Aikenhead, 2006; Regmi & Fleming, 2012; Zinyeka, 2013). Incorporating indigenous knowledge means bringing an audible native voice (Acton, Salter, Lenoy, & Stevenson, 2017) to the digital technology world. Indigenous knowledge, thus, can be said to be "a body of knowledge built up by a group of people through generations of living in close contact with nature" (Johnson, 1992, p.23). It is important to note that such knowledge evolves in the local environment so that it is specifically adapted to the requirements of local people and conditions. It is also creative and experimental, constantly incorporating outside influences and innovations to meet new conditions. It is usually a mistake to think of indigenous knowledge as 'old-fashioned,' 'backward,' 'static,' or 'unchanging (Handayani et al., 2018).

From the African perspective, indigenous knowledge can be conceptualized as indigenous systems that are localized in the African systems developed over long periods and whose patterns are based upon local knowledge systems and expressed in local languages (Gumbo, 2020). It would also be conceptualized as an indigenous system that is in balance with the local environment or would have sought such balance (Gumbo, 2020). Moreover, it would be viewed as systems that influence innovations that emerge from within themselves and from other indigenous systems to comply with the dictations of the international systems (Gumbo, 2020). To comply with the international systems means that indigenous knowledge systems are essentially African in origin with African attributes but comply with foreign attributes. This, therefore, implies that indigenous peoples would be those regarded as indigenous on account of their descent from the populations inhabiting the countries at the time of the conquest, colonization, or establishment of present state boundaries (Gumbo, 2020). However, these people have retained their own social, economic, cultural, and political institutions. Such a conceptualization encompasses major groupings and their uninterrupted indigenous skills. A significant criterion of indigenous would reflect African origination (Matowanyika, 1994).

Historically, indigenous knowledge has been labeled as primitive, backward, savage, rural, and unscientific. At the same time, it is the foundation of sustainable development as it informs the nature of education that benefits society (Kiggundu, 2007, p. 49). For this reason, colonialists from the West and mentally colonized non-Western scholars have viewed African indigenous knowledge as primitive and often accused it of lacking universality, a concept ascribed only to Western science (Kiggundu, 2007, p. 49). However, this argument is baseless because historical literature shows that education started in Africa and thus was built on the African indigenous knowledge in the pre-trans-Atlantic slave trade. Brush and Stabinsky (1996) described indigenous knowledge as being culture-specific, whereas formal (Western scientific) knowledge is de-cultured." In the academe and research, indigenous knowledge systems have been dismissed as archaic, old, and symptomatic of backwardness (Knudston and Suzuki, 1992, p. 11). Indigenous people's way of life has in the academia and other research-oriented and scholarly circles been tendered as simplistic, naïve, and even primitive, "reflective of an earlier, and therefore, inferior stage in human cultural progress" and consequently of no relevance to the highly advanced and technologically oriented needs of modern society (Knudston & Suzuki, 1992, p. 1).

CONCEPTUALIZING INDIGENOUS KNOWLEDGE THROUGH ONTOLOGY

The concept of knowledge comes from ancient times (ancient Egypt) and the Greek philosophers (Russel, 1972; Russel, 2009). Socrates noted that human beings are the measure of all things, the hypothesis that led them to the conclusion that "perception, then, is always something that is, and, as being knowledge, it is infallible" (Russel, 1972, p. 149). Perception is the process through which we see the nature around us and as it is. It generates representations of the physical objects in our mind, which is knowledge. Knowledge ontology became helpful in understanding indigenous knowledge management, acquisition, and transfer (Nonaka & Takeuchi,1995). Ontological framework helps guide knowledge managers throughout knowledge management from indigenous knowledge to education (Haron & Hamiz, 2014). It helped to share a common understanding and sensible structures of information, to make domain hypothesis explicit, to provide categorization structure, and to enable the reuse of domain knowledge (Haron & Hamiz, 2014).

The ontology is an example of knowledge modeling, representing knowledge in a manner that a computer can facilitate (Vassev, Hinchey & Gaudin, 2012). Using the knowledge ontology model helped develop a clear path to adopting knowledge management systems to support knowledge management and sharing. As Almeida and Barbosa (2009) pointed out, a declarative approach to ontology is thus needed for knowledge preservation because ontology is a method where the domain is represented in a structured manner and may benefit those who implement it. Knowledge management systematically influences knowledge exchange, application, and creation, hence creating knowledge value (Kozhakhmet & Nazri, 2017). Scholars argued that ontology in knowledge management processes aids in knowledge creation, acquisition, storage, transfer, application, and performance improvement (Sassson & Douglas, 2006; Haron & Hamiz, 2014).

African Indigenous Through Knowledge Creation

Knowledge creation is based on and informed by indigenous knowledge, which is embedded in the indigenous culture in every society. Von Krogh, Nonaka and Rechsteiner (2012) describe the creation of knowledge as an ongoing procedure by which knowledge comes into existence through cooperation or individual effort and is refined and enhanced within a corporate system. Grimsdottir and Edvardsson (2018) stated that new knowledge is frequently created or engendered by innovative concepts or urgent needs arising within the organization or from external market pressures. The knowledge creation process thus evaluates the stages of producing innovative knowledge, such as the application of figurative terms in which to render external knowledge (Grimsdottir & Edvardsson, 2018). Knowledge is also created through education, interaction, practice, and collaboration as the different types of knowledge are shared (Frost, 2010). Nonaka and Takeuchi (1995) noted that knowledge can be created through socialization, guidance, imitation, and observation. One can see that an amalgamation of the above combines codified knowledge sources such as documents that give information to create new knowledge. This implies that knowledge is internalized, thus modifying the user's tacit knowledge.

Maruta (2014) presents a mental model of how an individual acquires and creates knowledge. The comprehension ability depends on the knowledge framework existing already within everyone. "Thus, new knowledge needs to be linked to other knowledge in the repository, possibly becoming part of a hierarchy of knowledge assets, a case of aggregation" (Sherif & Xing, 2006, p. 531). Knowledge is a strategic resource and becomes unique, valuable, and difficult for those who want to imitate it and must have no substitution for knowledge (Johnson, Whittington & Scholes, 2011). Thus, African indigenous knowledge promoted knowledge creation because it became the mother of all civilizations. It became the father of science and technology because its outcomes satisfy technological requirements. The first requirement was to make Africans become innovators and be able to solve problems that surrounded them through innovation (Christensen, 2003). This means they choose between an incremental and disruptive innovation strategy that best fits the company's vision (Bratianu, 2018).

African Indigenous Through Knowledge Acquisition

Knowledge acquisition was crucial in spreading African indigenous knowledge to other parts of the continent and world. Dalkir (2005) defines knowledge acquisition as the stage at which knowledge is contextualized to be understood. It is the process of accepting knowledge from external sources to use it in the organization (Pacharapha & Ractham, 2012). For this reason, key scholars have argued that knowledge is to be constructed in a cooperative modeling effort between the domain expert(s) and knowledge engineer(s) and may not have explicitly 'existed' anywhere before (Clancey, 1993; Gaines, 1993). Acquiring knowledge is achieved by extracting, interpreting, and transferring knowledge to improve and update existing organizational knowledge. Shongwe (2016) stated that an acquisition process thus involves accessing, gathering, locating, and capturing knowledge from suppliers, participants, employers, and other knowledge sources. In this way, African indigenous knowledge was acquired from repositories, learning from others, and learning from experiences.

Moreover, acquiring knowledge is an essential activity for intellectual growth and innovation. It consists of elicitation, collection, analysis, modeling, and validation of knowledge (Tomei, 2009, p. 134). Locke (2001) and Parker (2004) argued that the birth state of the human mind is a blank slate or tabula rasa. Mack and Meadowcroft (2009) argued that people are born without knowledge and that we acquire knowledge only through experiences. Knowledge acquisition could be conceptualized as learning through experiences and experiments (Mathew, 1985). It is about grasping, integrating, adapting, and confirming knowledge for concept formation, clarification, formulating questions, understanding the problem to be solved, or reaching conclusions (Mathew, 1985). This makes digital technology knowledge rooted in African indigenous knowledge as it is the outcome of learning or knowledge acquisition (Al-Khatib, 2011). This is because knowledge evolves and grows. In this process, it went through processes of learning and acquisition about something new (Chisholm, 1982). Psychologists have argued that knowledge acquisition has three stages: behaviorism, cognitivism, and humanism.

Kaba and Chennupati (2020) noted that behaviorists believe the human mind is like a "black box." They did not try to analyze the inner activities of the mind, which include thoughts, feelings, or motivation. According to this school of psychology, knowledge acquisition is a visible change in one's behavior and can be measured (Graham, 2010; Behaviorism, 2011). The proponents of this school noted that a knowledge seeker starts seeking knowledge from a transparent state and responds to environmental incentives (Kaba & Ramaiah, 2020). It is important to note that those responses could be changed through positive and negative reinforcement, depending on the environment. Kaba and Ramaiah (2020) noted that the cognitive school sees the learning process as the active acquisition of new knowledge and the development of adequate mental constructions. Scholars in this school see knowledge seekers as the focus of control in knowledge acquisition and not just passive participants (Kaba & Ramaiah, 2020). Scholars have found that in the knowledge acquisition process, an individual tries to open his mind's "black box" and explain complex cognitive processes and architecture (James, 1890; Rock, 1990). Humanist perspectives on knowledge acquisition suggest that acquiring knowledge is a natural desire, a means of self-actualization, and the development of personal potential. According to them, the importance of acquiring knowledge lies in the process, not the outcome. As a result, learners should have more control over the learning process based on observing and exploring. They believe that, as a teacher, you must be a good example to your students and give those reasons and motivations for the learning process. Proponents of this approach associate the process of learning and seeking knowledge with the need to achieve self-actualization (Kramlinger & Huberty, 1990).

African Indigenous Through Knowledge Storage

Knowledge storage utilizes technical infrastructure such as modern informational hardware and software and human processes to identify the knowledge in an organization and then to code and index the knowledge for later retrieval (Nonaka & Takeuchi, 1995; Armstrong, 2009). Knowledge storage refers to the existence and identification of information in the company's database (Shongwe, 2016). It codifies existing knowledge and know-how into organizational memory (Dalkir, 2005). It is, therefore, necessary to store knowledge in a safe place for future purposes since knowers take their knowledge with them when they withdraw from the organization. However, knowledge may be lost, particularly if it is tacit, improperly managed, or preserved. Shongwe (2016) added that knowledge could be carried digitally or stored manually in minutes of meetings, reports, policies, and many other physical and organizational documents, while it can be stored electronically in organizational databases, portals, and emails.

African Indigenous Through Knowledge Transfer

Gaura, Hongjia and Baoshan (2019) define knowledge transfer as the design and transmission of knowledge within an organization or between organizations to enhance learning and productivity of workers, which is essential in the overall success of the organization. It is the process in which knowledge is shared or communicated to other individuals or groups within the organizations through workshops, seminars, conferences, classrooms, meetings, face-to-face interactions, or using technology (conferencing software, emails) (Likalu et al., 2010). Sagsan (2009) stated that knowledge transfer requires the prerequisites of knowledge-sharing mechanisms that allow the sharing of tacit and explicit knowledge through technological and social communication infrastructure channels. Social communication means informal working settings and it helps in transferring tacit to tacit knowledge while technological communication infrastructure is useful for structuring data and transferring knowledge timely (Sagsan, 2009). However, effective knowledge transfer may not be complete without the use of knowledge. Knowledge transfer thus involves donation and collection of knowledge, and it is possible that donated knowledge that is not put into proper use may not yield any positive benefit (Adeyemi, Uzamot & Modupe, 2022). Knowledge managers thus need to develop strategies for effective knowledge transfer that will aid efficient knowledge use.

African Indigenous Through Knowledge Application

Knowledge application refers to the actual use of knowledge captured and stored in organizational databases or the knowledge in people's heads (Shongwe, 2016). Knowledge application can help to transform knowledge from being a potential power tool into actual innovations or inventions, which can enhance the overall performance of organizations (Matin & Mirraboopa, 2003). Knowledge use is essential to knowledge recipients who employ knowledge for innovation and consequently improve organizational performance (Carvalho & Gomes, 2017). Gottschalk (2007) also emphasizes the necessity of using knowledge in organizational practices, processes, and policies. Knowledge application is when available knowledge is used for decision-making, problem-solving, and performing tasks through direction and routines (Becerra-Fernandez Gonzalez & Sabherwa, 2004). Direction refers to the process through which the individual possessing the knowledge directs the action of another individual without transferring to that individual the knowledge underlying the direction (Becerra-Fernandez et al., 2004). Routines involve utilizing knowledge embedded in procedures, rules, norms, and processes that guide future behavior. Trans-Atlantic slave trade, colonialism, and contemporary coloniality have prevented Africa's indigenous knowledge production, transfer, and application.

CATEGORIES OF AFRICA'S INDIGENOUS KNOWLEDGE

While indigenous knowledge research initially emphasized indigenous technical knowledge of the environment, it is now accepted that indigenous knowledge goes beyond this narrow interpretation (Emery, 1996). Indigenous knowledge is understood to be cultural knowledge and, in its broadest sense, includes all the social, political, economic, and spiritual aspects of a local way of life. Engineering, sustainable development, and economic growth researchers, by and large, argued that indigenous knowledge is made of four categories of particular interest. The first category is resource management knowledge (Emery, 1996). The tools in this category involve techniques, practices, and rules related to pastoralism, agriculture, agroforestry, water management, and wild food gathering (Emery, 1996). The second category is the classification systems for plants, technologies, environment, animals, soils, raw minerals, natural resources, water, and weather. The third category of indigenous knowledge is empirical knowledge about flora, fauna, and inanimate resources and their practical uses (Emery, 1996). The fourth category is the worldview, which is the manner in which the local group or indigenous people perceive indigenous knowledge's relationship to the natural world (Emery, 1996).

While research may focus on a particular category or type of indigenous knowledge, any indigenous knowledge under investigation must be viewed in terms of the overall cultural context. indigenous knowledge is embedded in a dynamic system in which spirituality, kinship, local politics, and other factors are tied together and influence one another. Researchers should be prepared to examine other aspects of a culture that may be essential in shaping indigenous knowledge. For example, religion is an integral part of indigenous knowledge and cannot necessarily be separated from technical forms of knowledge. Spiritual beliefs about nature may influence how resources are managed, and people are willing to adopt new resource management strategies (IIRR, 1996). The research in indigenous knowledge research has covered the following topics:

- *Local organization, controls, and enforcement*: institutions for resource management; standard property management practices; decision-making processes; conflict management practices; traditional laws, rights, taboos, and rituals; and community controls on harvesting.
- *Social networks:* Kinship ties and their effect on power relations, economic strategies, and allocation of resources.
- *Local classification and quantification:* a community`s definitions and classification systems for plants, animals, soils, water, air, and weather; and indigenous methods of counting and quantifying.
- *Learning systems:* indigenous methods of imparting knowledge, indigenous approaches to innovation and experimentation, and indigenous specialists.
- *Pastoral systems:* herd movement; range evaluation and monitoring; animal breeding and production; traditional fodder and forage species and their specific uses; animal diseases and traditional ethnoveterinary medicine.
- *Agriculture:* farming and crop systems; indigenous indicators to determine favourable times to prepare, plant, and harvest gardens; land preparation practices; ways to propagate plants; seed storage and processing; crop planting, harvesting and storage practices; food processing and marketing; and pest management systems and plant protection methods.
- *Agroforestry:* the management of forest plots and trees; the knowledge and use of forest plants and animals; and the interrelationships between trees, crops, herds and soil fertility.
- *Water:* traditional water-management and water conservation systems; traditional techniques for irrigation; and use of specific species for water conservation.
- *Soil:* soil conservation practices; the use of specific species for soil conservation; and soil fertility enhancement practices.
- *Plants:* as a source of wild food, building material, household tools, personal uses (dyes, perfumes, soaps), fuel wood and charcoal, medicinal purposes.
- *Wildlife:* animal behavior, habitats, uses.
- *Worldview:* Views of the universe and humanity`s place, the relationship between humans and nature, myths, beliefs, and customs (Grenier, 1998).

FEATURES OF INDIGENOUS KNOWLEDGE

Indigenous knowledge is the knowledge that is composed of knowledge from previous generations. Indigenous knowledge is a comprehensive knowledge that incorporates technologies and practices used by native people for their continuation, survival, and adaptation to a change of environment (Kasanda et al., 2005; Onwu & Mosimege,

2004). Battiste (2002:2) emphasized the holistic nature of indigenous knowledge and proposed that native knowledge should contain the complex set of technologies improved and sustained by the indigenous community. Ogawa (1995) defines indigenous knowledge as culturally dependent collective intellectual perceiving reality where collective means are to be held in an adequately similar form by people to provide effective communication but independent of a particular mind. All people involved in indigenous knowledge, from moderate to high levels in a community, are experts. They are actors of their knowledge. Indigenous knowledge is manifested in practices and communicated orally and sometimes through copying, illustration, painting, and other artifacts (Zinyeka et al., 2016).

Indigenous knowledge is learned and identified by communities and people within a cultural context (Fields, 1991; Bebbington, 1991). The knowledge base thus uses a specific language, sayings, and belief processes. This allows for cultural interaction and acceptance not identifiable in other situations or cultural contexts (Bebbington, 1991). Rocheleau (1991) argued that indigenous knowledge is the science of survival where indigenous people combine social, economic, and ecological strategies to manage high-risk production patterns (Materer, Valdivia & Gilles, 2002). The pattern of indigenous knowledge is like the collective thinking of a place or region based on natural phenomena that incorporate human and non-human thought integration, such as the scientific knowledge rooted in the local culture (Alessa et al., 2016). Indigenous knowledge relates to the people's cognitive and wise legacy because of their interaction with nature in the joint region (Hart, 2010). The ethnographical background of indigenous knowledge reveals compromised efforts, strengths, and preservation in a day-to-day continuation and generation of experimental living with nature (Chandra, 2014). It provides the foundational teaching and lessons on how people are linked to the rest of their lives.

Indigenous science describes how the local environment runs through a scientific process that includes objective observation of natural phenomena and classifies and solves problems encompassed in all perspectives of native culture (Snively & Corsiglia, 2000). Characteristics of indigenous science combine local development and application, such as testing hypotheses, experiments, and problem-solving related to sociocultural dynamics. The body of knowledge, equivalent to the peer-reviewed study, consists of a continuous living awareness of nature such as agriculture, architecture, mathematics, climatology or climate change, astrology, medicine, plant varieties, etc. (Alessa et al., 2016; Kasanda et al., 2005; Onwu & Mosimege, 2004). Indigenous knowledge is both factual and the existing practice of indigenous people needed for survival and adaptation in diverse conditions through natural and socio-cultural synergies with the environment (Regmi & Fleming, 2012). Although factual and practical knowledge are different, they influence one another.

SIGNIFICANCE INDIGENOUS KNOWLEDGE

There are two basic reasons why researchers need to consider indigenous knowledge when carrying out research projects. First and foremost, incorporating indigenous knowledge into research projects can contribute to local empowerment and development, increasing self-sufficiency and strengthening self-determination (Thrupp, 1989). Utilizing indigenous knowledge in research projects and management plans gives it legitimacy and credibility in the eyes of local people and outside scientists, increasing cultural pride and thus motivation to solve local problems with local ingenuity and resources (Thrupp, 1989). Local capacity-building is a crucial aspect of sustainable development, and researchers and development specialists should design approaches that support and strengthen appropriate indigenous knowledge and institutions.

Second, indigenous people can provide valuable input about the local environment and how to manage its natural resources effectively. Outside interest in indigenous knowledge systems has been fuelled by the recent worldwide ecological crisis and the realization that its causes lie partly in the overexploitation of natural resources based on inappropriate attitudes and technologies. Scientists now recognize that indigenous people have managed the environments in which they have lived for generations, often without significantly damaging local ecologies (Emery, 1996). Many feel that indigenous knowledge can thus provide a robust basis for developing alternative ways of managing resources. indigenous knowledge technologies and know-how have an advantage over introduced forms in that they rely on locally available skills and materials and are thus often more cost-effective than introducing exotic technologies from outside sources (IIRR, 1996). Local people are also familiar with them, so they do not need specialized training (ibid.).

Numerous features of indigenous knowledge are relevant to the advancement and growth of digital technology. The first feature presents indigenous knowledge as locally appropriate (Dewalt, 1994). At this feature, indigenous knowledge represents a way of life that has evolved with the local environment, so it is specifically adapted to the requirements of local conditions, and to enable local citizens to achieve and realize their survivals. The second feature presents indigenous knowledge as a restraint in resource exploitation (Attwood & Markus,1999). At this feature, technology serves as a tool to produce subsistence needed for the local communities, and only what is needed for immediate survival is taken from the environment. The third feature of indigenous knowledge is diversified production systems (Attwood & Markus,1999). This means that at this feature, technology does not promote overexploitation of a single resource or raw material. Risk is often spread by utilizing several subsistence strategies to produce goods that cater for the needs of the people in society.

The fourth feature of indigenous knowledge is the respect for nature. This means that indigenous knowledge in the use of raw material to utilise technology to achieve good and services needed by the people, it is conscious of the impact of pollution on the environment. This means that at this feature, there is a serious conservation ethic for the environment which shows that the protection of the environment and its role in promoting digital technology is recognized. The natural resources, raw materials and land are considered sacred; humans depend on them for survival, and all species are interconnected. The other feature is the fact that indigenous knowledge is flexible and can adapt at any environment (Dewalt, 1994). At this feature, indigenous knowledge can adapt to new conditions and incorporate outside knowledge. It incorporates social responsibility which relies on the strong family and community ties, and with them, feelings of obligation and responsibility to preserve the land for future generations (Dewalt, 1994).

Indigenous Knowledge and Education

In recent years, a requirement for a holistic understanding of the sustainability aspect has regularly appeared, resulting in a need and condition for determining a current perspective of education (Fedosejeva et al., 2018). The increase in awareness of indigenous sustainability is filled by enthusiasm for nature conservation, social justice, and the development of science in classes based on cultural identity. Aikenhead and Ogawa (2007) stated the reasons that stimulated the native research: the scholars would like to develop the containing domain of science, a chance to improve the supremacy and cultural continuation of indigenous people. They would like to raise several educators who are aware of the cultural impact on learners' science achievement. Education is not only strongly tied to formal schooling but also is a comprehensive process. People must learn, observe, remember, and develop a social, moral, and emotional response to their circumstances (Herusatoto, 2012; McInnes, 2017). What happens is pupils who represent different cultures, genders, education, ages, communities, and interests (McInnes, 2017) bring various ideas into the classroom based on their expertise and diverse backgrounds that also frequently represent different science concepts. Their mind is full of experience and knowledge, and their consciousness has evolved through assimilation and accommodation, known as preconceptions, which have not been realized (Mestre & Touger, 1989). Learning should emphasize beginning a lesson based on students' previous knowledge and experience. Everything the learner has experienced, or thought can be a source of scientific knowledge. They should be encouraged to strengthen their cultural identities and be interpreted in a meaningful practice.

In education, the most efficient way of strengthening indigenous knowledge is integrating the knowledge into school science (Meyer & Crawford, 2011; Regmi &

Fleming, 2012; Zinyeka, 2013; Zinyeka, Onwu, & Braun, 2016). It is a challenge for educators and researchers to promote lessons and curricula to stay in synergy with the times' demands without abandoning local values. Teachers must recognize that indigenous knowledge and classroom scientific knowledge can synchronize and support each other (Regmi & Fleming, 2012). The addition of indigenous knowledge in the curricula is an essential element of contemporary science education. Indigenizing curriculum refers to integrating indigenous knowledge into the school science curriculum (Moyo & Kizito, 2014), incorporating an audible native voice (Acton, Salter, Lenoy & Stevenson, 2017). The introduction of indigenous knowledge into science school might present science more appropriate for a student in a culturally different classroom (De Beer & Whitlock, 2009), could have a definite impact on students' enthusiasm for science (Kasanda et al., 2005), and help them appreciate as well as maintain alive the native knowledge (Ngíasike, 2011). Indigenous knowledge affords the conceptual improvement of pupils in science school since this cultural knowledge frequently supports the formal curriculum idea adequately (Cronje, de Beer & Ankiewicz, 2015).

Science is a way of knowing about the phenomena that occur in nature by using methods and ways of systematic thinking (Howe, 2002). It is an objective perception of reality. Science refers to conceptual constructs established by rational empiricism behind observation and experiment (Snively & Corsiglia, 2000). Chiappetta and Koballa (2009) are convinced that science is a systematic effort to create, construct, and organize knowledge of natural phenomena that begins from the human nature of curiosity, which is then followed up by an inquiry to investigate the most straightforward and consistent explanations and predictions of the phenomena. Science is regarded with testable phenomena and studies the universe as knowable (Ogunniyi, 2011). Science is essential and includes:

- a process: an inquiry procedure that includes real phenomena.
- products: facts, concepts, principles, laws, and theories that interpret and predict the phenomena.
- attitude: curiosity about nature that is studied through persistence, honesty, and openness to new opportunities (Lawson, 1995).

Science consists of facts, laws, and theories and involves human activities, such as investigations, processes, attitudes, and beliefs. Science in the domain of knowledge must be factual, conceptual, procedural, and metacognitive (Krathwohl, 2002; Suwarto, 2010). Factual knowledge is an essential component that pupils must recognize in the discipline. The core elements are given following academic science, which is easy to understand and systematically arranged. Conceptual knowledge connects, associates, and combines essential elements in a systematic and shared common structure. This knowledge can be a scheme, a mental model, or explicit and implicit theories in different cognitive psychology. Procedural knowledge emphasizes how to make discoveries and sets criteria for using skills, algorithms, techniques, and methods collectively. Metacognition implies awareness and responsibility for one's learning and thinking. Sandoval and Reiser (2004) explain that the epistemology of science comprises not only sources of scientific knowledge and the value utilized to sustain that knowledge but also the way of knowing handled by the community to accept scientific knowledge.

Contemporary African indigenous education involves the inculcation of traditional societal norms and values in the young generations through a systemic socialization process that involves the training of the mind, the body, and the soul (Sifuna, 2008). It emphasizes moral values and societal responsibilities among indigenes and, as such, aligns with worldwide educational agendas associated with 21st century skills and social, emotional, and ethical learning.

INDIGENOUS KNOWLEDGE AND DIGITAL TECHNOLOGY NEXUS

There is a huge nexus between indigenous knowledge and digital technology. Technology is "the use of knowledge, skills, values, and resources to meet people's needs and wants by developing practical solutions to problems, taking social and

Figure 1. The interaction between indigenous knowledge and education

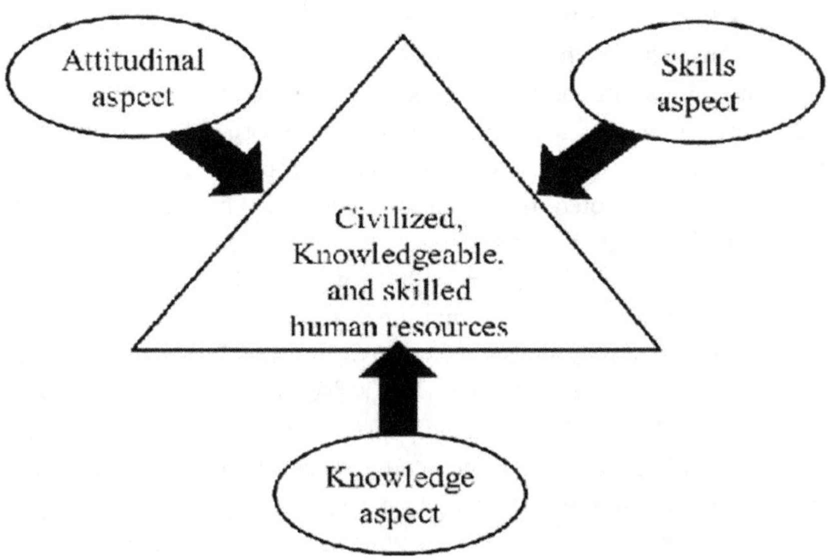

environmental factors into consideration" (Department of Basic Education, 2011, p. 8). As stated in the introduction, Department of Basic Education (2011, p. 8) defines technology as "the use of knowledge, skills, values and resources to meet people's needs and wants by developing practical solutions to problems, taking social and environmental factors into consideration." To encourage the promotion of indigenous technology, this definition should have been expanded to include the cultural perspective of technology since indigeneity raises issues of culture—for example, the marginalization of certain cultures through hegemonic aggrandizement. The cultural view of technology is that technology is any human-made or culture-generated devices, formulations, or organizations utilizable for the purpose of producing or creating needed goods and services (Obikeze, 2011). This means that any technology, whether digital or manual, is based on indigenous knowledge since this knowledge is housed by culture and is human-made knowledge.

The concepts of culture and design embedded in indigenous knowledge complement each other, and one is inconceivable without the other (Moalosi, Popovic, Hudson & Kumar, 2005; Ogungbure, 2011). It is vital to note that culture is a phenomenon that undergirds all material and non-material expressions of a people (Ogungbure, 2011). These expressions are tangible or intangible devices, formulations, and techniques that fulfill some need or provide some service for humankind in a particular environment (Obikeze, 2011). As Ogungbure (2011) argued, culture within indigenous knowledge can be viewed as the pattern of behavior that enables people to live in social groups and learn, create, and share. The argument that indigenous knowledge creates connotes technological activities. Cluster (1995, p. 223) thus argued that indigenous knowledge technological artifacts, married to culture, are wonderful and diverse cultural expressions. In this sense, artifacts within the indigenous knowledge become creations that represent, inspire, and define a given culture, thus inspiring further advancement in technological tools. It means the need to free our minds from only confining the understanding of technology as form being a distinguishing criterion to how the values, priorities, and needs of various cultures take form through the creative energy of their people (Custer, 1995, p. 224).

As Williams (2000) noted, this linear approach does not always represent how technology is applied. Indigenous learners "invent a process as they proceed toward task completion" (Williams, 2000, p. 49). Rodil et al. (2011) alluded to the knowledge that has sustained mainly rural communities in Southern Africa for generations, evidenced in animal husbandry, plant cultivation and harvest, processing, and conservation of local resources. Common ways of sharing such knowledge, according to Ngara (2007), Rodil et al. (2011), and Owusu-Ansah and Mji (2013), include oral participation in ordinary activities and rituals, some of which include storytelling, proverbs, folktales, recitation, demonstration, sport, epic, poetry, reasoning, riddles, praise songs, word games, puzzles, tongue twisters, dance, music. However, Rodil et al. (2011) argued further that tempering indigenous knowledge systems by various changes, which range from education to transport, threatens indigenous knowledge systems with extinction as elders pass on these systems without passing them on to the young. In digital technology education, this threat is perpetuated by the lackadaisical attitude informed by Western orientations to deny indigenous knowledge perspectives a space. This problem is exacerbated by the modern technology and lifestyles that the youth adopt, especially in cities, and from the education that they receive, which contrasts with indigenous environments (Rodil et al., 2011). This can be seen, for example, in written literacy versus land literacy and cell phones versus local social communication protocol systems.

The predominantly Western technologies, which are orientated and tailored for urban life, steal away the values, logic, and literacies which are the bedrock of indigenous knowledge and communities (Rodil et al., 2011). This, therefore, prevents Africa's contributions to the digital technology world. Understanding the effects of colonialism and coloniality on African indigenous knowledge and the role of indigenous knowledge in digital technology education are two of the aims of this chapter. It means that it is necessary to understand African indigenous knowledge promoted African innovators, led to the building of sophisticated pyramids in Ancient Egypt which modern architecture cannot phantom. However, with the event of trans-Atlantic slave trade and the advent of colonialism which colonial education, African indigenous knowledge and African innovators died away and was buried. The search for African indigenous knowledge would promote digital technologies that better align and reconcile the contributions of Africa to the world of digital technology.

The starting point is to redefine digital technology to include the indigenous perspective in education. It is important to note that this redefinition must not formulated through the lenses of colonial or Western educational perspective. To perpetuate this practice promotes the dominance of perceiving technology from a Western perspective, is biased, denies indigenous contexts space in offering the alternative, conditions teachers and learners in a cocoon of inferiority and unrelatedness of digital technology to their contexts (Heptinstall, 1998: 4). From a western perspective, digital technology denotes a human activity that can be traced to time immemorial, such as wheel and axle; melting of metals, the boat (De Vries, 1996; Heptinstall, 1998; Laufenberg, 2009). From Indigenous knowledge, digital technology implies the way in which people modify the natural world to achieve services and goods that suit their purpose and survival. Gumbo (2003) draws from the work of Williams (1996) and maintains intact the definition of digital technology that associates technological tools with Western education. This understanding omits African indigenous knowledge and its contributions as if Western indigenous knowledge is the basis for all aspects of technological education (Department of Basic Education, 2011). Technology has failed to capture the contributions of

African indigenous knowledge in its definition and conceptualization. Moreover, the curriculum inherited from the Western education system provides minimal recognition of African indigenous technology as part of Africa's contributions to the wave of digital technology. This signifies that technology is colonized and needs to be decolonized to welcome African indigenous knowledge contributions to digital technology.

To build on the indigenous perspective of technology from the perspective of indigenous knowledge and education, technology can be defined as covering complex fields in which technological designs and devices are self-expressive (Hoppers, 2001). This can be understood from woven baskets and handicrafts to the technologies such as looms, textile, jewelry, and brasswork manufacture, and technological knowledge in agriculture, fishing, forestry, resource exploitation, atmospheric management techniques, knowledge transmission systems, architecture, medicine, and pharmacy (Hoppers, 2001). Technology forms a body of knowledge developed by a culture that provides methods or means to control the environment, extract resources, produce goods and services, and improve the quality of life (Cheek, 1992, p. 32), thus covering the crucial aspect of sustainable development. This means that any definition of technology needs to situate digital technology within cultural lenses. Ogungbure (2011, p. 87) argued that "every technology within a social praxis is a product of culture." This means that technology and culture inform one's reasoning that undergirds a population's material and non-material expressions (Ogungbure, 2011, p. 87).

Obikeze (2011) argued that these expressions come in tangible or intangible devices, formulations, and techniques that fulfill some need or provide some service for humankind in a particular environment. For this reason, this chapter provides two categories of these technologies: material and social. While material technology covers engineering fields and means using raw materials to fabricate steels, social technology covers other fields in social science (Etzemüller, 2009; Knobloch, 2006). Both material and social technology means using social techniques, social pedagogy, administrative technique, technocracy, socio-technique, political science engineering, planned society, efficiency engineering, and social (economic) planning to achieve common goals (Leibetseder, 2011).

Standard in all the terms is the focus on the social, which should be dealt with and be acted upon. Social technique describes sociological knowledge (theories, methods, or expertise) in solving practical problems. In other words, the knowledge of sociology is used to provide applied expertise for specific decisions or actions. Social technology applies methods and theories to obtain a scientific-based analysis for a purpose, which is then used for political decisions (Leibetseder, 2011). Technique refers to applying specific evidence in everyday life, unlike technology, which is described as a system of evidence (Büschges, 2002). Materials technology, also known as materials science, examines the structure and properties of a material, how it is made, and how it can be used. For example, materials are the umbrella term for valuable products like metals, plastics, and ceramics, which typically have different properties. Social technology serves the purpose of controlling society for a better world.

This means that technology and culture go hand in hand and can be organized in terms of goods and services (material goods, social goods/services, and intellectual goods). This cultural perspective of technology enables one to understand and define technology as any human-made or culture-generated device. Technology captures those formulations or organizations utilizable to produce needed goods and services. Thus, African Indigenous knowledge has a role to play and has contributed hugely as it has provided for the needs of precolonial Africans.

The cultural perspective of technology, then, portrays culture as being characterized as the totality of the way of life evolved by people in their attempt to meet the challenges of living in their environment (Ogungbure, 2011.p. 88). It evolved from the people's initiative to hold on instinct to question the unknown to find a rational explanation for the way things are in the universe (Ogungbure, 2011, p. 88). This means that it evolves from the effect to seek understanding of the environment, harness the resources therein, make use of the earth's possessions that have been the principal force in the development of humankind and the society (Ogungbure, 2011, p. 88). Ogungbure (2011, pp.89-96) takes the definition further by looking at the intellectual and educational of technology:

Intellectually, technology harnesses nature in its entirety for humankind's development and sustenance as man constantly thinks about the best way to do things to suit his needs while drawing inspiration from the cultural mindset within a social praxis. Educationally, the "-ology" in the term technology relates to the systematic study and development of techniques for making and doing things, i.e., machines (products and byproducts), methods, and processes that involve knowledge, skills, and resources in the environment inhabited by humankind. Man, systematically applies his indigenous technological knowledge to produce goods and provide services to achieve perceived socioeconomic systems.

Furthermore, indigenous technology from the education perspective means the use of indigenous methods and techniques, which are locally available resources, to produce essential goods for purposes of sale or consumption by predominantly designing and making devices that help facilitate the production processes (Amara, 1987).

Chahine and Kinuthia (2013, p.10) list examples from Amara processing and preservation of foodstuff, production of chemicals such as salt, metal work, charcoal production, preparation of herbal medicines, production of local brews, trap making, arts and crafts. Chahine and Kinuthia (2013, p.11) also understand indigenous technologies as "the embodiments comprising information codes that reflect potential sources of implicit and explicit narratives and modes of strategic thought most valued by culture and which are communally accepted as emulations of the people's cultural heritage and social identity." Technology influences values and practices and becomes part of the culture (Custer, 1995; Chahine & Kinuthia, 2011).

Understood this way, technology brings a different perspective into understanding what it is. From an indigenous perspective, one needs to incorporate culture to it, add commonality to it, think about tangible and non-tangible dimensions, and relate it to economic growth (Handayani et al., 2018). Technology's cultural aspect is crucial for consideration, particularly in the teaching and learning situation (Handayani et al., 2018). If all that learners are asked to think about, talk about, see, and do in digital technology education is a portrayal of Western knowledge, which denies African indigenous inventors space to showcase what Africa has contributed to a wave of digital technology. This is because wherever indigenous knowledge exists, numerous forms of technologies are at work, and innovative thinking is producing services and goods for the population. The treatment of the definition of technology here does not suggest that teachers and learners should reject completely Western conceptualizations of technology (Handayani et al., 2018). This means that African indigenous knowledge and Western knowledge have many similarities that can be drawn. These similarities include the use of resources and raw materials needed for the technology to achieve goods and services. Resources and raw materials, the ingredients of technology, show, from a cultural perspective, that indigenous knowledge is needed when using technology to achieve goods and services for the population.

The aspect of context that could bring some difference in defining technology cannot be ruled out. Technology is the technical application of tangible and intangible aspects of human cultural products to create favorable life conditions in the natural and unnatural (human) world while ensuring sustainable economic growth (Handayani et al., 2018). Sustainable and economic growth allows a space to include indigenous knowledge, perspectives, and cultures. Sustainability is about nature (earth, biodiversity, ecosystems), life support (ecosystem services, resources, environment), and community (cultures, groups, places) (Handayani et al., 2018). Economic growth enables citizens to achieve sustainable livelihoods that boost their survival, life expectancy, and quality education. it also means that citizens can achieve a sustainable economic income that enables them to be productive in society. The diagram below emphasizes the above assertions and enables an understanding of the nexus between indigenous knowledge and education.

Elaborating Indigenous Knowledge in Digital Technology

Figure 2.
Sources: Handayani et al., 2018.

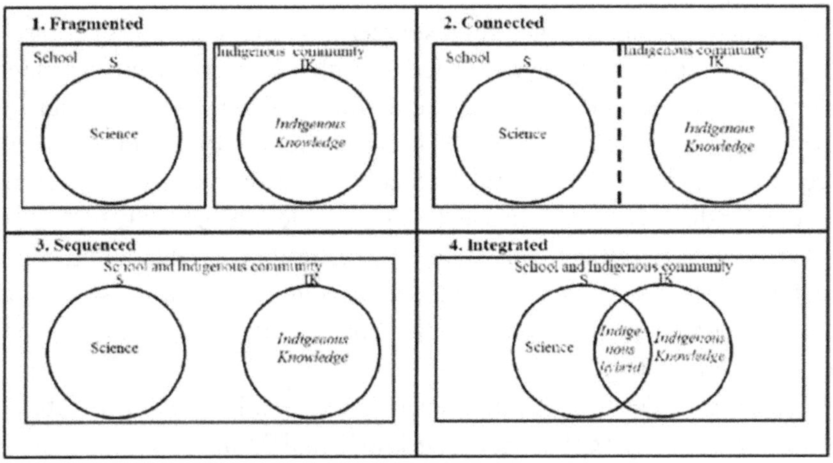

EFFECTS OF COLONIALISM ON INDIGENOUS KNOWLEDGE

African indigenous knowledge has suffered for decades from several strategies of disinformation embedded in Western-centric, colonial, and post-colonial education and Western religion, science, and technology. Western-centric technology that inhibits the growth of digital technology has affected African technological inventors and innovators, thus hampering the growth of non-European achievements, inventions, and technologies in academic tasks in Africa. Often, data on indigenous knowledge is distorted to confirm the hypothesis of non-Africanist scholars (Eyong, 2007, p. 131). The greatest of these challenges, however, was the incursion of the Europeans in the contextual development of African indigenous knowledge and practices. A significant challenge here is that African indigenous knowledge was developed and contextualized by non-Africans and outside the spheres and shores of Africa for Africans (Adigbuo, 2017).

Many individuals exclusively blame colonialism for the perceived destruction of African Indigenous education and ways of knowing (Shizha, 2016; Emeagwali & Shizha, 2016; Diop, 2012). Colonialism in Africa was an event full of brutal subjugation and humiliation of indigenous peoples. Colonialism has left the continent devastated and is still held up in the status underdeveloped, poverty and bad governance issues which are the legacies that have prevailed any treatments (Blakemore, 2019). Blakemore (2019) defines colonialism as control by one power over a dependent area or people, and it occurs when one nation subjugates another, conquering its population and exploiting it, often while forcing its own language and cultural values upon its people. Colonialism colonized and exploited Africa's rich in oil, copper, diamonds, gold, and many other resources upon which technology development is based.

Africa's raw materials and natural resources have made European nations blind and cruel to the African people and started to exploit them in the most violent ways possible (Blakemore, 2019). European nations claimed they were in Africa to boost local livelihood, whereas they were in Africa for monetary colonization and shipping resources for their troubled economy. Numerous African countries were compelled to import oil for their own use because their economies are narrowly tied to exports, and higher oil prices hit them. The forces of cultural genocide, colonization, and colonial policy perpetuated over the last several centuries by successive occupying settler governments are responsible for the current state of indigenous knowledge systems (Simpson, 2004).

Colonialists regarded indigenous knowledge as primitive, uncivilized, backward, superstitious, and savage (Briggs & Sharp, 2004), and this knowledge was viewed as irrelevant to development and an obstacle rather than a force of change (Blakemore, 2019). Sillitoe (2006) noted that the premise used to confirm indigenous knowledge's insignificance was drawn from modernity and dependency models. Hobart (2002) argued that the dependency theory is anchored on exploiting the resources and labor of the local people, creating inequalities among people. The prevalence of colonial legacies in Africa has led to coloniality, which is much uglier and deadlier than colonialism. Vázquez (2011) argued that coloniality is the darker side of modernity, and this is because former colonialists have used democracy, globalization, and liberalization as their tools to keep exploiting Africa's raw mineral and natural resources to enrich their digital technological advancements. The effect of colonialism on indigenous knowledge has included environmental degradation, economic instability, ethnic rivalries and human rights violations issues that can outlast one group's colonial rule (Blakemore, 2019).

Colonialism is, therefore, the current threat to indigenous knowledge, and it continues advancing the oppression of the world's indigenous nations. Morgan (2003) argued that as indigenous knowledge is still suppressed, the Western system also allowed the appropriation and exploitation of this knowledge for the benefit of the colonizers. Briggs (2005) observed that integrating indigenous knowledge into the education system dominated by the Western worldview occupies a lesser position than the other. This is because indigenous knowledge must meet the standards set by science to be accepted, yet Western systems control this standard. The education system continues to favor Western knowledge over indigenous knowledge (Waziyatawin, 2004). Morgan (2003) concurs that Western worldviews are still dominant in higher education despite various groups' efforts to push for recognizing indigenous knowledge in such institutions. Morgan (2003) thus emphasized that those once colonized should initiate decolonization, a process that allows the revaluing and recovering of the lost. This means that for Africa's contributions to the wave of digital technology to be seen and recognized, the need to decolonize education and incorporate African indigenous knowledge would play a vital advantage.

Through coloniality, the colonial powers are still using brutal policies and devious methods to subjugate the African indigenous knowledge to inhibit any contributions of the African continent to the world of technology, and yet what is making digital technology to fly are the raw materials and natural resources taken out from Africa. The policies and methods to prevent such contributions have been inferiorizing African indigenous knowledge and cultures and concerted efforts to erase existing systems of knowledge and their replacement with Western-driven belief and knowledge systems. These policies and methods have successfully culminated in the absolute submission of the communities and stigmatisation of their knowledge systems. The repercussion of this have been weakening African indigenous knowledge and the production of mentally enslaved scholars and political leaders who would perpetuate their own subjugation (Progler,1999: 1). The tools for this mental enslavement have been Western education, Christianisation, and degeneration of relatively self-sufficient economies into dependent consumers (Eyong, 2007: 131). Moreover, the colonial design succeeded to produce the economic imbalances essential for the growth and domination of European capitalism and imperialism (Eyong, 2007, p. 131).

It has been noted that African indigenous knowledge is almost buried to death and does not exist in Africa's post-colonial states (Progler, 1999, p. 1). Zeleza (2006:196) argued that post-colonial African systems of production and dissemination of academic knowledge reflect the Western hegemony. Zeleza (2006, p.197) further argued that the provisions of intellectual exchange are distinctly unequal. African Studies, including African indigenous knowledge, constitute a marginal part of the academy, while European epistemology remains central in African Studies. Therefore, the Western domination of knowledge and marginalization of African knowledge systems continue to be an academic challenge that calls for a comprehensive evaluation, rigorous planning, and watchful implementation of policies. This would ensure the recognition and provision of space for the locals in the existing political, economic, cultural, and pedagogical domains. Only then would African indigenous knowledge be established and gradually contextualized within these domains, thus overcoming all forces that inhibit Africa's contributions to the wave of digital technology.

Boahen (1980) identifies African elites and their role in the diffusion of Western lifestyles as threatening African indigenous knowledge and ways of knowing. Coe (2020) blamed African elites for stripping African indigenous knowledge and practices of their deep and authentic meanings. One can see that colonialism has rendered traditional practices and indigenous knowledge irrelevant in Africa. The most potent legacy of the colonial system is the African elites, who continue to reproduce colonial attitudes and oppressive systems (Dalleo, 2012; Rickford, 2016). Fanon (1967) makes a similar argument, noting that an inferiority complex took root among the colonized, who assimilated metropolitan culture and rejected their Blackness in their quest to be seen as equals to the Whites. Fanon (1967) argues that this tendency is a legitimate coping mechanism in response to the violence to which Black people have been subjected.

Coloniality power has made technology education in Africa void of the indigenous technological perspective, thus exacerbating by the global western knowledge (mostly referred to as science) and technological developments at the expense of other knowledge forms such as indigenous knowledge. O'Riley (2001: 5) argued that modern science and technology has dislocated African indigenous knowledge, destroyed traditional cultures, and played havoc with the environment in Africa. It has also replaced a way of knowing, which is multi-dimensional and based on synthesis, in African societies with a linear, clinical, inhuman, and rationalist mode of thought (O'Riley, 2001: 5). Western science and technology have systematically plundered African countries in the name of scientific rationality. O'Riley (2001), the West portrays the technologies that fall outside the mechanical model of reality and technologies associated with non-Western cultures as antiquated. There is also even alienation from the knowledge itself. This further influences the moral, social, cultural, and natural crises that have caused the humanitarian crisis (Herusatoto, 2012; McInnes, 2017).

EFFECTS OF COLONIAL EDUCATION ON INDIGENOUS KNOWLEDGE

Since the end of colonialism, Western knowledge has been driving digital technology education in African curriculum developments, showing that Africa's primitivity is to be addressed by Western curriculum. It is necessary to recognise the role of African indigenous knowledge and include its principles, such as indigenous knowledge systems, in the curriculum of education. For example, however, in South Africa, the technology education curriculum has attempted to include indigenous technology so pupils can be introduced to indigenous knowledge (Department of Basic Education, 2011, pp. 7-8). Learners should be aware of different coexisting knowledge systems wherever possible, learn factors that make indigenous cultures use specific materials and processes to satisfy needs, and become aware of indigenous intellectual property rights.

However, "where possible" shows that indigenous knowledge has not been taken seriously. The statement does not convey a commitment to integrating indigenous technology and making it compulsory. It shows that students can either go for it or not. This means that Western educational systems are still in the people's psyche, making it a big challenge to overcome. This shows how deep colonialism was and is. Moreover, how colonial education based on Western education systems is a tool for coloniality. Barnhardt and Kawagley (1996, pp.1-2) and Barnhardt and Kawagley (1999) argued that as indigenous people globally have begun reasserting their indigenous rights to self-determination, they have faced many challenges. Given the long history of sidelining indigenous knowledge in Africa, one can see that the statement makes indigenous technology more vulnerable to exclusion. It can be argued making learners aware and teaching them about indigenous technology should enable learners to understand indigenous knowledge systems and its role in technological advancement.

The lack of commitment to integrate indigenous technologies can be observed in teacher development workshops and teacher practice, which do not incorporate indigenous perspectives. Gumbo and Williams (2014) showed that African teachers are ignorant of the role of indigenous knowledge and about integrating indigenous technologies in digital technology. Gumbo and Williams (2014) further noted that teachers' pedagogical content knowledge is Western-oriented and does not reflect African indigenous knowledge. They found that indigenous perspectives are almost absent in both contexts in the learning and teaching materials and the planned teaching and learning activities in Africa (Gumbo & Williams, 2014). The teachers assumed they were integrating indigenous technologies when it was evident in their practice, and from the records, they were integrating and spreading Western-centric education. This strengthens the case for integrating indigenous technologies at the practice level and more studies that focus on incorporating indigenous knowledge systems (Gumbo, 2015). The void in the indigenous perspective in technology education is attributed mainly to the current dominant Western knowledge (Williams, 1996a; Williams, 1996b).

There has been less, if no, commitment whatsoever to let indigenous people participate in the economic and technological development of the country (Rakometsi, 2008), let alone even recognizing indigenous technologies. Coloniality forces are still sidelining indigenous people educationally through separate development. Western knowledge, which is more visible through science as stated above, promotes a scientific approach towards an inquiry into a phenomenon, which manifests through the definition of the problem, a gathering of information, formulation of a hypothesis, observation, testing of hypotheses, and drawing of conclusions (Williams, 2000, p. 49). This scientific approach has essentially cloned itself in digital technology and has refused to accommodate and recognize the contributions of African indigenous knowledge.

The term science is a Western term that perpetrates Eurocentrism (Herman, Vizina, Augustus & Sawyer, 2008). When talking about science from an indigenous point of view, some authors prefer to use indigenous knowledge to indigenous science due to the inseparability of indigenous knowledge from culture (Aikenhead, 2006a). Aikenhead (2006b: 3) denoted that "science is a rational, empirically based description explanation of nature." His understanding of science embraces both the Western and indigenous points of view surrounding science (Aikenhead, 2006b: 3). Herman et al. (2008:35), and Owusu-Ansah and Mji (2013) argued that there are huge differences between Western science and indigenous science.

Thus, the reconceptualization of science needs to include indigenous perspectives. However, the focus of this chapter is not on science as such. Hence, the author did not have to go to the extent of contributing the definition.

The above table and diagram show the difference between indigenous knowledge systems and Western knowledge in Africa. From a digital technology perspective, indigenous knowledge is more physical and metaphysical with practical and experimental skills than Western knowledge by assuring the integrity of locally situated cultural knowledge and skills and by critiquing the learning processes by which such knowledge is acquired and utilized.

Table 1. Differences between Western science and Indigenous science

Western Science Concepts	Indigenous Science Concepts
Science is a subset of Euro-western culture	Physical and spiritual are inseparable
People are separate from the world around them	Holistic, communal and contextual
What counts as science is what can be measured	Inclusive, cooperative, peaceful
Focus is on physical world; absence of the sacred	Focuses on natural laws and land
Scientific method is a compartmentalized knowledge	Respect and reverence for all life
Fragmented worldview, linear, hierarchical	Reciprocity and interdependence
Mechanistic, reductionist, rational	Balance and wholeness; inner and outer harmony
Impersonal mathematically idealized, decontextualized	Nourishing the learning spirit
Exploitive, materialistic, ideological, elitist	Sustainable ways of being

Source: (Gumbo, 2020, p. 7).

Figure 3. The 'ethical space' between Indigenous and Western knowledge
Adapted from Barnhardt and Kawagley (2005)

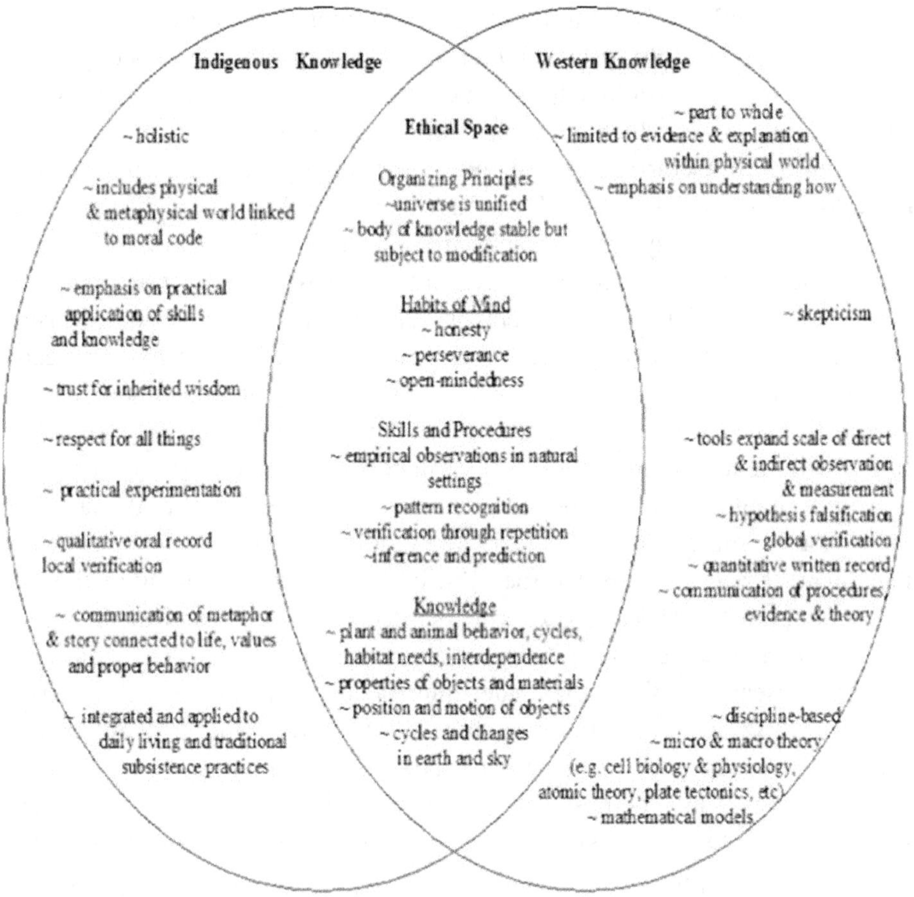

DECOLONIZING INDIGENOUS KNOWLEDGE

Decolonization is recovering from colonial impact and restoring indigenous people's identities, languages, and experiences (Datta, 2018). This means indigenous communities can disentangle themselves from the oppressive control of colonizing state government through policy and decolonized strategies (Simpson, 2004). Denzin et al. (2008) further described decolonization as a continuous anti-colonial struggle that honors indigenous approaches to technology and recognizing indigenous sovereignty and autonomy in science and technology. Decolonial thinking has been used to critique the colonial-modern function of assumptions and knowledge forms deeply embedded in the discipline and the broader field of Western social science (Seth, 2013). Smith (1999) defines decolonization as a process for conducting research with indigenous communities that places indigenous voices and epistemologies in the center of the research process. For example, interest in research on African indigenous vegetables by Agriculture Research Council (ARC) has been regarded as decolonization process. Smith (1999) highlighted that the process of decolonization of indigenous research would help regain control over indigenous ways of knowing and being, and ways in which research can be used for social justice.

Colonialism and coloniality had led to the subjugation and suppression of indigenous ways of life (Heleta, 2018), and this was because the colonialists considered indigenous knowledge system as backward and not worthy of any technological development. There has also been a general disregard of indigenous knowledge system amongst academics and scientists, and as a result, the value of primary knowledge was strategically rejected among academics (Mji, 2019). Even indigenous researchers who were aware of the benefits and superiority of indigenous techniques remarked that they were afraid to admit an interest on this sphere of knowledge, basing this on fear of being ridiculed by Western peers (Mji, 2019). Researchers thus requires contextualized research processes that are relevant to the challenges of indigenous communities and contribute to their development, using acceptable indigenous research methods and theories. Indigenous decolonizing methods and theories suitable to the African contexts include decolonizing research methodologies.

Decolonizing research methodology is an approach that is used to challenge the Eurocentric research methods that undermine the local knowledge and experiences of marginalized population groups (Smith, 1999). The Ubuntu concept expresses the African philosophy of humanness, which states that a human being is a human through the other human beings. This shows the culture of communalism of having things in common, that people should live in common recognizing their commonality. The Afrocentric is a philosophical and theoretical perspective that suggests cultural and social immersion as the best approach to understanding African phenomena instead of scientific distance (Mkabela, 2005). Furthermore, Goduka et al. (2013) argued that for research to be relevant and improve the quality of life of indigenous people, it needs to be rooted in indigenous worldviews, cultural values, and languages.

CONCLUSION

Indigenous knowledge, including intangible heritage, can contribute immensely to the wave of digital technology if value and a modicum of respect are accorded to them. The chapter presented that the relevancy of indigenous knowledge and intangible heritage to the contribution of digital technology can neither be overlooked nor underestimated. Eurocentric discourse has greatly influenced research, teaching and learning of social sciences in higher education in Africa: the principal focus of developments in social thought continues to originate from the work of American, British, French and German scholars. The implication of this is that higher educational institutions in Africa have reduced themselves to the reproduction of the intellectual outputs of social thinkers of western countries, including their theories and methodologies of selection of research problem priorities. As a result of this, there is little attention given to African indigenous literary and philosophical traditions because they tend to be viewed as primitive and unscientific and hence not proper sources for social theory and research development. This is accompanied by the inability of African social scientists to generate their own indigenous concepts, definitions, theories and methods which could guide the intellectual development in their research and academic fields. African indigenous learning can, therefore, be viewed as the core foundation for education self-reliance in modern education (Hamilton-Ekeke & Dorgu, 2015). Chawinga and Zozie (2016:4) This explains the relevance of technology-enhanced teaching/learning, which ODL represents in the re-expression of AIK, particularly in Nigeria. According to UNESCO (2002:8), the term "open and distance learning reflects both the fact that all or most of the teaching is conducted by someone removed in time and space from the learner and that the mission aims to include greater dimensions of openness and flexibility, whether in terms of access, curriculum, or other elements of structure." The need for Indigenous knowledge re-creation is crucial in creating Indigenous technology that will serve the local people.

REFERENCES

Acton, R., Salter, P., Lenoy, M., & Stevenson, R. (2017). Conversations on cultural sustainability: Stimuli for embedding indigenous knowledges and ways of being into curriculum. *Higher Education Research & Development, 36*(7), 1311-1325. . doi:10.1080/07294360.2017.1325852

Adeyemi, I. O., Uzamot, W. O., & Modupe, F. (2022). Knowledge Transfer and Use as Predictors of Law Firm Performance: Nigerian Lawyer's Perspectives. *International Journal of Knowledge Management, 18*(1), 1–17. doi:10.4018/IJKM.291097

Adigbuo, E. R. (2017). African Indigenous Knowledge. *The Challenges of Industrialization International Journal of Humanities Social Sciences and Education, 4*(5), 76–84.

Aikenhead, G. (2006a). *Science education for everyday life: Evidence-based practice*. The Althouse Press.

Aikenhead, G. (2006b). Towards decolonizing the Pan-Canadian Science Framework. *Canadian Journal of Science, Mathematics, &. Technology Education, 6*(4), 387–399.

Aikenhead, G. S., & Ogawa, M. (2007). Indigenous knowledge and science revisited. *Cultural Studies of Science Education, 2*(3), 539n620. https://doi.org// doi:10.1007

Al-Khatib, S. A. (2011). *General psychology*. Bright Horizon.

Alessa, L., Kliskey, A., Gamble, J., Fidel, M., Beaujean, G., & Gosz, J. (2016). The role of Indigenous science and local knowledge in integrated observing systems: moving toward adaptive capacity indices and early warning systems. *Sustainability Science, 11*(1), 91-102. doi:10.1007/s11625-015-0295-7

Almeida, M. B., & Barbosa, R. R. (2009). Ontologies in Knowledge Management Support: A Case Study. *Journal of the American Society for Information Science and Technology, 60*(10), 2032–2047. doi:10.1002/asi.21120

Amara, J. (1987). Indigenous technology of Sierra Leone and the science education of girls. *International Journal of Science Education, 9*(3), 317–324. doi:10.1080/0950069870090308

Armstrong, M. (2009). *Handbook of Human Resource Management Practice* (11th ed.). Kogan Page Ltd.

Attwood, B., & Markus, A. (1999). *The struggle for aboriginal rights: A documentary history*. Allen & Unwin.

Bah & Scott. (2022). *Evaluation of the UNESCO's Local and Indigenous Knowledge Systems (LINKS) programme*. Academic Press.

Barnhardt, R., & Kawagley, A. O. (1999). Education Indigenous to Place: Western Science Meets Indigenous Reality. In *Ecological Education In Action* (pp. 117 – 140). State University Of New York Press. http://www.ankn.uaf.edu/Curriculum/Articles/BarnhardtKawagley/EIP.html

Barnhardt, R., & Oscar Kawagley, A. (2005). Indigenous Knowledge Systems And Alaska Native Ways Of Knowing. *Anthropology & Education Quarterly, 36*(1), 8–23. doi:10.1525/aeq.2005.36.1.008

Battiste, M. (2002). *Indigenous knowledge and pedagogy in first nations education a literature review with recommendations*. Retrieved from https://www.afn. ca/uploads/files/education/24._2002_oct_marie_battiste_indigenousknowledge andpedagogy_lit_review_for_min_working_group.pdf

Bebbington, A. (1991). Indigenous Agricultural Knowledge Systems, Human Interests, and Critical Analysis: Reflections on Farmer Organizations in Ecuador. *Agriculture and Human Values*, *8*(1-2), 14–24. doi:10.1007/BF01579652

Beccerra-Fernandez, I., Gonzalez, A., & Sabherwa, R. (2004). *Knowledge management: challenges, solutions and technologies*. Pearson Prentice Hall.

Bettina, L. (2011). A Critical Review On The Concept Of Social Technology. *Socialinės Technologijos Social Technologies*, *1*(1), 7–24.

Blakemore, E. (2019, February 19). Colonialism facts and information. *National Geographic*. Available at: https://www.nationalgeographic.com/culture/article/colonialism

Boahen, A. A. (1980). *Topics in West African history*. Longman.

Bratianu, C. (2015). *Organizational Knowledge Dynamics: Managing Knowledge Creation, Acquisition, Sharing, and Transformation*. IGI Global. doi:10.4018/978-1-4666-8318-1

Briggs, J. (2005). The Use of Indigenous Knowledge in Development: Problems and Challenges. *Progress in Development Studies*, *5*(2), 99–114. doi:10.1191/1464993405ps105oa

Briggs, J., & Sharp, J. (2004). Indigenous knowledges and development: A postcolonial caution. *Third World Quarterly*, *25*(4), 661–676. doi:10.1080/01436590410001678915

Bruchac, M. M. (2014). Indigenous Knowledge and Traditional Knowledge. In C. Smith (Ed.), *Encyclopedia of Global Archaeology* (pp. 3814–3824). Springer Science and Business Media. doi:10.1007/978-1-4419-0465-2_10

Brush, S., & Stabinsky, D. (1996). *Valuing Local Knowledge: Indigenous Peoples and Intellectual Property Rights*. Island Press.

Büschges, G. (2002). *Wörterbuch der Soziologie. 2. völlig neubearbeitete und erweiterte Auflage*. Lucius & Lucius.

Carvalho, N., & Gomes, I. (2017). Knowledge Sharing between Enterprises of the Same Group. *International Journal of Knowledge Management*, *13*(1), 34–52. doi:10.4018/IJKM.2017010103

Chahine, I., & Kinuthia, W. (2013). Juxtaposing form, function and social symbolism: An ethnomathematical analysis of indigenous technologies in the Zulu culture. *Journal of Mathematics & Culture*, *7*(1), 1–30.

Chandra, D. V. (2014). *Re-examining the importance of indigenous perspectives in the Western environmental education for sustainability: iFrom tribal to mainstream*. Academic Press.

Chisholm, R. M. (1982). *Foundations of Knowing*. University of Minnesota Press.

Chiwanza, K., Musingafi, M., & Mupa, P. (2013). Challenges in preserving indigenous knowledge systems: Learning from past experiences. *Information and Knowledge Management*, *3*(2), 19–25.

Christensen, C. (2003). *The innovator's dilemma. The revolutionary book that will change the new way you do business*. HarperBusiness Essentials.

Churchill, W. (1996). I Am Indigenist: Notes on the Ideology World. In *A Native Son: Selected Essays on Indigenism, 1985-1995*. South End Press.

Clancey, W. (1993). The knowledge level reinterpreted: Modeling socio-technical systems. *International Journal of Intelligent Systems*, *8*(1), 33–50. doi:10.1002/int.4550080104

Cronje, A., de Beer, J., & Ankiewicz, P. (2015). The development and use of an instrument to investigate science teachersi views on indigenous knowledge. *African Journal of Research in Mathematics, Science and Technology Education, 19*(3), 319-332. . doi:10.1080/10288457.2015.1108567

Custer, R. L. (1995). Examining the dimensions of technology. *International Journal of Technology and Design Education*, *5*(3), 219–244. doi:10.1007/BF00769905

Dalkir, K. (2005). *Knowledge management in theory and practice*. Elsevier.

Dalleo, R. (2012). Ngũgĩ wa Thiong'o's *Wizard of the Crow* and Postcolonial Pedagogy. *Research in African Literatures*, *43*(2), 138–154. doi:10.2979/reseafrilite.43.2.138

Datta, R. (2018). Decolonizing both researcher and research and its effectiveness in Indigenous research. *Research Ethics*, *14*(2), 1–24. doi:10.1177/1747016117733296

David, M. (2001). Truth as epistemic goal. In M. Steup (Ed.), *knowledge, truth, and duty: Essays on epistemic justification, responsibility, and virtue*. Oxford University Press. doi:10.1093/0195128923.003.0010

De Beer, J., & Whitlock, E. (2009). Indigenous knowledge in the life sciences classroom: Put on your de Bono hats! *The American Biology Teacher*, *71*(4), 209–216. doi:10.2307/27669414

De Vries, M. J. (1996). Technology Education: Beyond the "Technology is applied science" paradigm. *Journal of Technology Education, 8*(1), 7–15. doi:10.21061/jte.v8i1.a.1

Department of Basic Education. (2011). *Curriculum and Assessment Policy Statement Grades 7-9. Technology*. Government Printers.

DeWalt, B. R. (1994). Using indigenous knowledge to improve agriculture and natural resource management. *Human Organization, 53*(2), 123–131. doi:10.17730/humo.53.2.ku60563817m03n73

Diop, S. (2012). African elites and their Post-Colonial legacy: Cultural, political and economic discontent–by way of literature. *Africa Development. Afrique et Developpement, 37*(4), 221–235.

Ellen, R., & Harris, H. (1996). Concepts of indigenous environmental knowledge in scientific and development studies literature: A critical assessment. East-West Environmental Linkages Network Workshop.

Emeagwali, G., & Shizha, E. (Eds.). (2016). *African Indigenous knowledge and the sciences*. Sense Publishers. doi:10.1007/978-94-6300-515-9

Emery, A. (1996). *The Participation of Indigenous Peoples and Their Knowledge in Environmental Assessment and Development Planning (draft)*. Centre for Traditional Knowledge.

Etzemüller, T. (2009). Social engineering als Verhaltenslehre des kühlen Kopfes. In Die Ordnung der Moderne. Social Engineering im 20. Jahrhundert. Transcript Verlag.

Eyong, C. T. (2007). Indigenous knowledge and sustainable development in Africa: Case study on central Africa. *Tribes and Tribals. Special Volume*, (1), 121–139.

Ezinwa Nwagwu, W. (2007). Africa's indigenous knowledge Creating science and technology information databases for developing and sustaining sub-Saharan Citations Creating science and technology information databases for developing and sustaining sub-Saharan Africa's indigenous knowledge. *Journal of Information Science, 33*(6), 737–751. doi:10.1177/0165551506077374

Fanon, F. (1967). *Black skin, white masks*. Grove Press.

Fedosejeva, J., BoEe, A., Romanova, M., Ilisko, D., & Ivanova, O. (2018). Education for sustainable development: The choice of pedagogical approaches and methods for the implementation of pedagogical tasks in the anthropocene age. *Journal of Teacher Education for Sustainability, 20*(1), 157-179. . doi:10.2478/jtes-2018-0010

Fields, L. (1991). Tools for Indigenous Agricultural Development in Latin America: An Anthropologist's Perspective. *Agriculture and Human Values*, *8*(1-2), 85–92. doi:10.1007/BF01579659

Gaines, B. (1993). Modeling practical reasoning. *International Journal of Intelligent Systems*, *8*(1), 51–70. doi:10.1002/int.4550080105

Gaura, A. S., Hongjia, M. A., & Baoshan, G. E. (2019). MNC strategy, knowledge transfer context, and knowledge flow in MNEs. *Journal of Knowledge Management*, *23*(9), 1885–1900. doi:10.1108/JKM-08-2018-0476

Goduka, N., Madolo, Y., Rozani, C., Notsi, L., & Talen, V. (2013). Creating spaces for eZiko Sipheka Sisophula theoretical framework for teaching and researching in higher education: A philosophical exposition. *Indilinga*, *12*(1), 1–12.

Graham, G. (2010). Behaviorism. *Stanford Encyclopedia of Philosophy*. Retrieved February 26, 2011, from https://plato.stanford.edu/entries/behaviorism/

Grenier, L. (1998). *Working With Indigenous Knowledge: A Guide For Researchers*. IDRC.

Grimsdottir, E., & Edvardsson, I. R. (2018). Knowledge Management, Knowledge Creation, and Open Innovation in Icelandic SMEs. *International Journal of Knowledge Management*, *14*(4), 1–13.

Gumbo, M. T. (2003). *Indigenous technologies: Implications for a Technology Education curriculum* [Unpublished PhD thesis]. Vista University.

Gumbo, M. T. (2015). Indigenous technology in Technology Education curricula and teaching. In P. J. Williams, A. Jones, & C. Buntting (Eds.), *Contemporary issues in Technology Education: The future of Technology Education* (pp. 57–75). Springer.

Gumbo, M. T. (2017). An indigenous perspective on Technology Education. In P. Ngulube (Ed.), *Handbook of research on Indigenous Knowledge Systems in developing countries* (pp. 137–160). IGI.

Gumbo, M. T., & Msila, V. (Eds.). (2017). African Voices on Indigenisation of the Curriculum: Insights from Practice. Wandsbeck: Reach.

Gumbo, M. T., & Williams, J. P. (2014). Investigating Grade 8 Technology teachers' pedagogical content knowledge in the Tshwane District of Gauteng Province of South Africa. *International Journal of Educational Sciences*, *6*(3), 479–488. doi:10.1080/09751122.2014.11890159

Gumbo Mishack, T. (2020). Indigenising Technological Pedagogical Content Knowledge in Open Distance Learning, Africa. *Education Review*, *17*(4), 72–86. doi:10.1080/18146627.2020.1868073

Handayani, I.D., Wilujeng, I., & Prasetyo, Z.K. (2018). *Elaborating Indigenous Knowledge*. Academic Press.

Haron, H., & Hamiz, M. (2014). An Ontological Framework to Preserve Malay Indigenous Health Knowledge. *Journal of Computational and Theoretical Nanoscience*, *20*(1), 226–23.

Hart, M. A. (2010). Indigenous worldviews, knowledge, and research. *The development.*

Heleta, S. (2018). Decolonizing Knowledge in South Africa: Dismantling the pedagogy of big lies. Ufahamu. *Journal of African Studies*, *40*(2), 47–65.

Herman, M., Vizina, Y., Augustus, C., & Sawyer, J. (2008). *Learning indigenous science from the place*. Retrieved 23 October, 2015 from https://iportal.usask.ca/docs/Learningindigenousscience.pdf

Herusatoto, B. (2012). *Mitologi Jawa* [The Javanese mythology]. Oncor Semesta.

Hoppers, C. A. (2001). *Decolonising the curriculum, indigenous knowledge systems and globalisation*. HSRC.

Howe, A. C. (2002). *Engaging children in science* (3rd ed.). Prentice Hall.

IIRR (International Institute of Rural Reconstruction). (1996a). *Recording and Using Indigenous Knowledge A Manual. IIRR: Silang, Philippines. IUCN/UNEP/WWF 1991 Summary - Caring for the Earth: A Strategy for Sustainable Living*. IUCN/UNEP/WWF.

James, W. (1890). *The Principles of Psychology*. Holt.

Johnson, G., Whittington, R., & Scholes, K. (2011). *Exploring strategy. Text and case* (9th ed.). Financial Times/ Prentice Hall.

Kaba, A., & Ramaiah, K. C (2020). Measuring Knowledge Acquisition and Knowledge Creation: A Review of the Literature. *Library Philosophy and Practice (e-journal)*. 4723.

Kasanda, Lubben, F., Gaoseb, N., Kandjeo-Marenga, U., Kapenda, H., & Campbell, B. (1823). The role of everyday contexts in learner centred teaching: The practice in Namibian secondary schools. *International Journal of Science Education*, *27*(15), 1805. doi:10.1080/09500690500277854

Kiggundu, J. (2007). Intellectual property law and the protection of indigenous knowledge. In I. Mazonde & P. Thomas (Eds.), *Indigenous knowledge systems and intellectual property in the twenty-first century; perspectives from Southern Africa* (pp. 26–47). Dakar Codesria.

Knobloch, H. (2006). *Sozialtechnologie, Soziologie und Rethorik*. http://www. as.tu-berlin.de/v-menue/mitarbeiter/prof_ hubert_knoblauch/publikationen/

Knudston, P., & Suzuki, D. (1992). *Native and Scientific Ways of Knowing about Nature*. Mountain Man Graphics.

Kozhakhmet, S., & Nazri, M. (2017). Governing Knowledge Sharing Behavior in Post-Soviet Kazakhstan. *Journal of Workplace Learning*, *29*(3), 1–18. doi:10.1108/JWL-06-2016-0053

Kramlinger, T., & Huberty, T. (1990). Behaviorism versus humanism. *Training and Development Journal*, *44*(12), 41–45.

Laufenberg, V. (2009). *About Technology*. Retrieved May 15, 2015 from mhtml:file://B:\TE KIT\AboutTechnology.mht.

Likalu, M., Abdullah, R., Selamat, M. H., Ibrahim, H., & Nor, M. Z. M. (2010). A framework of collaborative knowledge management system in open-source software development environment. *Computer and Information Science*, *3*(1), 81–90. doi:10.5539/cis.v3n1p81

Locke, J. (2001). *An essay concerning human understanding*. Batoche Books.

Mack, E., & Meadowcroft, J. (Eds.). (2009). *Major Conservative and Libertarian Thinkers: John Locke*. Continuum.

Martin, K., & Mirraboopa, B. (2003). Ways of knowing, being and doing: A theoretical framework and methods for indigenous and indigenist research. *Journal of Australian Studies*, *27*(76), 203–214. doi:10.1080/14443050309387838

Maruta, R. (2014). The creation and management of organizational knowledge. *Knowledge-Based Systems*, *67*, 26–34. doi:10.1016/j.knosys.2014.06.012

Materer, S. M., Valdivia, C., & Gilles, J. L. (2002). *Indigenous Knowledge Systems: Characteristics and Importance to Climatic Uncertainty*. Working or Discussion Paper.

Mathew, R. M. (1985). Social analysis of information producation and consumption: the new challenges and tasks of Third World Countries. In A. I. Mikhalov (Ed.), *Theoretical problems of informatics: social aspects of modern informatics* (pp. 37–47). All Union Institute for Scientific and Technical Information.

McInnes, B. D. (2017). Preparing teachers as allies in indigenous education: Benefits of an American Indian content and pedagogy course. *Teaching Education, 28*(2), 145-161. doi:10.1080/10476210.2016.1224831

Mestre, J., & Touger, J. (1989). Cognitive research — What's in it for physics teachers? *The Physics Teacher*, *27*(6), 447–456. doi:10.1119/1.2342824

Meyer, X., & Crawford, B. A. (2011). Teaching science as a cultural way of knowing: Merging authentic inquiry, nature of science, and multicultural strategies. *Cultural Studies of Science Education*, *6*(3), 525–547. Advance online publication. doi:10.1007/s11422-011-9318-6

Mji, G. (2019). *Opting for a walk without limbs: the avoidance of Stellenbosch University to be an African University.* Stellenbosch University Library Auditorium.

Mkabela, Q. (2005). Using the Afrocentric Method in Researching Indigenous African Culture. *Qualitative Report, 10*(1), 178–189.

Morgan, D. L. (2003). Appropriation, Appreciation, Accommodation: Indigenous Wisdoms and Knowledges in Higher Education. *International Review of Education, 49*(1/2), 35–49. doi:10.1023/A:1022937003545

Moyo, P. V., & Kizito, R. (2014). Prospects and challenges of using the argumentation instructional method to indigenise school science teaching. *African Journal of Research in Mathematics, Science and Technology Education, 18*(2). https://www.tandfonline.com/doi/abs/10.1080/10288457.2014.912831

Mustonen, T. (2014). *2021 Towards Inclusion of Indigenous Knowledge and Local Knowledge in Global Reports on Climate Change*. Available from: https://www.researchgate.net/publication/352440686_2021_Towards_Inclusion_of_Indigenous_Knowledge_and_Local_Knowledge_in_Global_Reports_on_Climate_Change

Mustonen, T., Harper, S. L., Rivera Ferre, M., Postigo, J., Ayanlade, A., Benjaminsen, T., Morgan, R., & Okem, A. (Eds.). (2021). 2021 Compendium of Indigenous Knowledge and Local Knowledge: Towards Inclusion of Indigenous Knowledge and Local Knowledge in Global Reports on Climate Change. Snowchange Cooperative.

Nadaraj, Ronicka, & Angela. (2016). *Engaging Indigenous Knowledge Holders in Teaching Preservice Teachers in IKS Food Production and Practices: Implications for Higher Education*. Available at: https://www.semanticscholar.org/paper/Engaging-Indigenous-Knowledge-Holders-in-Teaching-Govender-Mudaly/f99a4ebcd3839b60b4344699a80c56c573caa2e8

Ngiasike, J. T. (2011). Turkana children's rights to education and indigenous knowledge in science teaching in Kenya. *New Zealand Journal of Teacher's Work, 8*(1), 55-67. Retrieved from https://ir-library.ku.ac.ke/handle/123456789/9530

Ngulube, P. (2002). Managing and Preserving Indigenous Knowledge in the Knowledge Management Era: Challenges and opportunities for information professionals. *Information Development, 18*(2), 95–101. doi:10.1177/026666602400842486

Ninnes, P. (2000). Representation of Indigenous Knowledge in Secondary School Science Textbooks in Australia and Canada. *International Journal of Science Education, 22*(6), 603–617. doi:10.1080/095006900289697

Nonaka, I., & Takeuchi, H. (1995). *The knowledge-creating company. How Japanese companies create the dynamics of innovation*. Oxford University Press. doi:10.1093/oso/9780195092691.001.0001

Nozick, R. (1981). *Philosophical explanations*. Harvard University Press.

O'Hara, K. (2001). *Plato and the Internet*. Icon Books.

O'Riley, P. (2001). *A different storytelling of technology education curriculum revisions: A storytelling of difference*. Retrieved May 31, 2015 from https://scholar.lib.vt.edu/ejournals/JTE/v7n2/pdf/oriley.pdf

Obikeze, D. S. (2011). *Indigenous knowledge systems and the transformation of the academy in Africa: The CULPIP model*. Retrieved February 11, 2015 from https://citeseerx.ist.psu.edu/viewdoc/download?doi=10.1.1.164.9357&rep=rep1&type= pdf.

Ogawa, M. (1995). Science education in a multiscience perspective. *Science Education, 79*(5), 583–593. doi:10.1002/sce.3730790507

Ogunbure, A. A. (2011). The possibilities of technological development in Africa: An evaluation of the role of culture. *The Journal of Pan African Studies, 4*(3), 86–100.

Onwu, G., & Mosimege, M. (2004). Indigenous knowledge systems and science and technology education: A dialogue. *African Journal of Research in Mathematics, Science and Technology Education, 8*(1), 1-12. doi:10.1080/10288457.2004.10740556

Owusu-Ansah, F. E., & Mji, G. (2013). African indigenous knowledge and research. *African Journal of Disability*, *2*(1), 1–5. doi:10.4102/ajod.v2i1.30 PMID:28729984

Pacharapha, T., & Ractham, V. (2012). Knowledge acquisition: The roles of perceived value of knowledge content and source. *Journal of Knowledge Management*, *16*(5), 724–739. Advance online publication. doi:10.1108/13673271211262772

Pardi, P. (2012). *What is knowledge?* Philosophy News. Retrieved March 18, 2012 from: https://www.philosophynews.com/post/2011/09/22/What-is-Knowledge.aspx

Parker, K. I. (2004). *Biblical Politics of John Locke*. Wilfrid Laurier University Press.

Petter, G. (2007). *Knowledge Management Systems: Value Shop Creation*. Idea Group Publisher.

Progler, Y. (1999). *Towards defining Western research on Indigenous Knowledge*. Muslimedia.

Rakometsi, M. S. (2008). *The transformation of black school education in South Africa, 1950-1990: A historical perspective* [Unpublished PhD thesis]. Bloemfontein: University of Free State.

Regmi, J., & Fleming, M. (2012). Indigenous knowledge and science in a globalized age. *Cultural Studies of Science Education, 7*(2), 479-484. https://doi.org// doi:10.1007

Rocheleau, D. E. (1991). Gender, Ecology, and the Science of Survival: Stories and Lessons from Kenya. *Agriculture and Human Values*, *8*(Winter-Spring), 156–165. doi:10.1007/BF01579669

Rocheleau, D. E. (1991). Gender, Ecology, and the Science of Survival: Stories and Lessons from Kenya. *Agriculture and Human Values*, *8*(Winter-Spring), 156–165. doi:10.1007/BF01579669

Rock, I., & Plamer, S. (1990). The legacy of gestal psychology. *Scientific American*, *263*(6), 48–61. doi:10.1038/scientificamerican1290-84 PMID:2270461

Rodil, K., Winschiers-Theophilus, H., Bidwell, N. J., Eskildsen, S., Rehm, M., & Kapuire, G. K. (2011). *A new visualization approach to re-contextualize indigenous knowledge in rural Africa*. Paper presented at INTERACT2011, Lisbon, Portugal. 10.1007/978-3-642-23771-3_23

Russel, B. (1972). *A history of Western philosophy*. Simon and Schuster.

Russel, B. (2009). *Human knowledge. Its scope and limits. With an introduction bu John G. Slater*. Routledge. doi:10.4324/9780203875353

Sagsan, M. (2009). Knowledge management discipline: Test for an undergraduate program in Turkey. *Electronic Journal of Knowledge Management*, *7*(5), 627–636.

Sassson, J.R., & Douglas, I. (2006). A conceptual integration of performance analysis. *Knowledge Management and Technology: From concept to prototype*, 81-99.

Schumpeter, J. A. (1912). The Theory of Economic Development. Oxford University Press.

Senanayake, S. G. J. N. (2006). Indigenous knowledge as a key to sustainable development. *The Journal of Agricultural Sciences, 2*. Available at: https://www.researchgate.net/publication/265197993

Seth, S. (2013). Once Was Blind but Now Can See': Modernity and the Social Sciences. *International Political Sociology*, *7*(2), 136–151. doi:10.1111/ips.12014

Shava, S. (2013). The representation of indigenous knowledges. In R. B. Stevenson, M. Brody, J. Dillon, & A. Wals (Eds.), *International Handbook of Research on Environmental Education*. Routledge.

Shava, S., & O'Donoghue, R. (2014). *Teaching Indigenous Knowledge and Technology. Fundisa for Change Programme*. Environmental Learning Research Centre, Rhodes University.

Sherif, K., & Xing, B. (2006). Adaptive processes for knowledge creation in complex systems: The case of a global IT consulting firm. *Information & Management*, *43*(4), 530–540. doi:10.1016/j.im.2005.12.003

Shizha, E. (2016). African Indigenous Perspectives on Technology. In G. Emeagwali & E. Shizha (Eds.), *African Indigenous Knowledge and the Sciences*. Anti-colonial Educational Perspectives for Transformative Change. SensePublishers. doi:10.1007/978-94-6300-515-9_4

Shongwe, M. M. (2016). An Analysis of Knowledge Management Lifecycle Frameworks: Towards a Unified Framework. *Electronic Journal of Knowledge Management*, *14*(3), 140–153.

Sifuna, D. N. (2008). The case of primary education in Africa from colonialism to globalisation. *African Traditional Herbal Research Clinic*, *10*(3), 20–33.

Sillitoe, P. (2006). Introduction: Indigenous Knowledge in Development. *Anthropology in Action : Newsletter of the British Association for Social Anthropology in Policy and Practice (BASAPP)*, *13*(3), 1–12. doi:10.3167/aia.2006.130302

Simpson. (2004). Anticolonial Strategies for the Recovery and Maintenance of Indigenous Knowledge. *The American Indian Quarterly, 28*(3&4), 373-384. doi:10.1353/aiq.2004.0107

Sithole, J. (2007). The challenges faced by African libraries and information centres in documenting and preserving indigenous knowledge. *IFLA Journal, 33*(2), 117–122. doi:10.1177/0340035207080304

Smith, L. (1999). *Decolonizing methodologies: Research and indigenous peoples*. Zed Books.

Snively, G., & Corsiglia, J. (2000). Discovering indigenous science: Implications for science education. *Science Education, 85*(1), 6-34. . doi:10.1002/1098-237X(200101)85:1<6::AID-SCE3>3.0.CO;2-R

Tharakan, J. (2017). Indigenous knowledge systems for appropriate technology development. *IntechOpen*, *7*, 124–134. doi:10.5772/intechopen.69889

Thrupp, L. A. (1989). Legitimizing local knowledge: From displacement to empowerment for third world people. *Agriculture and Human Values*, *6*(3), 13–24. doi:10.1007/BF02217665

Tomei, L. (2009). *Lexicon of online and distance learning*. Rowman & Littlefield Education.

Vassev, E., Hinchey, M., & Gaudin, B. (2012). Knowledge Representation for Self-Adaptive Behavior. *Proceeding of Conference on Computer Science & Software Engineering*.

Vázquez, R. (2011). Translation as Erasure: Thoughts on Modernity's Epistemic Violence. *Journal of Historical Sociology*, *24*(1), 27–44. doi:10.1111/j.1467-6443.2011.01387.x

Von Krogh, G., Nonaka, I., & Rechsteiner, L. (2012). Leadership in organizational knowledge creation: A review and framework. *Journal of Management Studies*, *49*(1), 240–277. doi:10.1111/j.1467-6486.2010.00978.x

Warren, D. M. (1991). The role of indigenous knowledge in facilitating the agricultural extension process. Paper Presented at *International Workshop on Agricultural Knowledge Systems and the Role of Extension*, Bad Boll, Germany.

Warren, D. M., Slikerveer, L. J., & Brokensha, D. (Eds.). (1995). *The Cultural Dimensions of Development: Indigenous Knowledge Systems*. Intermediate Technology Publications. doi:10.3362/9781780444734

Waziyatawin, A. W. (2004). Introduction: Indigenous Knowledge Recovery Is Indigenous Empowerment. *American Indian Quarterly*, *28*(3/4), 359–372. https://www.jstor.org/stable/4138922

Weitzman, M. L. (1998). Recombinant growth. *The Quarterly Journal of Economics*, *113*(2), 331–360. doi:10.1162/003355398555595

Welbourne, M. (2001). *Knowledge.* Acumen. doi:10.1515/9780773585294

Williams, A. (1996). An introduction to Technology Education. In J. Williams & A. Williams (Eds.), Technology education for teachers (pp. 1-26). Macmillan.

Williams, J. (1980). *The educated and professional elite in the Gold Coast and Sierra Leone, 1885-1914.* University of California.

Williams, P. J. (1996). *International developments in technology education: Lessons for South Africa.* Paper presented at the International Conference on Technology Education for Development in South Africa, Cape Town, South Africa.

Williams, P. J. (2000). The only methodology of Technology? *Journal of Technology Education, 11*(2), 48–60. doi:10.21061/jte.v11i2.a.4

Wilson, V., & Harris, M. (2003). Designing the best: A review of effective teaching and learning of Design and Technology. *International Journal of Technology and Design Education, 13*(3), 223–241. doi:10.1023/A:1026133808945

World Bank. (1998). W*orld development report 1998/1999: Knowledge for development.* Retrieved 11th March, 2024, from https://www.worldbank.org/afr/ik/ikrept.pdf

Zeleza, P. T. (2006). The disciplinary, interdisciplinary and global dimensions of African Studies. *International Journal of African Renaissance Studies*, *1*(2), 195–220. doi:10.1080/18186870608529717

Zinyeka, G. (2013). Onwu and Mosimege on indigenous knowledge systems and science and technology education: A dialogue in some remaining issues. *Greener Journal of Educational Research, 3*(9), 432-437.

Zinyeka, G., Onwu, G. O. M., & Braun, M. (2016). A truth-based epistemological framework for supporting teachers in integrating indigenous knowledge into science teaching. *African Journal of Research in Mathematics, Science and Technology Education, 20*(3), 256-266. doi:10.1080/18117295.2016.1239963

Chapter 2
Africa at the Center of Digital Technology

Cleophas Gwakwara
https://orcid.org/0000-0003-3380-4825
University of Johannesburg, South Africa

ABSTRACT

Digital technology is defined as electronic technology that produces, stores, and processes data. Digital technology has evolved over time with several global players, including those in Africa, making varied but significant contributions within their capacities and digital spaces. A critical question that has been asked is to locate and clarify the myths and misconceptions surrounding the contribution of African Indigenous knowledge (AIK) to modern digital technology. Voronkova et al. assert that the development of digital technology was dominated by the United States (US) and Germany during the second industrial revolution following the widespread use of electricity from the second half of the 19th century to the early 20th century in that order.

INTRODUCTION

Digital technology is defined as electronic technology that produces, stores and processes data, Moore and Czerwinska (2019:3). Digital technology has evolved over time with a number of global players, including those in Africa making varied, but significant contributions within their capacities and digital spaces. A critical question that has been asked is to locate and clarify the myths and misconceptions surrounding the contribution of African Indigenous Knowledge (AIK) to modern digital technology. Voronkova et al. (2023:734) assert that the development of digital

DOI: 10.4018/978-1-6684-7851-6.ch002

Copyright © 2024, IGI Global. Copying or distributing in print or electronic forms without written permission of IGI Global is prohibited.

technology was dominated by the United States (US) and Germany during the second industrial revolution following the widespread use of electricity from the second half of the 19th century to the early 20th century in that order. Shizha and Emeagwali (2016:47) argue that Africa's development, particularly in the field of science and technology, is judged and adjudged from the viewpoints of its former colonizers, with a deliberate bias on Africa (negatively) as not having possessed any form of science and technology before Europeans set foot on the continent. Consequently, it is the duty of this study to interrogate such assertions.

Hegel in his narrative asserts that the Black man cannot conceive neither a history nor a science because he acts on basic instincts most of the time, which are devoid of civilization (Kienon-Kabore, 2017:50). Whilst, the narrative that 'modern' should replace 'traditional' is also misplaced (Battiste, 2017), Kaya and Seleti (2013:36) note that the Eurocentric view perpetuates the misconceptions and stereotypes narrative by articulating that AIK is obsolete and static. These sentiments are also echoed by Khumalo and Baloyi (2017:4) who proclaim that the failure to properly position AIK can also be ascribed to biases and stereotypes brought about by colonialism and its negative perceptions. Hountondji (2017) is of the opinion that much of the existing research on African communities is still dominated by Euro-centric biases and lenses which view Africa in the negative trajectory.

Denial and a lack of research on how technical and scientific AIK influenced the development of digital technology in Sub-Saharan Africa has been put forward by Kienon-

Kabore (2017). Other scholars question indigenous knowledge's accuracy including (Smith, 2012); on credentials of community experts (Apffel-Marglin, 2011); its ultimate replication (Ellen et al., 2000:17) and the consequential lack of documentation (Kienon-Kabore, 2017). It is, thus, a cosmopolitan of views regarding the birth, evolution and growth of African Indigenous Knowledge Systems as they relate to digital technology development and consequent widespread use.

Utsua (2015) posits that many modern high-school-level concepts in mathematics were first established, developed and typified in Africa, as the first method of counting. The Dogon people of Mali amassed and digested a wealth of detailed astronomical observations (Adams, 1983). It is critical to note that the Kingdom of Mali, being in existence in West Africa from the 13th to the 16th century, was well known for its wealth, its highly advanced political institutions and centres of learning and excellence, such as Timbuktu. The city was a major hub for trade and manuscripts from Timbuktu are littered with evidence of advanced knowledge in areas such as astronomy, mathematics and medicine. Scholars argue that Africa created its own indigenous technology using its own scientific knowledge (Shizha and Emeagwali (2016:47) data to build structures (Shishima, 2014); the important use fire (Shillington, 2005); expertise in the construction of boats and sailing (Utsua, 2015:15); and carried out mathematical and numerical computations (Sydella, 2013), long before the advent of colonisation. In this regard, Africa's contribution to digital technology has not received due consideration. The researchers argue that despite the myths and misconceptions around Africa's role in the advancement of digital technology, Africa remains at the centre and core of digital technology and digitization. There should be a deliberate effort to decolonize and demystify modern digital technology so as to recognize Africa's contributions to its development. Using an exploratory, narrative and descriptive approach the study seeks to use documentary evidence in its investigation to rightfully position the role of AIK within the background of the Fourth Industrial Revolution (4IR).

The next section presents misconceptions and stereotypes on the central thesis that Africa is at the centre of digital technology, while section 3 addresses African indigenous knowledge, with section 4 illuminating Africa's rich technological heritage, against the impact of colonisation in section 5. Finally, section 6 concludes this research on Africa's role in propelling the use of digital technology.

MISCONCEPTIONS AND STEREOTYPES

Conservatively, the African continent is mostly left out written out of discussions on technology. In circumstances where Africa and its interplay with technology, there is a deliberate portrayal of the development of technology being influenced by the West with Africans not being active actors in growth (Mavhunga, 2014, 2017). However, some current work on race and technology have been quite helpful in bringing to the fore the role that the African people have played in the production of digital technology (McIlwain, 2020).

Munyoki (2019:1) asserts that there is a notion that the developed world is the prime mover of innovation and for many decades, it has been proffered that no meaningful technological innovations came from the African Continent. It is thus, falsely held that most inventions that we know today came from the developed world and tends to perpetuate the fallacies that Africa lagged behind in contributing to technological advancements including the digital divide. Shizha and Emeagwali (2016:47) argue that Africa's development, particularly in science and technology, is judged from the viewpoints and lenses of its former colonizers, with a deliberate bias on Africa as not having possessed any science and technology before Europeans set foot on the continent. Such a notion tends to ignore the notable strides that Africa made in contributing to science and technology prior to colonization and subjugation.

Hegel, the philosopher and scientist, has also influenced the collective mentality which does not give due credit to the contributions of Africa to the development of science (Kienon-Kabore,2017:50. According to Hegel's "...congenital inferiority of the Black", the black man cannot have known technique and science as his thinking is mostly influenced by instinct which is characteristic of "wild societies", contrary to what transpires in the civilised societies. This same biased view is also spread by Kienon-Kabore (2017:50) who argues that the African is not capable of technique and science. This tends to undermine the role and importance of AIK in the development of digital technology.

The conversation that portrays Africa at the centre of digital technology also has to be put in context by changing the mindset, attitudes and practices of researchers and extension workers working in African local communities as advanced by Battiste (2017).

The supposition that that the "modern" must replace the "traditional" should not be entertained by researches (ibid), as it perpetuates biases. Local people, as "insiders", should give importance to their own local knowledges system as it forms the basis of their collective livelihoods and development trajectory (ibid). Researchers, as "outsiders" should accept this line of thinking (ibid). Battiste cautions that if natives do not properly position their knowledge and wisdom, it will gradually vanish and could even be stolen (ibid).

Another stereotype that is often thrown around is that African civilizations, cultures, and traditions are often assumed by outsiders to be static and unchanging (Falola and Fleming, n.d.). This, in earnest, defies logic and is tainted with bias. Falola and Fleming also disputes this assertion, arguing that it is untrue as African "tradition" has constantly been in a state of flux. A scholarly discussion on how AIK lost its influence revealed some surprising reactions. Some scholars argue that society sees IK as immaterial because it lacks supportive theory (ies) (Kaya and Seleti, 2014).

The Eurocentric view is that African indigenous knowledge is outdated and motionless

(Kaya and Seleti, 2013:36). Apparently, this contradicts the view on AIK that the communities hold and is dear to them. To them, AIK is part of their culture and livelihoods, whether one is educated or not, lives in the city or countryside (ibid). Moreso, this community view, held by the natives, supports the notion that knowledge is dynamic, developing, and collectively created by people to find solutions to their survival needs and difficulties (ibid). In addition, the erroneous view that AIKs is community-based results in it lacking universal usage (Ibid). Kaya and Seleti (2013) contend that this is a simplistic view of the nature of knowledge production and use because all knowledge systems are initially created and utilised locally, but with the potential to be used universally. In support, Nkondo (2012) claims that all knowledge is local but becomes universal through processes of conquest and colonialism. Further, Kaya and Seleti (2013:36) refute the Eurocentric view that Africa was a *tabula rasa* before colonisation, that is, it is a continent with no history of civilization. It is on the basis of this Eurocentric view that African Indigenous ways of knowing and knowledge production are considered unscientific and primitive (Ibid; McCarthy; Moodie, 2017) and, therefore, incapable of producing human capital capable of meeting the challenges of modern science and technology associated with globalization (McCarthy, 2004).

Hountondji (2017) is of the opinion that existing researches related to African communities, which by extension (includes (AIKs)), is dominated by Eurocentric views. African students and researchers are trained with western programmes and are guided by western schema and bench mark/contextual information (Knudston & Suzuki, 1992). This tends to distort and undermine the relevance and positioning of AIK in advancing digital technology in the global context. The language used in AIK research is also another area of concern. Hountondji (2017) notes that in Africa, most academic and research activities are conducted in the languages of the colonisers, especially English, French and Portuguese which consequently weakens researches on AIK due to mismatch on native theoretical bases and patterns. 5

There are also concerns regarding the role of indigenous knowledge to the growth of the broader knowledge. Some scholars have questioned the degree to which indigenous peoples should be allowed to exert control over scientific research (Nicholas 2010; Smith & Wobst 2005). Other scholars have also questioned the accuracy of indigenous knowledge systems, or challenged the credentials of community experts who have no apparent codified scientific training (Apffel-Marglin 2011; Augustine 1997; Smith 2012). Another group wonder erroneously whether native knowledge generated in one locality could be useful outside that locality (Ellen et al., 2000:17). AIK remains masked by current science discoveries. Tauli-Corpuz (2017) argues that contemporary global thinking continues to take African political, economic, cultural and knowledge systems as regressive, impractical and idealistic. Such prejudices and doubts on AIK by some individuals and organisations in technologically advanced countries find it problematic to take that any respectable science knowledge can come from Africa (Communique, 2017).

The influence of Sub-Saharan Africa to the digital divide has also been a bone of contention. This view is expressed by Kienon-Kabore (2017) who notes that the influence to the technical and scientific level of sub-Saharan Africa to the technical and scientific legacy of humankind remains unfamiliar because there is a denial and a lack of research on this theme. Kaya and Seleti (2013) implore African intellectuals to help Africa close the yawning gap created by over four hundred years of domination and marginalisation of African Indigenous Knowledge Systems (AIKs), by rejecting the utilisation of the main western world view of knowing and knowledge production as the only way of knowing. Ngugi wa Thiong'o (1986) also raise similar sentiments in his inspiring book titled

"Decolonizing the mind", where he avers that AIK should not be construed as an alternative knowledge system but rather as a body of the knowledge systems in its own right, with its own framework and unique world view.

Nkondo (2012) adds his voice by reiterating that African social scientists are not helping to advance AIK by failing to generate ideas, explanations, schemes and approaches which resonate with their indigenous knowledge systems thus informing the academic progress in their research work. Smith (2002) adds that this tends to lower confidence among African academics, as western research ideas, explanations, schemes, and approaches are implemented without due scrutiny and used in African societies which are usually poverty stricken, rendering them inappropriate to their settings.

Indigenous sub-Saharan Africa metallurgy is loaded with imagery, ritual, and taboos (Bandama, 2020:1). This results in it being ridiculed by Western science as magical and, consequently, undeserving of suitable scientific study (Ibid). This biased thinking has since been condemned by sound research, but the ripple effects still remain and archaeo -metallurgical investigation is still generally underfunded when compared to other elements of anthropological inquiry (Ibid).

AFRICA'S INDIGENOUS KNOWLEDGE

Warren (1991) broadly defines Indigenous knowledge (IK) as the knowledge that is used by people in a particular local environment for their livelihood. Johnson (1992) also defines Indigenous knowledge as a frame of knowledge that is developed by a group of people, spanning generations of interaction with nature. Both definitions underscore the importance and nature of AIK being contextual, environment specific (local), and long-time span. The interplay of these factors determines the nature of the knowledge generated. Contrary to the notion usually advanced that AIK is static, Langill (1999) argues that AIK is also dynamic, and incorporates creativity and experimentation elements, as dictated by internal and outside influences and inventions. It is generally a mistake, therefore, to label indigenous knowledge as 'outdated,' 'retrogressive,' 'static' or 'rigid.' (Langill, 1999).

AIK has often been misunderstood with the proponents of the Western view claiming that the promoters of AIK cannot even provide a plausible definition and conceptualisation of "knowledge" (Kaya and Seleti, 2013:31). Chavunduka (1995) does not agree with these critical sentiments towards AIK, arguing that western worldview of "knowledge" has since its introduction in Africa and other non-western societies, lacked an understanding of the comprehensive nature and process of non-western ways of knowing and knowledge making. Nkondo (2012) agreeing with Chavhunduka's stance, argues that the western perception of AIK as mere repetition of practices without any theory to explain them is a representation of western cultural and intellectual pride.

Kimwaga (2010) shares a similar view that, African societies have for quite some centuries come up with their own sets of practices and ways of ways of explaining aspects related to their own environment. This line of thinking resonates with the contention advanced by Kay (2014) that the basis of all knowledge systems is homegrown. Kay further notes that due to unequal power relations emanating from colonialism and other forms of imperialism, other nations and cultures have generally forced their knowledge systems (wa Thiong'o, 1986; Timothy, 1998; Schutte, 1999; Walter, 2002; Smith, 2002). Africa is a victim of these machinations.

Matike (2008) elaborating, notes that learning is a product of how social reality is shaped by the different cultures. This, to some extent, explains the distortions and misrepresentations being flouted on AIK and consequently their contribution to the digital age.

Wane (2005:31) outlines the following IK characteristics:

- trusts in invisible powers;
- upholds dependency among things;
- has regard for power of personal interactions in strengthening relationships;
- upholds that people who hold these customs can pass them on; and
- is created within societies, is position and values specific, verbal, not systematic, all-inclusive and emphasises trustworthiness.

Africa, using the narrative advanced by Shizha and Emeagwali (2016:47), created its own indigenous technology using its own scientific knowledge. Each African society, from the East to the West and from the North to South, for instance, had its own scientific knowledge, which was put into practice and functioned for the good of its people. The tools with which societies worked and the manner in which they organized their labour are both important indices of social and technological development (Ibid).

AIK has not been accorded its rightful place in the development discourse due to numerous factors. Khumalo and Baloyi (2017:6) argue that the viewpoint that guides AIK is different from the rationalist western current systematic knowledge, and should, therefore, embrace pluralism (Afisi, 2016:63), anchored on grounded theories (Khan, 2014:224) and interpretivism (Gemma, 2018:4).

AIK is mostly oral, undocumented (Kienon-Kabore, 2017), rooted in culture, exclusive to a given location, inadequately used (Wayne, 2017), ignored in developmental projects (World Bank Group, 2017), and least analysed, considered or understood for its influence to knowledge creation or even to the existence of communities precisely when related to other bodies of knowledge (Wane, 2017).

IK is distinctive, that is, the knowledge is passed on to the child by its parents, or it could be descendants communicating to one through dreams, incoherent language to many except the recipient only (Sithole, 2017), making it difficult to document.

In order to understand AIK it is important, therefore, to engage IK practitioners who are well versed with it as advised by Khumalo and Baloyi (2017. Lengthy, qualitative, discourses with IK practitioners would help to understand AIK better (Wane,2017). However, there could be a hitch in terms of the secretive nature and myths harboured by the African traditional knowledge practitioners. Generally, there is some fear, in most African societies, that if this knowledge is shared it would lose its importance and relevance, as the spiritual realm would have been angered by this move.

Shillington (2005:4) explains that, modern man first developed in the Great Rift Valley of Africa, the first development of tools is found there as well: The Homo habilis, residing in East Africa, developed the first tool making industry, the Olduwan, around 2.3 million BC. Much progress in metallurgy and tool making, to include the steam engines, metal chisels and saws, copper and iron tools and weapons, nails, glue, carbon steel and bronze weapons and art was made across the entire old Africa (Van Sertina, 2013; Brooks,2013). The Archeulean stone tool industry, precisely hand-axes, were developed by Homo ergaster, in Africa, 1.5 million BC. Blatch (2013) asserts that progress in metallurgy and tools in Tanzania, Rwanda and Uganda between 1500 and 2000 years ago exceeded that of Europeans then and surprised the Europeans when they knew of it.

Around 90,000 to 60,000 BC, modern humans fashioned bone tools and the back blade, which ultimately became the foundation of the later stone tool trades (ibid).

Energy is also an important factor in developing digital technology, especially in the assembling of hardware. The Homo erectus innovatively began to use fire. Ancient Tanzanian furnaces could reach 1,800°C — 200 to 400°C warmer than those of the Romans (Shore, 2013). This further affirms the thesis of this chapter that Africa is at the centre of digital technology. Abstract art also appeared during this period. The oldest abstract art in the world is the shell necklace dated 82,000 years in the Cave of Pigeons in Tforalt, eastern Morocco (ibid). The second oldest abstract art and the oldest rock art is found in the Blombos Cave at the cape in South Africa, dated 77,000 years (Ehret, 2002:22).

In 1987 the third oldest canoe in the world and the oldest in Africa, dated 8000 years, and made of African mahogany, was discovered in Nigeria by fulani herdsmen, near the Yobe river, in the village of Dufuna (Utsua, 2015:15). It was dated 8000 years, cut out of African mahogany. Based on brilliant "stylistic sophistication", the tradition of canoe building must have gone further back in time, noted one archaeologist (Shishima, 2012:114). There were some sophisticated structures that were built by some African societies. These include the puzzlingly raised pillars and the more than 80 pyramids that were engineered by the Egyptians (Utsua, 2015). Later, in the 12th century and further to the south, there were hundreds of great cities in Zimbabwe and Mozambique (ibid).

AIK also made significant contributions to modern day medicines although they used unprocessed herbs. Imotep, who came from African Egypt and was an accomplished scientist, can assertively be introduced as the father of medicine (Adigbuo, 2017:77). Blatch (2013) observes that before Europeans invaded Africa, medicine in present day Egypt, Nigeria and South Africa, among others, was more advanced than European medicine. Examples of such medicines cited by Sertima (2013) include use of plants to treat pain (with salicylic acid), diarrhoea (as in Kaopectate), and some extractions that were shown in the 20th century to destroy Gram positive bacteria (Van Sertima, 2013). Even modern day, there are varied remedies found in the African context well know for its efficacy.

African scientists seem to be dragging in advancing the AIK contribution to digital technology narrative. This has been aptly observed by Khumalo and Baloyi (2017:2) who note that African scientists are not advancing AIKs as they appear not to care much about localized native knowledge medication and research or examining how African herbs can cure local illnesses, among other things. In order to put AIK in its true pedestal of scholarship, Naamwintome and Miller (2015), advocate for African scholars, who in most cases are Western schooled, to reposition themselves, give credit and advance their places in the broader world of science. They should ask provoking questions that address how the way that Africans know relates to the claims of reasonableness, impartiality and acceptance of western sciences (Naamwintome and Miller, 2015). This conforms to the current increase in interest on AIK.

Battiste (2017) posits that when people neglect their own knowledge and insight, it gradually gets lost and can easily be stolen. There is need for African scientists to vigorously push the AIK narrative in modern day digital technology development conversations and discourses. Khumalo and Baloyi (2017:2) also argue that Indigenous Knowledge (IK) systems continue to be in use informally in some communities. This, however, should not demean the contributions to science and technology that were made during this era and space. Oral traditions, whether communicated as historical narratives or mythical stories, constitute a form of traditional knowledge that can teach, carry, and reinforce other knowledges (Bruchac, 2014).

Funding is another factor that limits the exposure of AIK's contribution to development generally and digital technology, specifically. Brock-Utne (2000) regrets the continual reliance on external funding for education as it limits the contribution of AIK. This is mainly because the funders, usually Western in built and spread, will tend to advance their own Western knowledge systems into the curricula and research agendas thus side lining AIKs (Ezeanya, 2017). AIK benefits can only be fully appreciated in Africa if they are formalized. Currently, IK practices are informally kept lending them to manipulation (Ocholla and Dlamini, 2017). If IK were to be useful it has to be stored for recovery, hence the need for formalized databases (Khumalo and Baloyi, 2017:6). Le Roux (2003) The World Bank has made some efforts in developing IK databases and further initiatives have encompassed research into IK publication designs by Ocholla and Onyancha (2017).

However, Ocholla and Onyacha (Ibid) regret that some of those IK contents and records may not be dependable as not much IK research has been conducted. To that end, recording IK, creating laws, rules and structures to accommodate for IK go a long way in positioning IK (Khumalo and Baloyi, 2017:6). Okello et al. (2010: 1-2) warns that the recording of African traditional plants requires urgency following the speedy loss of natural habitats, native community life, cultural variety and knowledge of therapeutic plants. To allow knowledge on medicines to be transmitted orally, from generation to generation of healers, would be quite damaging to IK databases, as it would eventually disappear (Weldegerima, 2017).

Thrupp (1989) cautions that IK, just like scientific knowledge, should not be accepted blindly, by local people, as it may be incorrect or even damaging, in some cases. Practices based on, for example, misguided views, defective investigations, or erroneous evidence can be risky and may even be a blockade to improving the welfare of indigenous people (Langill, 1999).

COLONIALISM

Colonialism came about between 1884 and 1885, following the signing of the General Act of the Conference of Berlin, that sought to partition Africa among the colonising countries, under the guise of White supremacy, deeply entrenched anti-Black feelings, subjugation of Africans and unrestricted colonial authority (Stelios and Elias, 2016). In effect, African nations, in accordance with the Berlin Act, had no sovereignty and no legal claim over their state, land, or resources (Ibid).

Colonialism, according to Ocheni and Nwankwo (2012:46) refers to the complete control of one country by another with state power being in the hands of an external authority. Ocheni and Nwankwo (2012:46) assert that colonialism seeks to achieve two basic goals, that is, political control and ability to exploit the occupied country, hence Thondlana and Garwe (2021) liken it to imperialism. Raw materials looted from the colonies were meant to feed into the new ways of producing goods and services ushered in by the industrial revolution.

"Whites" had a belief, (erroneous though), that Africans were originally backward and immature and that Europeans had to act as trustees of their governance until such time when they had matured enough to govern themselves (Danielle, 2016:419).

Such sentiments show how the colonisers sought to advance their own knowledge systems and relegating AIK to the dustbin.

THE IMPACT OF COLONIZATION

Mawere (2020: 31-43) argues that colonialism is fundamentally centred on destroying all knowledge contrary to Eurocentric ideologies, opinions and views. Mawere further notes that during colonial wars of subjugation, for instance, the colonisers from the West instituted what he refers to as "epistemicide", which, according to him, essentially sought to destroy African knowledge and systems and impose a Western colonial epistemology (Ibid).

Colonialism has also been blamed for distorting the AIK through the education system. Several individuals entirely blame colonization for the apparent eradication of African Indigenous education and ways of knowing (Diop, 2023; Emeagwali & Shizha, 2016; Olufemi, 2023; Ocheni and Nwankwo, 2012).

Historical accounts of White (1959) in Angola; Abubakar (1980) in Nigeria; Schapera (1970) in Botswana; Ashton (1967) in Lesotho; Boone (2003) in West Africa; and Mamdani (1996) in Sub-Saharan Africa, Gennaioli and Rainer (2005) commonly conclude that pre-colonial Africa was disturbed in its growth path but continued to do so in spite of the disturbances (Shizha and Emeagwali, 2016:48).

In addition, European colonisation is said to have delayed and devastated the social relationships of production and the development of indigenous technological advancement that were in place (Ibid), with Basil Davidson (1992) arguing that colonialism prevented the potential for the natural maturation of pre-colonial African institutions. Colonisation also deprived Africa of resources that could have stimulated indigenous technological advancements. Onipede (2010:86), aptly notes that the dawn of colonial rule and by extension 'imperialism' laid the basis for Nigeria's industrial underdevelopment, as it shifted to the metropolitan states the assets of the underdeveloped nations, thereby exploiting them through capital and human deprivation.

The whole of Africa was victim to colonial exploitation. Post-colonial Africa, observes Shizha and Emeagwali (2016:48), suffered from the loss of skills and disorders in the course of development, grossly caused by the stifling colonial regimes. The consequences of colonialism and imperialism in corrupting native development and technologies in Africa were previously discussed at length in Walter Rodney's (1982) classic 'How Europe Underdeveloped Africa', which rationally and judgmentally examines the relationship between Europe's economic development and Africa's underdevelopment. In this seminal book, Rodney argues that European exploitation, through mechanisms such as the slave trade and resource extraction, not only led to the economic underdevelopment of Africa but also disrupted its social and political structures including technological innovation capacities. He contends that the development of Europe was intricately linked to the underdevelopment of Africa, creating a global economic system that perpetuated and continues to perpetuate inequality and dependency. Rodney's analysis is rooted in historical materialism, providing a comprehensive assessment of the historical processes that shaped the economic disparities between Europe and Africa which consequentially include the digital divide.

Kresge (2011) argues that despite suffering through the horrific system of slavery, sharecropping and the Jim Crow era, early African-Americans made immeasurable contributions to science and technology. Due to decades of colonization, cultural expansionism and the power of multi-national medicinal industry, traditional doctors and traditional medications have been downgraded and their value to communities understated (Richter, 2017).

AFRICA'S RICH TECHNOLOGICAL HERITAGE

Technology is defined as scientific knowledge that is put to practical ends (Peter, 2015:15). This knowledge mainly originates from mathematics and the physical sciences, and it is used in designing machinery, materials and industrial processes (Ibid).

Africa has over the years made some contributions to knowledge towards digital technology. In fact, mathematics and other numerical computations were done in Africa for thousands of years (Sydella, 2013). Ahmes, the great mathematician, was African and is renowned for writing the first textbook in mathematics (Aniakor, 2011:15-16). A number of modern high-school-level ideas in mathematics first were developed in Africa, as was the first method of counting (Blatch, 2013). More than 35,000 years ago, Egyptians scripted textbooks about mathematics that included division and multiplication of fractions and geometric formulas to calculate the area and volume of shapes (Woods,2013).

Eight thousand years ago, people in current Zaire formulated their own counting system, as did Yoruba people in what is now Nigeria (Blatch,2013). The Yoruba system was based on units of 20 (instead of 10) and needed an astonishing measure of subtraction to identify different numbers (Ibid). Distances and angles were also calculated, algebraic equations were solved and mathematically based predictions were made founded on the size of floods of the river Nile (Utsua, 2015).

The [1] theory of binary numeral system was also widely known throughout Africa before much of the world (ibid). This theory is said to have influenced western geomancy which led to the development of the digital computer (Eglash, 1999).

Similarly, Cook (1974: xiv) asserts that Pythagorean mathematics and modern-day science have their roots in Egyptian cosmogony and science. Mathematics with its origins in Africa, is therefore, a contributory element to digital technology as it is used in coming up with computer programs and systems used in capturing, sorting and analysing data.

According to Shillington (2005) modern man first developed in the Great Rift Valley of Africa, where the first development of tools is found. Specifically, The Homo habilis, who resided in East Africa, developed the first tool making industry, the Olduwan, around 2.3 million BC (Ibid). Homo ergaster developed the Acheulean stone tool industry, precisely hand-axes, in Africa, 1.5 million BC (Ibid). This tool industry is said to have spread to the Middle East and Europe around 800,000 to 600,000 BC. Homo erectus began to use fire while Homo sapiens or modern humans created bone tools and the back blade around 90,000 to 60,000 BC, in Southern and Eastern Africa (Ibid). The use of bone tools and back blades became characteristic of later stone tool industries (Ibid).

The appearance of abstract art also emerged during this period. The oldest abstract art in the world is a shell necklace dated 82,000 years in the Cave of Pigeons in Taforalt, eastern Morocco (Peter, 2015). The second oldest abstract art and the oldest rock art is found in the Blombos Cave at the cape in South Africa, dated 77,000 years. (Ehret, 2002)

Lithium is among minerals that are currently being used in the making of computer chips that help in the processing of data. Africans had an idea of what type of stone would be suitable for which activity long before the era of colonisation.

In the area of built environments, various past African societies made some significant contributions. An engineering accomplishment of note, is that built by Egyptians, with strangely raised pillars and built more than 80 pyramids. Imotep, the renowned Egyptian scientist, besides making significant contributions to medicine, also designed the Egyptian pyramid, which happens to be one of the wonders of the world (Adigbuo, 2017:77). The largest of the pyramids covers 13 acres and is made of 2.25 million blocks of stone (Shishima, 2014). Later, in the 12th century and much farther south, there were hundreds of great cities in Zimbabwe and Mozambique (Utsua, 2015:15).

Navigation is another contested area where AIK has not been properly 1 acknowledged and positioned. It is generally said and believed by a number of people that the Europeans were the first to sail to the Americas (ibid). To the contrary, numerous lines of [6] historic evidence point to ancient Africans sailing to South America and Asia hundreds of years before Europeans, in some cases, using small reed[1] -based vessels, sailboats and impressive structures with many compartments and cookery amenities (ibid). Iduring the same time that the ancient Africans were sailing to South America in the 13th century, these ancient persons also sailed sailed to China and back, carrying consignments of elephants (Ibid). The historical evidence of ancient Africans navigating vast oceans and engaging in transcontinental voyages to South America and Asia, including carrying elephants to China, underscores an extraordinary level of sophistication and maritime prowess. These insurmountable achievements speak to a depth of knowledge in navigation, shipbuilding and resource management that challenges stereotypes and misconceptions about the capabilities of ancient societies. The ability to undertake such mammoth tasks requires not only advanced technical skills but also a profound understanding of astronomy, geography and complex logistical arrangements and skillful planning. Recognizing these historical feats highlights the need to reevaluate our understanding of ancient African civilizations and their contributions to global exploration, emphasizing the remarkable levels of sophistication that existed in various aspects of their societies.

Africa has also made significant contribution to digital technology through creative arts. Azi (2012) has coined this phenomenon *Afrimation* (African Animation). Azi explains *Afrimation* as home-grown digital contents that directly or indirectly inclined to Africa's progress and characterized by one or more of the following factors: (a) cultural significance; (b) economic capability, and (c) social worth.

An article on six (6) modern 'indigenous' *Afrimation* projects on Kabongo; Tinga Tinga Tales; Zambezia; The Lion of Judah; Magic Cellar and Interactive Child Learning Aid Project (i-CLAP) Model, Azi sheds insights on their probable contribution to the digital technology conversations, in keeping with the African Renaissance agenda. Figure 1 below show the 2D-animated series that was developed by Pictoon studio, in Senegal, established by Pierre Sauvalle and Ndiaye Aida (Unwin, 2005), which underscores the need for Africans to sell their culture or else others will do it for them. The latter should be avoided as it is likely to result in distortions and/ or loss of intellectual property benefits.

In 2003, the animation on 'Kabongo and Golo his Singing Monkey' got international recognition on the Canal France International (CFI) satellite service for its grandeur, motivation, and Africanness (Riggs, 2000).

In the right picture of Figure 1, Pictoon studio portrays 'The general assembly of diseases' which seeks to inform the African viewers on the menaces od diseases (Azi, 2012). These are confirmation of the contribution that Africa made to the digital technology thus helping to support the thesis of this chapter.

RECOMMENDATIONS ARISING OUT OF THE STUDY

In view of the above assertions, the following observations are critical in repositioning Africa as a global player in the context of harnessing AIK, for the benefit of mankind

Figure 1. Kabongo and Golo, his singing monkey (left). The general assembly of diseases by Pictoon (right)
Sources: www.wipo.int/wipo_magazine/en/2005/05/article_0004.html
https://wn.com/The_General_Assembly_of_Diseases_Scenerios_from_the_Sahel
(Original Idea by Mamadou Macki Bah, aged 17, from Kayes, Mali) (Greenleaf, 2011)

thereby rebutting the false impression that African has not played a part in the development of the digital divide;

1. **Rapid Invest in African-led research:** There is need to prioritize funding and resources for research initiatives led by African scholars and practitioners to explore and document the depth and breadth of African Indigenous Knowledge (AIK) in shaping digital technology. This ensures that perspectives from within the continent are central to understanding and harnessing its contributions to the use of the technologies.
2. **The need to Foster collaborative partnerships**: An urgent appeal has to be made to facilitate partnerships between local scholars, international researchers and indigenous communities to ensure a holistic and culturally sensitive approach to studying AIK. Collaborative efforts can lead to more accurate representation and utilization of indigenous knowledge systems within the digital and other technology sectors.
3. **Incorporate AIK in educational curricula**: There is need to integrate teachings and perspectives on African Indigenous Knowledge into formal education systems at all levels. By incorporating AIK into curricula, future generations can gain a deeper appreciation for the continent's historical contributions to digital technology and be better equipped to leverage indigenous knowledge for innovation and problem-solving in advancement of the Fourth Industrial Revolution and collorary industrialization drives.

CONCLUSION

In conclusion, it is shown in the above discussion that AIK is a major player in the development of digital technology despite the criticisms, misconceptions and stereotypes on it being inaccurate, oral, non-replicable, uncivilised, anti-Eurocentric, instinctive and static. Such negative views on AIK are perpetuated by colonisers and researchers who deliberately do not conduct researches that correctly position AIK. Specifically, AIK contributed towards developments in mathematics, numerical computations, engineering, energy, tool manufacturing, animation and abstract thinking, which were quite instrumental in shaping the modern digital age. Going forward there is need for a paradigm shift in acknowledging and celebrating Africa's role in shaping the digital age and global governance so as to harness indigenous knowledge for global problem-solving in the 4IR.

It is important to push for continued research on AIK, preferably by African researchers who best understand the context and issues around AIK, to strengthen the argument that the birth of digital technology is firmly rooted in Africa. In the process, African practitioners need to be consulted within their spaces and contexts, as they are the repositories of AIK. This does not come cheap. It requires substantial funding which must be mobilised from African states, the developed countries and partners.

While the contributions of African Indigenous Knowledge (AIK) to mathematics, numerical computations, engineering, energy, tool manufacturing and abstract thinking are evident, further exploration and documentation is necessary to fully understand the extent of Africa's influence on the development of digital technology. This research should involve collaborative efforts between local scholars, international researchers and indigenous communities to ensure a comprehensive and accurate representation of Africa's historical and contemporary contributions to this critical sector.

Moreover, as Africa strives to create a strong economic bloc, exemplified by initiatives like the African Continental Free Trade [7] Area (AfCFTA), there is a special window of opportunity to assert and celebrate the continent's pivotal role in the birth and growth of digital technologies. This paradigm shift is not only essential for acknowledging Africa's historical contributions but also for harnessing indigenous knowledge in addressing global challenges within the context of the Fourth Industrial Revolution (4IR). By recognizing and valuing Africa's role in shaping the digital age, the continent will continue to actively contribute to global problem-solving, innovation and sustainable development in the years to come. In actual fact, Africa can reposition itself as the foremost global player in the development and use of digital technologies.

REFERENCES

Adams, H. (1983). African Observers of the Universe: The Sirius Question. *Blacks in Science: Ancient and Modern Science*.

Adigbuo, E. R. (2017). African indigenous knowledge: The challenges of industrialization. *International Journal of Humanities Social Sciences and Education*, *4*(5), 76-84. www.arcjournals.org doi:10.20431/2349-0381.0405009

Aniakor, C. C. (2011). Strategies for human capital development in Nigeria: A perspective. In The humanities and human capital development. The Humanities, Delta State University.

Apffel-Marglin, F. (2011). *Subversive spiritualities: How rituals enact the world*. Oxford University Press.

Augustine, S. J. (1997). *Traditional aboriginal knowledge and science versus accidental science*. Paper prepared for the Biodiversity Convention Office of Environment Canada. http://www.nativemaps.org/?q=node/1399

Azi, J.I. (2012). Appraising the role of Afrimation (African-Animation) in promoting Africa's rich cultural heritage in a digital age. *International Journal of computer graphics and animation (IJCGA), 2*(2/3).

Bandama, F. (2020). *Preindustrial mining and metallurgy in Africa*. Available at: https://www.researchgate.net/publication/345918784

Battiste, M. (2017). Protecting indigenous knowledge and heritage. In *African indigenous knowledge: An underutilized and neglected resource for development. Library Philosophy and Practice (e-journal).* 1663. https://digitalcommons.unl.edu/libphilprac/1663

Blatch, S. (2013). *Great achievements in science and technology in ancient Africa*. ASMB. https://www.asbmb.org

Brock-Utne, B. (2000). Whose education for all: The recolonization of the African mind. In *African indigenous knowledge: An underutilized and neglected resource for development. Library Philosophy and Practice (e-journal).* 1663. https://digitalcommons.unl.edu/libphilprac/1663

Brooks, L. (2013). African achievements: Leaders, civilisations and culture of ancient. *Africa*.

Bruchac, M. M. (2014). Indigenous knowledge and traditional knowledge. In C. Smith (Ed.), *Encyclopedia of Global Archaeology*. Springer Science and Business Media. doi:10.1007/978-1-4419-0465-2_10

Cook, M. (1974). *The African origin of civilization: Myth or reality*. Lawrence Hill & Company.

Danielle, C. (2019). Digital colonialism: The 21st century scramble for Africa through the extraction and control of user data and the limitations of data protection laws. *MICH. J. RACE & L, 24*(2), 417. https://repository.law.umich.edu/mjrl/vol24/iss2/6

Diop, S. (2012). African elites and their post-colonial legacy: Cultural, political and economic discontent–by way of literature. *Alter Native, 19*(1), 71-79. https://digitalcommons.unl.edu/libphilprac/1663

Eglash, R. (1999). *African fractals: Modern computing and indigenous design*. Rutgers University Press.

Ehret, C. (2002). *The Civilizations of Africa*. University of Virginia Press.

Ellen, R., Parker, P., & Bicker, A. (2000). *Indigenous environmental knowledge and its transformations: Critical anthropological perspectives*. Harwood Academic.

Emeagwali, G., & Shizha, E. (Eds.). (2016). *African Indigenous Knowledge and the sciences*. Sense Publishers. doi:10.1007/978-94-6300-515-9

Ezeanya, C. (2015). Research, Innovation and Indigenous Knowledge in Africa. In *African indigenous knowledge: An underutilized and neglected resource for development. Library Philosophy and Practice (ejournal).* 1663. https://digitalcommons.unl.edu/libphilprac/1663

Fayola, T., & Fleming, T. (n.d.). Afican civilisations: From the pre-colonial to the modern day. *Encyclopedia of Life Support Systems (EOLSS).* Available at: https://www.eolss.net/ebooklib/sc

Gemma, R. B. (2018). *Introduction to positivism, interpretivism and critical theory*. https://www.researchgate.net/publication/323811451

Greenleaf, V. (2011). *'Lion of Judah' Animated Movie Premieres in June*. Charisma News Online. http://www.charismamag.com/index.php/news/31147-lion-of-judah-animated-movie-premieres-injune

Hountondji, P. J. (2017.) Knowledge appropriation in a post-colonial context. In *African indigenous knowledge: An underutilized and neglected resource for development. Library Philosophy and Practice (e-journal).* 1663. https://digitalcommons.unl.edu/libphilprac/1663

Jana, P. (2001). African Renaissance and the millennium action plan. *Quest, 15*, 41–46.

Johnson, M. (1992). *Lore: Capturing Traditional Environmental Knowledge*. IDRC.

Kaya, H., & Seleti, Y. (2014). African Indigenous knowledge systems and relevance of higher education in South Africa. *International Education Journal: Comparative Perspectives, 12*(1), 30–44.

Kaya, H. O., & Seleti, Y. N. (2013). African indigenous knowledge systems and relevance of higher education in South Africa. *The International Education Journal: Comparative Perspectives, 12*(1), 30–44.

Khan, S. N. (2014). Qualitative research method. *International Journal of Business and Management, 9*(11), 224–233. doi:10.5539/ijbm.v9n11p224

Khumalo, N. B., & Baloyi, C. (2017). African indigenous knowledge: An underutilized and neglected resource for development. *Library Philosophy and Practice (e-journal).* 1663. https://digitalcommons.unl.edu/libphilprac/1663

Kienon-Kabore, T. H. (2017). History of traditional science and technology in Sub-Saharan Africa: Problematic and methodology of approach. *International Journal of Academic Research and Reflection, 5*(5), 47–59.

Kimwaga, S. (2010). *African indigenous psychology and eurocentricism*. Unpublished Manuscript. Dar es Salaam, Tanzania: College of Business Education.

Knudston, P., & Suzuki, D. (2017.) The Wisdom of the Elders. Toronto, Ontario: StoddartLePena. In *African indigenous knowledge: An underutilized and neglected resource for development. Library Philosophy and Practice (e-journal).* 1663. https://digitalcommons.unl.edu/libphilprac/1663

Kresge, N. (2011). *A history of black scientists*. ASBMB Today.

Langill, S. (1999). *Indigenous knowledge: A resource kit for sustainable development researchers in dryland Africa*. IDRC.

Matike, E. (2008). *Knowledge and Perceptions of Educators and Learners in the Incorporation of IKS into School Curriculum.* Unpublished BA (Honours) thesis. North-West University, South Africa.

Mavhunga, C. (2014). *Transient workspaces: Technologies of everyday innovation in Zimbabwe*. MIT Press. doi:10.7551/mitpress/9780262027243.001.0001

Mavhunga, C. (2017). *What do science, technology, and innovation mean from Africa?* MIT Press. doi:10.7551/mitpress/10769.001.0001

Mawere, J. (202)0. Decolonising legal education in South Africa: A review of African indigenous law in the curriculum. *Pretoria Student Law Review, 14*(1).

Mbeki, T. (1996). *I am an African.* Statement on Behalf of the African National Congress on the occasion of the adoption by the Constitutional Assembly of the Republic of South Africa Constitution Bill.

McCarthy, S. (2017). In globalization and education. In *African indigenous knowledge: An underutilized and neglected resource for development. Library Philosophy and Practice (e-journal).* 1663. https://digitalcommons.unl.edu/libphilprac/1663

McIlwain, C. D. (2020). Black software: The internet and racial justice. Oxford University Press.

Moodie, T. (2017). Alternative ways of knowing–doing justice to non-Western intellectual traditions in a postmodern era. In *African indigenous knowledge: An underutilized and neglected resource for development. Library Philosophy and Practice (e-journal).* 1663. https://digitalcommons.unl.edu/libphilprac/1663

Moore, N., & Czerwinska, K. (2019). *Understanding the use of digital technology in the career development sector.* The University of Derby.

Munyoki, J. M. (2019). A Paradigm Shift for Innovation and Creativity in Africa in the 21st Century. *European Journal of Business and Management, 11*(21). Advance online publication. doi:10.7176/EJBM

Naamwintome, B. A., & Millar, D. (2015). Indigenous knowledge and the african way forward: Challenges and opportunities. *OAlib*, *2*(3), e1295. doi:10.4236/oalib.1101295

Ngugi, T. (1986). Decolonising the mind. The International Education Journal: Comparative Perspectives, 12(1), 30–44.

Nkondo, M. (2012). *Indigenous African knowledge systems in a polyepistemic world: The capabilities approach and the translatability of knowledge systems.* Paper presented at the Southern African Regional Colloquium on Indigenous African knowledge systems: methodologies and epistemologies for research, teaching, learning and community engagement in higher education. Howard College Campus: University of KwaZulu-Natal.

Ocheni, S., & Nwankwo, B. C. (2012). Analysis of colonialism and Its impact in Africa. *Cross-Cultural Communication, 8*(3), 46–54. doi:10.3968/j.ccc.1923670020120803.1189

Ocholla, D. N., & Dlamini, P. (2017). *Exploring the use of ICT tools for capturing, preserving, and disseminating Indigenous knowledge among IK owners and users: A preliminary study of IK users in the KwaZulu Natal Province, South Africa.* Paper Presented at the 3rd Moi University Annual International Conference, Moi University, Kenya. https://digitalcommons.unl.edu/libphilprac/1663

Ocholla, D. N., & Onyancha, O. B. (2017). The marginalized knowledge: An Afrocentric analysis of indigenous knowledge publications 1990-2004. In *African indigenous knowledge: An underutilized and neglected resource for development. Library Philosophy and Practice (e-journal).* 1663. https://digitalcommons.unl.edu/libphilprac/1663

Richter, M. (2017). Discussion paper prepared for the Treatment Action Campaign and AIDS Law Project. In *African indigenous knowledge: An underutilized and neglected resource for development. Library Philosophy and Practice (e-journal).* 1663. https://digitalcommons.unl.edu/libphilprac/1663

Riggs, F. W. (2000). *Globalization key concepts*. COCTA Panel on Globalization, World Sociology Congress, Montreal, Canada.

Rodney, W. (1982). How Europe Underdeveloped Africa. In African indigenous knowledge and the sciences. Sense Publishers.

Schutte, G. (1999). Colonialism and Eurocentrism. *Revista de Antropologia, 22*(2), 120–134.

Shillington, K. (2005). *History of Africa* (Rev. 2nd ed.). Palgrave Macmillan. Available at https://www.africaworld.net/afrel

Shishima, S. D. (2012). Challenges to religious freedom in Nigeria: The case of African religion. Department of Religion & Philosophy, Benue State University, Makurdi.

Shishima, S. D. (2014). African religions. Makurdi: Obeta Continental Press.

Shizha, E., & Emeagwali, G. (Eds.). (2016). *African indigenous knowledge and the sciences*. Sense Publishers.

Shore, D. (2013). Steel-making in ancient Africa. In *Great achievements in science and technology in ancient Africa.* ASMB. https://www.asbmb.org

Sithole, J. (2017). *The challenges faced by African libraries and information centres in documenting and preserving indigenous knowledge*. Paper Presented at the World Library and Information Congress: 72nd IFLA General Conference and Council, 20-24 August 2006, Seoul, Korea. https://digitalcommons.unl.edu/libphilprac/1663

Smith, A. (2002). Power and hierarchy of knowledge. *Geoforum*, *40*(1), 230–248.

Smith, L. T. (2012). *Decolonizing methodologies: Research and indigenous peoples* (2nd ed.). Zed Books.

Stelios, M., & Elias, P. (2016). The long-run effects of the scramble for Africa, 106 AM. *Economic Review (Kansas City, Mo.)*, *1802*, 1807.

Sydella, B. (2013). *Great achievements in science and technology in ancient Africa*. ASBMB.

Tauli-Corpuz, V. (2017). Cultural diversity: The right of indigenous people to remain different and diverse. In *African indigenous knowledge: An underutilized and neglected resource for development. Library Philosophy and Practice (e-journal)*. 1663. https://digitalcommons.unl.edu/libphilprac/1663

Thondlana, J., & Garwe, E. V. (2021). Repositioning of Africa in knowledge production: Shaking off historical stigmas — introduction. *Journal of the British Academy*, *9*(s1), 1–17. doi:10.5871/jba/009s1.001

Thrupp, L. A. (1989). Legitimizing local knowledge: From displacement to empowerment for third world people. *Agriculture and Human Values*, *6*(Summer Issue), 13–24. doi:10.1007/BF02217665

Timothy, R. (1998). Views from the South: Intellectual hegemony and postmodernism in developing societies. *Revista de Antropologia*, *20*(1), 61–78.

Unwin, T. (2005). *Partnerships in development practice: Evidence from multi-stakeholder ICT4D Partnership Practice in Africa*. UNESCO Publications for the World Summit on the Information Society.

Utsua, T. P. (2015). The Evolution of African Indigenous Science and Technology. *Historical Research Letter, 16*. Available at: http://www.iis.te.org

Van Sertima, I. (2013). The lost sciences of Africa: An overview. In *Great achievements in science and technology in ancient Africa*. ASMB. https://www.asbmb.org

Voronkova, V., Nikitenko, V., Oleksenko, R., Andriukaitiene, R., Kharchenko, J. & Kliuienko, E. (2023). Digital Technology Evolution of the Industrial Revolution From 4G to 5G in the Context of the Challenges of Digital Globalization. *TEM Journal, 12*(2), 732-742.

Walter, D. (2002). Colonial and post-colonial discourses in social sciences: A cultural critique of colonialism. *Latin American Research Review*, *28*(3), 120–134.

Wane, N. N. (2017). African Indigenous Knowledge: Claiming, writing, storing, and sharing the discourse. *Journal of Thought, 40*(2), 27-46. https://digitalcommons.unl.edu/libphilprac/1663

Warren, D. M. (1991). *Using indigenous knowledge for agricultural development*. World Bank discussion paper 127.

Weldegerima, B. (2017). Review on the importance of documenting ethno pharmacological information on medicinal plants. In *African indigenous knowledge: An underutilized and neglected resource for development. Library Philosophy and Practice (e-journal).* 1663. https://digitalcommons.unl.edu/libphilprac/1663

Woods, G. (2013). Science in Ancient Egypt. In *Great achievements in science and technology in ancient Africa*. ASMB. https://www.asbmb.org

Chapter 3
Effects of the Trans-Atlantic Slave Trade and Colonialism on Indigenous Knowledge for Digital Technology in Africa

Eric Blanco Niyitunga
University of Johannesburg, South Africa

Logic L. Modibedi
Johannesburg Business School, South Africa

ABSTRACT

Over time, Indigenous knowledge has preserved distinctive understandings rooted in cultural experience and creativities that guide relations among human, non-human, and other-than-human beings in specific ecosystems. These understandings and relations constitute a system broadly identified as Indigenous knowledge, also called traditional or Aboriginal knowledge. Archaeologists conducting excavations in Indigenous locales may uncover physical evidence of Indigenous knowledge (e.g., artifacts, landscape modifications, ritual markers, stone carvings, faunal remains).

INTRODUCTION

Over time, Indigenous Knowledge has preserved distinctive understandings rooted in cultural experience and creativities that guide relations among human, non-human, and other-than-human beings in specific ecosystems. These understandings and relations constitute a system broadly identified as Indigenous knowledge, also

DOI: 10.4018/978-1-6684-7851-6.ch003

Copyright © 2024, IGI Global. Copying or distributing in print or electronic forms without written permission of IGI Global is prohibited.

called traditional or Aboriginal knowledge. Archaeologists conducting excavations in Indigenous locales may uncover physical evidence of Indigenous knowledge (e.g., artifacts, landscape modifications, ritual markers, stone carvings, faunal remains). Still, the meaning of this evidence may not be evident to non-Indigenous or non-local investigators. Researchers can gain information and insight by consulting Indigenous traditions; this localized knowledge contains crucial information that can explain and contextualize scientific data. Archaeologists should, however, strive to avoid interference with esoteric pieces of knowledge, sacred sites, ritual landscapes, and cultural property. Research consultation with local Indigenous knowledge bearers is recommended to ensure ethical practice and avoid unnecessary harm to sensitive sites and practices.

This chapter aims to investigate and analyze the effects of the trans-Atlantic slave trade and colonialism events on the development of indigenous knowledge and explain how these effects have prevented the contributions of Africa to the wave of digital technology in the global system. The chapter seeks to explain how these two events that occurred in Africa demolished and buried indigenous knowledge to death and erased some of the technological advancement the continent has achieved, and how the legacies of these events are still rampant in Africa, thus preventing any contributions of the continent to be remarkable in the world of digital technology. The chapter argues that while trans-Atlantic demolished local ideology and traditional knowledge that were the basis of technological advancement, colonialism erased all the evidence that justified the history of digital technology in Africa, led to colonial education and governance that kept in the darkness Africans as far the continent contributions to the wave of digital technology in the world. This has led Africans and the rest of the world to fail to see that digital technology has its origins in Africa, and its advancement depends on Africa's raw materials, natural resources, and indigenous knowledge that served as a starting principle. This is because scholars have not yet understood that Africa, as the mother of all civilizations (Ben-Jochannan, 1997), the mother of science, technology, mathematics, education, and the origin of mankind, has much to contribute to the world of digital technology. The argument is that Africa, the origin of mankind, automatically becomes the origin of technology and its advancement because you cannot separate mankind and technology. To understand these realities, one needs to assess the effects of the trans-Atlantic slave trade and colonialism and the prevalence of their legacies on digital technology in Africa. Africa suffered more than any other continent because, after the abolition of the trans-Atlantic slave trade, another deadly event called colonialism was imposed that inhibited any efforts to address the effects and legacies of the previous event on the technology. The introduction of colonialism, which was eviler than the trans-Atlantic slave trade, buried to the grave the contributions of the African continent to the wave of digital technology such that through proxy governance, all the contributions are attributed to the colonial masters. Hence, the main research question the chapter seeks to answer is how the effects of the trans-Atlantic slave trade and colonialism on Africa's indigenous knowledge affected the continent's contributions to digital technology.

The analysis in this chapter is divided into six main sections. The first section defines slavery and explains the trans-Atlantic slave trade in Africa. The second section gives a brief history of the trans-Atlantic slave trade and explains its ramifications on economic growth, development, and African governance to affect Africa's contributions to the wave of digital technology. The Third section explains the impact and effects of the trans-Atlantic slave trade on Africa's indigenous knowledge and digital technology advancement. This is because the advancement of digital technology depends not only on indigenous knowledge but also on economic growth, financial abilities, and good governance. The fourth section defines colonialism and explains its characteristics in Africa. The section equally explains colonialism's effects on Africa's indigenous knowledge and explains factors that have made those factors that affected Africa's contributions to the wave of digital technology. In this section, the chapter explains those effects individually and shows how they affected Africa's contributions to the wave of digital technology. For example, the effects of the colonial monetary system and governance have weakened digital technological advancement in Africa. The fifth section explains the factors that have made the existing colonial governance in Africa have emanated from colonial collaborators and thus have affected political governance and stability, without which Africa's contributions to the wave of digital technology would not be made available and heard to the global world. The sixth section consists of the chapter's concluding remarks.

SLAVERY DEFINED

Slavery is an old social institution in human history. From its inception in the distant past, there was no conscious historical theorization of the term, which was then more or less reduced to a state of mind that was often learned from incidental remarks, especially based on the treatment given to slaves (Schlaifer, 1936: 1). According to Morgan (2005: 1), almost every society in the history of the world had at one time or the other experienced slavery. He equally opines that a slave society is one in which slaves play a servile role and form a significant proportion of about 20% of the total population. Though slavery had been in existence for a time immemorial, there had never been a clear-cut definition of the term. Scholars, however, made several efforts to define and conceptualize the term in different but related ways. According to Allain and Hickey (2012:1), there is no clear understanding of the meaning of 'slavery' even in modern international law.

The Slavery Convention defines "slavery as the status or condition of a person over whom any or all of the powers attaching to the right of ownership is exercised" (Slavery Convention, 25 September 1926, 60 LNTS 253). From this definition, one can understand that slavery is the process of a person owning another person, whether mentally or physically. This means that slavery requires that a person owns another person. The definition also shows that possession is one of the powers attached to the right of ownership (Allain, 2017:37). To exercise possession over a person is foundational and paramount to slavery. It is the hallmark of trans-Atlantic slaves; slaves from Africa were owned by European slave traders and exercised power over them. Allain (2017:38-39) argued that slavery can only be present if possession is present and control tantamount to possession is being exercised. This foundational possession is a hallmark of slavery, and only if possession is exercised can any or all the other powers attached to ownership be exercised (Allain, 2017:38-39). This is because it becomes impossible to sell something without possessing it and having power over it. Similarly, one cannot sell a person if they do not have power over them or control them in a manner equivalent to possession. Thus, the ability to sell a person indicates the presence of control equivalent to possession. Honoré (2012) noted that possession allows for the ability to sell. The ability to sell on the other indicates possession (Honoré, 2012).

Possessing somebody is associated with the violation of human rights because it makes a person lose their freedom and deciding ability over their lives. It is worth emphasizing that losing their liberty to somebody perceived as superior or to be in control shows the evidence of slavery. For example, slaves taken from Africa during the trans-Atlantic slave trade lost their liberty because they were bought and sold. They lacked control over their lives and lost their ability to make their own decisions. This means they were in the possession of other forces, which compelled them to do what they would not want to do. For this reason, torture and violations of their rights because a norm. Slaves, having been sold to another person and forced to move to unfamiliar places, were on their own against their will. Losing human will would mean losing the ability to become independent and control one's life. In cases of slavery, somebody is exercising control in such a manner as to deprive the enslaved of her or his individual liberty significantly. The person with the power to enslave dictates what the enslaved is to do and backs up this dictum with violence, either actual or latent (Honoré, 2012).

A second power attached to the right of ownership is the ability to use a person. Again, one person can use another, but this need not amount to slavery (Honoré, 2012). Nevertheless, such use may amount to slavery if the background relationship of control is present to such an extent that it is equivalent to possession. By using a person, what is meant is the deriving of benefit from their service or labor (Allain, 2017:38-39). In the case of slavery, such benefit might be the savings incurred because of paying little or no salary for labor or the gratification from sexual services (Allain, 2008:342). The power attached to the right of ownership manifest in the ability to manage the use of a person is closely associated with the use of a person. In general terms, it goes without saying that to manage a person is not to enslave them.

FORMS OF SLAVERY

Slavery points to the situation where one is deprived of personal freedom and is forced to serve another. There are different forms of slavery, but the chapter only identified and explained four forms. The first form of slavery is royal slavery. This form of slavery existed when a robust community of slaves emerged, holding certain political positions. This form of slavery also emerged during the Arab slave and trans-Atlantic slave trades (Stilwell, Hamza, and Lovejoy, 2001, p. 273). Stilwell et al. (2001) argued that the royal slaves were also considered elitists who wield administrative power, especially in the political and socio-economic life of the people. Scholars have argued that numerous royal slaves are made warlords, heads of guards, chiefs of mounted escorts, and heads of royal farmland, among other responsibilities (Lovejoy, Mahadi and Mukhtar, 1993, pp.5-6). They play a vital role because they gather a wide range of intelligence information about politics and political intrigues obtained in their lives (Stilwell, 1999: 12).

The second form of slavery is war slaves. This form of slaves was used by powerful kingdoms and empires to fight wars of conquest and protect the political boundaries of such states (Nasidi, 2022, p. 141). An example of this form of a slave is found in the events surrounding the Ottoman Empire and the Funj Sultanate of Sinnar (Fadle, 2005, p. 17). The third form of a slave is economical, also known as an industrial slave. This form of slavery points at those slaves that were enslaved basically to serve as an economic workforce, mainly to produce surplus goods (Nasidi, 2022, p. 142). An example of this form of slave would be the military slaves, who were also found within the hearts of kingdoms and empires (Nasidi, 2022, p. 142). This form of slavery is at the bedrock of the trans-Atlantic slave trade that is the focus of this chapter. Trans-Atlantic slaves were taken from Africa and forcefully transported to Latin America and the West Indies to work in the various European plantation fields (Nasidi, 2022, p. 142).

The fourth form of slavery is mental slavery. This form of slavery is deadly because its hidden forces are usually not seen, and they have tremendous impacts on the people. This is because, in this form, there is no use of force against one's will that is used. This means that no force is used, and the slaves are presumably considered 'free', sing, and talk about as well as celebrate their freedom while mentally bound or enslaved. This kind of slavery existed in Africa in this era of coloniality, and the slave asters achieved it through social engineering, manipulation of knowledge, and civilization. It is spread through education and embedded in the coloniality of power and knowledge (this is primarily explained in Chapter 6). Coloniality is the current program used by former European colonialists on the various African states in the name of the civilizing mission. It has produced mental slavery, which prevents Africa's contributions to the wave of digital technology. Former European colonialists have succeeded through this means in luring many Africans to love and appreciate their mental slavery by thinking that the only way to civilize is to westernize (Nasidi, 2022, p. 145). This is a terrible form of slavery with horrible effects. Many Africans, both citizens and political class, have succumbed to this form of slavery and, by extension, are referred to as slaves. This manifests when the former colonial forces enslave Africans, which in turn makes Africans end up enslaving each other. As former colonialists would not wish to see Africa rising above the first world in terms of digital technology, they dictate to their slaves in academia and political leadership to inhibit any light advanced in that regard. Thus, Africa's contributions to the wave of digital technology are thwarted, and its light is deemed.

CHARACTERISTICS OF SLAVERY

There exist different characteristics of slavery that help one understand the ugly faces of slave trans-Atlantic slave trade. The lack of personal ownership is the first characteristic of slavery. This indicates the lack of ownership and will to do what one proposes without compulsion. Boahen (2012, p.167) argued that the lack of personal ownership is embedded in the institutions "whose powers spread to their members, ensuring that slaves become just 'pieces of property, lacking any other social dimension." This characteristic has led to an extensive lack of freedom to pursue their lives as they saw fit. This phenomenon was at work during the trans-Saharan slave trade. Through this feature, approximately twelve million African people were forcefully enslaved and transported to the West (Basil, 1961; Curtin, 1971; Christopher, 2001). Even though slaves lack personal ownership, a certain degree of freedom can be accorded to them to the extent that the difference between slaves and free individuals becomes slender. These freedoms may include rights to own properties, marriage, and the enjoyment of social mobility by rising to occupy sensitive positions of authority, among others (Boahen, 2012, p. 168).

Commodification is the second characteristic of slavery. During the trans-Atlantic slave trade, the commodification of African people was considered a major characteristic of slavery. Commodification is an act or fact of turning something into an item that can be bought and sold. Thus, African people were commodified and ceased to become human beings and were seen as commodities. This granted a slave owner some unrestricted rights to buy or sell his slaves. It was practiced during the trans-Atlantic slave trade, which encouraged the buying and selling of slaves as another form of commodity. Oliver and Fage (1969: 268), quoted in Nasidi (2022), stated that "it was only with the growing European demand, from the sixteenth century onwards, that the large-scale slave trade developed." Boahen (2012, p.178) further asserted that the growth of European mercantile activities greatly influenced the economic fortunes of the privileged while stimulating the rapid expansion of slavery to various coastal societies." For instance, in Senegal and Gambia, the trans-Atlantic slave traders initiated the slave trade as early as the 1500s. They compelled local chiefs to contribute to the trade by selling to their citizens. The same compulsion was used in the Benin Kingdom, which became a primary slave trading state during trans-Atlantic slave slavery (Manning, 2006, p. 101).

The third characteristic of slavery is dehumanization. This is the process of denying someone their full humanity in others, along with the accompanying cruelty and suffering. It is the viewing and the treatment of other people as though they lack the mental capacities commonly attributed to human beings. Dehumanization is a psychological process whereby opponents view each other as less than human, thus not deserving of moral consideration (Maiese, 2003). Though the concept of dehumanization can escape the definitional boundaries of what constitutes slavery to issues of global human trafficking, for instance, the concept is associated with slavery. This was the reason why many slaves were maltreated, dehumanized, and even killed without any socio-political intervention (Nasidi, 2022). The dehumanization of slaves was more pronounced in the European model of slavery, where slaves were forced to walk and work even when they were ill. Failure to do so was tantamount to severe punishments, including death. Unique chains were also used to tie slaves like animals. Dodge (2013, pp. 77-90) found that between 2.5 and 5 million enslaved Africans died during their shipment to Europe and the Americas. These death rates were no doubt attributed to the dehumanizing treatment of such African slaves by their European enslavers. Despite a series of slave resistance and revolts in different parts of the European plantation fields, it was always brutally suppressed, ending with the killing of many slaves. A classic example was the maroon rebel slave communities in the Americas (Burnard, 2008, pp. 143- 144).

The fourth characteristic of slavery is the deprivation of someone's rights. Enslaved people are deprived of their rights to association, rights to own properties, and the rights to religion and food simply because they are considered by societal 'laws' as individuals who have completely lost their personal ownership. On this basis, therefore, they can be treated as mere chattels that can be used and thrown away. As human beings, too, slaves are deprived of their rights to partake in the decision-making process concerning their lives and well-being. Being deprived of rights made them vulnerable and weak, such that they were interchanged and sold to other slave traders without their consent. This is because they were dehumanized and reduced to a commodity, which made them entirely owned by someone (Nasidi, 2022).

THE TRANS-ATLANTIC SLAVE TRADE EXPLAINED

From the 15th century onwards, European states introduced a fearful form of enslavement that destroyed African values on which political structures were built (Niyitunga, 2017). African slaves were sent to Europe and their colonies in the Americas. Africans are very experienced people when it comes to labor. They heavily labored for their European masters. Thus, one can argue that Western digital technology that has promoted economic development owes much to slaves from the African continent. This was the time when the Industrial Revolution was still in its infancy. For example, millions of Africans were forced into slavery and the Trans-Atlantic slave trade and sent to European colonies, particularly in North and South America, from 1520 to 1860 (Niyitunga, 2017). The impact of the trans-Atlantic slave trade on Africa was immensely irredeemable; people who would have advanced their indigenous knowledge of the Industrial Revolution and promoted economic development were tortured, some killed, and others sent to work for the building of European economies.

Between 1400 and 1900, the African continent experienced four simultaneous slave trades. The largest and most well-known is the trans-Atlantic slave trade, where, beginning in the fifteenth century, slaves were shipped from West Africa, West-Central Africa, and Eastern Africa to the European colonies in the New World (Nunn, 2008, p. 141). The three other slave trades, the trans-Saharan, Red Sea, and Indian Ocean slave trades, were much older and pre-dated the trans-Atlantic slave trade (Nunn, 2008, p. 141). During the trans-Saharan slave trade, slaves were taken from the south of the Saharan desert, Northern Africa, inland of the Red Sea, and shipped to the Middle East and India (Nunn, 2008, p. 142). Slaves were also taken from a long Indian Ocean slave trade, Eastern Africa, and shipped either to the Middle East and India or to plantation islands in the Indian Ocean (Nunn, 2008, p. 142). How could Africa and its people prosper with this happening? No way, as well explained in the following:

In addition to the Africans captured and sold, many were killed during raids. About two-thirds of those taken were men between 18 and 30. Slave traders chose young, strong, healthy people, leaving few behind to lead families and villages. African cities and towns did not have enough workers. Family structures were destroyed (Pheko, 2001, p. 9).

Destruction of the family structures means that all the principles and structures on which community and family leadership were established were destroyed and undermined. The trans-Atlantic slave trade also left the continent hopeless in the sense that people who could advance economic development in Africa were taken away from the continent. Africans who were taken to the New World left on the continent a vast vacuum that remained unfilled for a very long time. During the trans-Atlantic slave trade alone, approximately 12 million slaves were exported from Africa (Nunn, 2008, p. 142). Another 6 million were exported in the other three slave trades (Nunn, 2008, p. 142). These figures do not include those who were killed during the raids or those who died on their journey to the coast. The total effect of the slave trades was that by 1850, Africa's population was only half of what it would have been had the slave trades not taken place (Manning, 1990, p. 171). Motsoko Pheko says that:

Over ten million Africans were taken into slavery for over 500 years; twenty percent of them died between the time of capture and embarkation; family on family were swept away, not even infants being spared. Millions upon millions of sons and daughters of Africa were sent to destruction as if they had been wild animals (Pheko, 2001, p. 9).

Moreover, it created many crises in Africa, such as deep-rooted poverty and land failure, and it also destroyed Communalism, Africa's local ideology, which previously identified African culture and used it to develop ways to improve social life and leadership on the continent. There is no non-African nation that does not feel superior to Africans economically, scientifically, and technologically. During this period, Africa again lost its power and past glory worldwide (Pheko, 2001, p. 10). It was underdeveloped by the Trans-Atlantic slave trade, slavery, and colonialism, which reduced its capacity and strength to develop (Pheko, 2001, p. 10). It is interesting to note that slavery and slave trade on the continent were made possible by the complicity of black local chiefs, especially on the coasts, who, without hesitation, decided to exchange their brothers for Arab and Western commodities (Njock, 1991, p. 42). African local chiefs present the most terrible self-imposed injustice that blacks perpetrated against their own progeny (Njock, 1991, p. 42).

The other question one can ask is why Africans could sell and exchange their fellow Africans to Western slave traders. The bottom line is the fear of being sold and the hatred that slave traders created among them. This proves the argument on royal slavery that the trans-Atlantic slave traders used to acquire slaves from the continent. This, therefore, became the leading root cause of slavery and the slave trade and the root cause of mental slavery and royal slavery that Africa still faces, which has hindered its contributions to the digital world. The trans-Atlantic slave trade is basically the trigger of African problems, such as underdeveloping the continent and deep-rooted poverty. Because violence and injustices feed on poverty, conflicts in Africa became a norm, thus preventing the rise of the African continent in digital technology.

The Trans-Atlantic slave trade, also known as the Euro-American slave trade, was a phenomenon that lasted from the 16th century to the mid-19th century. It was marked by the shipping and transportation of African slaves to the New World, particularly in the southern part of the continent (Occelli, 2015). With her failed and failing states caused by the lack of solid and equal institutions of political leadership and attendant ungoverned spaces pooled with increasing gaps between the haves and have-nots, problems have become endemic. The continent's endemic problems have caused many debates among scholars and beyond. Despite the exciting debates around digital technology, much has been said, yet the continent's contributions are left behind.

Lobakeng (2013) argued that the continent continues to face many complex problems ranging from issues of governance and poor socioeconomic development within its borders to growing terror attacks from extremist groups. The problems also include deep-rooted poverty, political instability that has ignited political insecurity, poor leadership, xenophobia attacks, heavy human rights violations, gender inequality, and domestic violence, among others. These problems have contributed to Africa's lack of visibility in digital technology advancement.

As the Trans-Atlantic slave trade became an established form of commerce, the process of capturing, stowing, torturing, and transporting African slaves was intricately developed (Muhammad & Patricia, 2003, pp. 883-947). Accordingly, the trans-Atlantic slave trade evolved into an accepted norm of the international community. Europeans, primarily the Portuguese during the fifteenth and sixteenth centuries, explored various islands and waterways of Africa in search of new sources of slaves (Higginbotham & Jacobs, 1992). The question that comes to mind is to understand or know whether Africa's indigenous knowledge was affected by the trans-Atlantic slave trade, thus inhibiting the contributions of Africa to the wave of digital technology. Answering this question would help one know that Africa has contributed considerably to digital technologies. Unfortunately, these contributions have not been acknowledged because of the lingering trans-Atlantic slave trade in Africa. This helps one understand that trans-Atlantic affected Africa's indigenous knowledge.

EFFECTS OF TRANS-ATLANTIC SLAVE TRADE ON AFRICA'S INDIGENOUS KNOWLEDGE OF DIGITAL TECHNOLOGY

The trans-Atlantic slave trade damaged the functionalities of states; some communities were destroyed, and their populations were decimated as rivals absorbed them. Millions of Africans were forcibly removed from their homes, and towns and villages were depopulated. Numerous energetic Africans were killed in slave wars while others remained enslaved on the continent. Moreover, trans-Atlantic traders caused numerous sufferings and miseries to many African communities. This kind of trade led to detribalization and deculturalization as people lost ties with their tribal/ groupings and others were left homeless. The trans-Atlantic slave trade created many problems that have affected the continent and the people to date.

The trans-Atlantic slave trade had immense effects and problems in Africa, which affected the continent from using its raw minerals and resources to contribute to digital technology. These effects have remained unsolved and continue to manifest on the personal, family, communal, and continental levels. They include the deterioration of domestic legal institutions, weakening of states, political and social fragmentation, economic incentives of warlords, divisions (in terms of ethnicity and borders), depopulation, destruction of African societies, leaders taking part in continuing the trade, and underdevelopment.

The first effect emerging from the trans-Atlantic slave trade that affected the growth of digital technology in Africa is the deterioration of Africa's domestic legal institutions. The trans-Atlantic slave trade in Africa destroyed and deteriorated domestic legal institutions. The deterioration of legal institutions is a significant effect that Africa still faces today, as it has led to the lack of rule of law in many African countries. The lack of rule of law has then led to heavy human rights violations that resulted in civil wars on the continent. The trans-Atlantic slave trade destroyed the African legal institutions as slave traders used false accusations to obtain slaves. Trans-Atlantic slave traders aggressively engaged in false accusations against individuals for community crimes like witchcraft. It is argued that before the trans-Atlantic slave trade, the African communities had established judicial and punishment penalties such as imprisonment for such crimes (Nunn, 2008, p. 153). However, the trans-Atlantic slave merchants converted such penalties into enslavement (Nunn, 2008, p. 153).

Nunn argued that some of the converted penalties into enslavement included ex-communication from the community, beatings, compensation, and exile (Nunn, 2008, p. 153). One can see that instead of the African communities subjecting their members to these penalties, they decided to enslave them for false accusations. The act of converting the penalties into enslavement was an active abuse of the judicial system. It can be noted that African leaders fearing for their lives decided to support this tribute to protect their communities and themselves against enslavement. With the deterioration of Africa's domestic legal institutions, it became impossible to have governance that would protect Africa's minerals and resources from impacting the wave of digital technology. This is because governance for digital technology is crucial for one to see clearly where the main contributions come from.

The second effect that hindered Africa's contributions to the wave of digital technology was the fragmentation of the African states, which weakened the states. Fragmentations and weakening of the states in Africa is a significant problem. This problem is still alive in Africa and has remained intractable, leading to the stagnation of digital technology in Africa. Curtin (1969, p.178) argued that the demand for slaves by the European powers brought about pressure on African societies and exacerbated internal conflicts within the communities. Curtin (1969, p.179) further presented that because of the level of insecurity that arose during the trans-Atlantic slave trade in Africa, communities on the continent had to look for means of defense. The Africans resorted to using light weapons such as knives, swords, firearms, and spears as they were the only ones available. The rise of internal conflicts in Africa during the trans-Atlantic slave trade was a result of the weakened and fragmented states on the continent. It is important to note that this problem has continued and is still practiced in Africa. As Africa is entangled in intra-state conflicts that require the use of modern weapons, the African political leaders have resorted to the use of the exchange of natural resources for weapons with the first world states and now China as well. This has led to the fragmentation of digital technology in Africa.

The third effect emerging from the trans-Atlantic slave trade that affected the growth of digital technology in Africa is that the trans-Atlantic slave trade sabotaged and restructured social relations and traditional values in Africa. Bühnen (1993, pp.67-68) argued that the trans-Atlantic slave trade resulted in the development of predatory regimes and stagnation or regression in Africa. It also led to confusion about relocation as many communities relocated as far from the slavers' route as possible for their safety. In the process of relocating and moving away from the slave trade routes, Africa's technological and economic development was hindered as many Africans put their efforts and energy into hiding and defending themselves from being traded. This act on its own did not only bring political insecurity to the continent but also retarded its development. Moreover, European slave traders intervened in the political process in Africa at that time to prevent the rise of the African centralized states that would have hampered their operations (Bühnen, 1993, pp. 57-110). The ties between villages were socially fragmented and weakened. This, in turn, hindered the formation of larger communities and broader ethnic identities (Occelli, 2015). Kusimba (2004, p. 66) argued that "insecurity confined people within ethnic boundaries constructing spheres of interaction." This is the problem that has continued in Africa. The trans-Atlantic slave trade is a significant factor that helps understand the current high level of ethnic fractionalization in Africa.

The other significant impact and negative effect of the trans-Atlantic slave trade on Africa that affected the growth of indigenous knowledge in digital technology is ethnic fragmentation. This impact is detrimental, given the fact that Africans are attacking their fellow Africans, causing harm to them, and forcefully torturing them. Even today, they see their fellows as animals, calling them cockroaches or any other name that degrades and dehumanizes human beings. This was the case in 1994 in Rwanda, where a whole community was called a cockroach. During the trans-Atlantic slave trade, villages and states engaged in the act of raiding one another and capturing slaves for trade, and this led to hostility among the Africans (Inikori, 2003, p. 117). The hostility led to conflicts between communities and societies, and social interaction was highly compromised following high levels of insecurity among various social groups, and this led to the confinement of populations within their respective ethnic boundaries and groups (Nunn,2008, p. 141). Due to the rising incidences of conflicts, suspicion, and banditry arising from the trans-Atlantic slave trade, there was the destruction of almost all the village networks and federations that had been created before in the 19th century (Nunn,2008, p. 141).

The fourth effect emerging from the trans-Atlantic slave trade that affected the growth of digital technology in Africa is the creation of mistrust among Africa. Mistrust is a terrible disease for technological development because people who would come together to promote technological innovation do not trust each other and thus fear each other. Their talents, which would have nurtured digital technology and economic development, are not put to work. The increased lack of trust or mistrust among Africans has its origin in the trans-Atlantic slave trade. This lack of trust has prevented the continent from having one voice and unity to form a continental body with good governance for digital technological advancement. It is through technological advancement that many of the problems, such as poverty and unemployment, are addressed, thus achieving economic development in Africa. It is important to note that the trans-Atlantic slave trade created mistrust and increased lack of faith among the Africans and their political leaders. Scholars have argued that the trans-Atlantic slave trade created a culture of mistrust within Africans (Nunn & Wantchekon, 2011, pp. 3221-3222). The other effect is the fact that the trans-Atlantic slave trade caused the displacement of many communities. Numerous communities were forced to migrate and flee their motherland due to fear for their safety (Nunn & Wantchekon, 2011, pp. 3221-3223). Engerman and Sokoloff (2000:63) said that many of these communities felt insecure about staying along or near the enslavement routes. This, therefore, affected the development of indigenous knowledge, which is the basis for digital technology.

The mistrust among Africans has become a long-term effect of the trans-Atlantic slave trade on Africa and continues to cause many more problems on the continent. The effects on Africa's values, norms, culture, beliefs, and rules of thumb are remarkable, all internal to the individual and transmitted from parents to children over time (Nunn & Wantchekon, 2011, pp. 3221-3252). This mistrust has persisted because it is rooted and entrenched in complementarities between cultural norms and domestic institutions (Nunn & Wantchekon, 2011, pp. 3221-3252). Tabellini (2008) explained that individuals inherit norms and traits of mistrust from their parents and make political choices that determine the quality of domestic institutions. It can be said that domestic institutions established by Africans today are not robust because they are informed by mistrust. It has been seen that even though a leader is elected by the people, because of the lack of trust in him/herself and his/her people, he/she still goes to the colonial master to get approval for his elections and to be dictated how the governance must look like. This has increased proxy leaders, which have created proxy wars that engulf contemporary Africa. Proxy wars on the continent have been a significant problem caused by the mistrust among Africans, which originated from the trans-Atlantic slave trade.

Nunn and Wantchekon (2011, pp. 3221–3252) have argued that the trans-Atlantic slave trade created in Africans a negative shock to internal norms of cooperation, which made the next generations that followed to be less trusting. Scholars have argued that mistrust made Africans choose institutions with weaker enforcement, resulting in poor behavior and low trust among future generations (Nunn & Wantchekon, 2011, pp. 3221–3252). This attitude has created a self-enforcing outcome by which low levels of trust and weak institutions persist among future generations (Nunn & Wantchekon, 2011, pp. 3221–3252). Muhammad and Patricia argued that the trans-Atlantic slave trade also produced a culture of mistrust that still prevails in all African states.

The fifth effect emerging from the trans-Atlantic slave trade that affected the growth of digital technology in Africa was the dehumanization and exploited Africans. The trans-Atlantic slave trade became the source of dehumanization, physical dependence, and psychological destruction of the Africans. It has been noted that the trans-Atlantic slave trade led to the death of very many Africans during the entire period (Doubleday & Co, 1976, p. 150). The trans-Atlantic slave traders devised a plan to ensure the physical dependence and psychological destruction of a whole people (Doubleday & Co, 1976, p. 150). The method that they used to achieve that goal was the amputation of slaves (Doubleday & Co, 1976, p. 150). This was practiced when a slave trader caught or recaptured a slave after he attempted to escape in search of freedom; the limbs parts of the slave would be amputated to use him as an example to instill fear in the other slaves not to escape (Higginbotham & Jacobs, 1992). Hugh further argued that the amputation of different parts of the slaves occurred during the Middle Passage (Hugh, 1997). Moreover, Hugh (1997, p.225) argued that "even pregnant slaves endured horrendous injuries as a result of the Trans-Atlantic slave trade, and slave traders valued pregnant slaves because they viewed the unborn babies as a new property to sell."

The sixth effect emerging from the trans-Atlantic slave trade that affected the growth of digital technology in Africa is depopulation. Depopulation has inhibited the growth of digital technology in Africa, which is depopulation. Depopulation was also caused indirectly by the European slave traders. The traders brought deadly diseases such as European strains of syphilis and smallpox, typhus, and tuberculosis (Inikori, 2003, p. 106). This is a crime and is punishable by the law. Many scholars have called it a crime against humanity (Inikori, 2003, p. 106). As over twelve million slaves were transported from Africa during the trans-Atlantic slave trade into European New World colonies, many slaves died during the process of transportation, both to the coast and during the actual raids (Inikori, 2003, p. 106). Some people were taken by force, while others were betrayed into the slave trade by their communities and societies in which they lived (Inikori, 2003, p. 106).

The trans-Atlantic slave trade had tremendous demographic effects on Africans in that over twelve million Africans were removed from Africa. This has an immense effect on the growth of digital technology and thus prevents Africa's contributions to the digital world. Rodney (1972, pp. 56-57) argued that "the export of so many people had been a demographic disaster and left Africa permanently disadvantaged compared to other parts of the world, and largely explains that continent's continued poverty." Rodney (1972, pp. 56-57) further argued that "Africa's population stagnated during this period, while that of Europe and Asia grew dramatically, and the slave trade disrupted all other areas of the economy as the top merchants abandoned traditional industries to pursue slaving and the slaving itself disrupted the lower levels of the population."

The seventh effect emerging from the trans-Atlantic slave trade that affected the growth of digital technology in Africa is corruption and corrupt arrangements. The rampant corruption and corrupt arrangements that we see in Africa originated from the trans-Atlantic slave trade. Anyone who had the opportunity to count the number of times African states have been accused of being "corrupt" in foreign media will lose count. Thus, without obscurantism, there is no doubt that all the charges are accurate, as there are both domestic and foreign investigations of corruption cases in Africa. The existence of mental slavery has reduced the capacity of African leaders and academics to deal with the issue of corruption. On this basis, the government depends on the citizens to carry out its constitutional functions and provide the needed platform for social engagement between the leaders and the citizens in addressing socio-economic poverty (Nnoli, 2003).

COLONIALISM: DEFINED

As the trans-Atlantic slave trade was a very barbaric practice, voices to stop it and its impact were echoed in many places in the world. From the 18th century, Christianity's beliefs to stop the slave trade started and firmly condemned it and any states involved in the slave trade (Niyitunga, 2017). By 1808, the USA, the United Kingdom (UK), and Denmark, who were the central leading states in the slave trade, became the first states to abolish it and make it illegal (Niyitunga, 2017). When Europeans saw that slavery and the trans-Atlantic slave trade were made illegal, and they could not continue benefiting from it, they did not lose interest in Africans at all. As European travelers and missionaries such as Dr. David Livingstone and Henry Stanley came to Africa, they discovered natural minerals like diamonds, gold, oil, etc., on the continent, which they reported to their home governments. With the desire to benefit from Africa's raw materials to advance digital technology, European governments responded by sending armed forces into Africa and establishing foreign governments, which translated into colonialism. As documented, European states concluded that if they wanted to build our economies, Africa could supply both "raw materials, such as minerals, and new markets for goods. Christianity helped abolish the trans-Atlantic slave trade but paved the way for European colonialism in Africa. Thus, European colonialism in Africa replaced the trans-Atlantic slave trade on the continent.

Colonialism is "the direct and overall domination of one country by another based on state power being in the hands of a foreign power" (Ocheni & Nwankwo, 1992, pp. 46-54). It was a system of rules that assumed the right of one person to impose their will upon another (Oba & Onyije, 2011). It involves both the material and intellectual subjugation of the colonized (Olukayode, 2014, pp. 154-164). Horvath says that colonialism is a form of temporally extended domination by people over others; as such, it is part of the historical universe of forms of intergroup domination, subjugation, oppression, and exploitation (Horvath, 1972, pp. 45-57). From a world-systems perspective, Sanderson argues that much of the history of the capitalist world economy is a history of colonialism, consisting of repeated and more or less successful attempts by the core to create a periphery to control it politically in order to exploit it economically (Sanderson, 2005, pp. 179-213).

The main objective of colonialists was political domination and the exploitation of the colonised country (Niyitunga, 2021). Settles says that political domination was used to exploit the physical, human and economic resources of the colonised state in order to benefit the colonising states (Settles, 1996). Niyitunga (2021) argued that the imposition of colonialism on Africa changed the nature and features of Africa's history forever. For example, the African modes of thought, patterns of cultural development and ways of life were forever impacted by the changes in political structures brought about by colonialism (Settles, 1996). Niyitunga (2021) further argued that Africa's image in the international system was changed through the processes of imperialism remarked in the political and economic policies that accompanied colonization. These processes have continued in the postcolonial African states, leading to endemic problems.

Colonialism had a devastating effect or impact on the African colonies. It is responsible for the present situation explained by Walter Rodney in his book *How Europe Underdeveloped Africa* (Niyitunga, 2021). It introduced a dichotomy between the center and the periphery nations. The center nations exploit the periphery nations. This constitutes its colonization and promotes poverty in peripheral states while promoting development in core states. The periphery nations produce raw materials which are expropriated by the center nations. Africa is a periphery continent, making it remain colonized and subjected to first-world dictations. African colonies produced raw materials, which the colonialists expropriated.

EFFECTS OF COLONIALISM ON INDIGENOUS KNOWLEDGE

The first effect of colonialism, which inhibited the growth of digital technology and industrialization in Africa, is the colonial monetary system, which was installed by colonialists and imposed on the Africans. This has a prominent effect on the development of digital technology and, unfortunately, has escaped the attention of scholars and political leadership classes. It continues to devalue the money system in Africa, thus making it impossible for the African states to industrialize and use the raw materials from Africa to press on with digitalization. Colonialists introduced a dual economic structure within the African economy that devastated the continent. It also brought about the disarticulation of the African economy, education, trade, market, transport, and currency institutions (Ocheni & Nwankwo, 2012, p. 51). While the trans-Atlantic slave trade introduced mistrust and dependency to Africans, "colonialists made African colonies dependent by introducing a mono-cultural economy for the territories. This dehumanized the African labor force and traders and forced Africans to work at very low wages" (Ocheni & Nwankwo, 2012, p. 51).

The monetization of the African economy was another effective instrument or strategy used by the colonialists to take direct control and political administration of the African territories (Ocheni & Nwankwo, 2012, p. 51). Before colonization, the African economy was essentially based on the barter system. Furthermore, even where a sort of currency was introduced, these currencies lacked general acceptability, were too heavy or bulky, and were hardly divisible into smaller exchange units. The currencies lacked the excellent qualities of a modern medium of exchange. The colonialists needed to monetize the African economy to integrate it into the world market and international trade to prevent the continent from growing and prospering. The currencies introduced in the African territories were those used by the colonialists back home. It was, therefore, easy for them to regulate the use and value of the currency to maintain effective control of the African economy and their administration (Ocheni & Nwankwo, 2012, p. 51).

The colonialists made the currency too difficult for Africans to obtain, thus imposing their own currency, which they controlled and valorized upon the continent. The way they did this was to make the prices of raw materials and agricultural products produced by Africans to be too cheap. On the other hand, the colonialists made the prices of goods manufactured by them to be too dear or high so that an African would spend all he had toiled for, for a year or more, to purchase a small quantity of foreign goods. The implication was that Africans kept working hard and making their labour service available to the colonialists to enjoy some of the foreign manufactured goods they required. The consequence was that while Africans became poorer, colonialist profits increased. Since the colonialists controlled the currency used in the African colonial territories, they determined the character and nature of the development of the African economy and political administration. The monetization of the African economy and the introduction of currency institutions were effective imperialistic instruments used by colonialists to maintain control and domination of African territories (Ocheni & Nwankwo, 2012, p. 51).

The second effect of colonialism on Africa's contributions to the wave of digital technology is the deterioration of Africa's economy. Rodney (1972, p. 244) asserts that "African development was blunted, halved, and turned back by colonialism without offering anything of compensatory value. The significant impact of colonialism in Africa is that it brought about the under-development of African territories, which hindered the process of industrialization in Africa (Ocheni & Nwankwo, 2012, p. 51). It is usually argued in favor of colonialism that it brought Western education and, hence, western civilization to the shores of Africa, which by implication is a positive contribution towards African development. Another significant impact of colonialism in Africa is the disarticulation of their economy. Colonialism distorted the African pattern of economic development in many ways. There was disarticulation in the production of goods, markets, traders, transport, provision of social amenities, pattern of urbanization, etc. The colonialists introduced a pattern of the international division of labor, which was to the disadvantage of Africans (Ocheni & Nwankwo, 2012, p. 51). There was also disarticulation in the type of goods produced by Africans. The colonialists compelled Africans to concentrate on producing goods meant for export. Africans were not encouraged to produce those goods required by the local population (Ocheni & Nwankwo, 2012, p. 51).

The third significant impact of colonialism that hindered Africa's contributions to the wave of digital technology was the disarticulation of their economy. Colonialism distorted the African pattern of economic development in many different ways. There was disarticulation in the production of goods, markets, traders, transport, provision of social amenities, pattern of urbanization, etc. The colonialists introduced a pattern of the international division of labor, which was to the disadvantage of Africans. They assigned to Africa the role of producing raw materials and primary products for their industries at home. Africans were not allowed nor encouraged to go into manufacturing. The only industries Africans were encouraged to build were those that would facilitate processing the raw materials for export. The African raw materials were bought at a very low price, while manufactured goods from abroad were sold at expensive prices. This situation accounted for the impoverishment of most Africans (Ocheni & Nwankwo, 2012, p. 51).

There was also disarticulation in the type of goods produced by Africans. The colonialists compelled Africans to concentrate on producing goods meant for export. Africans were not encouraged to produce those goods required by the local population. This made many Africans abandon the production of food items required to feed the teeming and growing population. The effect of this was food shortage and an escalation in food prices. The present-day situation where Africans now import their food is a carry-over from colonialism. The point being stressed here is that colonialism distorted the satisfaction of local needs regarding food production and other requirements in preference to production and satisfaction of foreign needs, especially the industries (Ocheni & Nwankwo, 2012, p. 51).

Colonialism also disarticulated African markets and trades. The traditional or original African marketing centers were distorted by colonialism. Most traditional African marketing centers or routes were formed based on local needs. When colonialism came and introduced a different need, this changed the original or traditional marketing centers to new marketing centers because it rendered them irrelevant. Colonialists created new marketing centers and routes where their required raw materials could be easily bought and evacuated back home. This led to the gradual decay or death of most original or traditional marketing centers, distorting the African development and urbanization pattern. As we know, most of these traditional African market centers constituted the traditional or original African centers. Colonialism also made African trade mainly export-import-oriented (Ocheni & Nwankwo, 2012, p. 51).

It prematurely integrated African trade and economy into the world market and international trade. It is a known fact that before a local economy fully integrates into the world economy or trade, it must have developed its internal dynamics and forces of production adequately. The consequences of premature integration are that the more advanced ones will hijack such an economy, and the vagaries in international trade will make the country a perpetual debtor. Furthermore, premature integration cannot absorb shock from the international market and will never enjoy a trade balance or comparative advantage. The export-import orientation pattern of the African economy introduced by colonialism does not allow for accelerator and multiplier effects necessary for economic advancement and development. The raw materials produced by Africans were not used by industries located in Africa but abroad. Therefore, there was no organic linkage between Africa's agricultural and industrial sectors. Consequently, the African economy could not move forward because the surplus profit appropriated from the economy by the colonialists was not plowed back or spent within the economy (Ocheni & Nwankwo, 2012, p. 51).

The colonialists distorted and disarticulated the development of a comprehensive transport system in Africa. The transport network developed was not to link different towns and rural areas for effective communication and development. The colonialists-built transport routes to quickly evacuate the raw materials from their sources or bases to the destination point where they could be effectively exported abroad. The transport network developed was essentially rails and seaports. There was no excellent effort to develop an organized road network that would help to improve the lives of the African people and their interaction with their relations in the different parts of the territories. The distorted, disjointed, and disarticulated transport system developed by the colonialists did not allow for effective agricultural and economic integration within the different parts of the African enclaves and territories. There was, therefore, an absence of economic integration and cooperation among the African territories during the period of colonialism (Ocheni & Nwankwo, 2012, p. 51).

Colonialism also brought about disarticulation in providing social amenities and the urbanization pattern in Africa. Most of the little social amenities provided during the colonial period were concentrated in a place. This made most people migrate from the rural areas where these amenities were virtually non-existent to colonial urban centers where they could be found. The consequence of this was the struggle and over-use of these amenities, the attendant overcrowding of the areas (cities), and the problems of urbanization. The consequent problems of disarticulation of provision of amenities and urbanization include rural-urban migration, overcrowding, filthy and slump environments, poor hygienic conditions, the spread of epidemic disease, social vices, tribal and ethnic problems, etc. Managing the above problems created by colonial distortion and disarticulation of amenities and urbanization in Africa has remained the most crucial problem confronting African states today (Ocheni & Nwankwo, 2012, p. 51).

The fourth effect of colonialism on digital technology in Africa is the emergence and institutionalization of classes and class struggle in the socio-economic and political life of the people. Colonialism aided the apparent emergence and development of classes in Africa. These classes include the comprador bourgeoisie, petty bourgeoisie, proletariat, and peasant. The African petty bourgeoisie served as the conveyor belt through which the colonialists exploited and siphoned the economy of African countries. There is an excellent harmony of interest between the African petty bourgeoisie and the European comprador bourgeoisie. This was why, during the period of political independence, it was the African petty bourgeois that got the mantle of leadership. The African petty bourgeois maintained the same relationship with the erstwhile colonial masters, which is why they ran their states' economy and political administration like the colonialists did (Ocheni & Nwankwo, 2012, p. 51).

Most of the African leaders or petty bourgeois maintain strong links with their erstwhile colonial masters. The African petty bourgeoisie maintained the long exploitation of the proletariat and the peasant classes. The rampant and complex nature of political instability and socio-economic malaise being experienced in most African states today has recourse to the nature and character of classes introduced in Africa by colonialism. Africa's economic and other resources are shared between the petty bourgeoisie and their European/colonial counterparts, even in this contemporary time. The nature of the political power struggle and the distribution of wealth and economic resources in the contemporary African state reflect the understanding and harmony of interest between the African petty bourgeoisie and their colonial partners/friends.

The severe impoverishment of most Africans by their petty-bourgeois leaders and marginalization, as well as oppression of the masses by those who have access to state power, are an offshoot of colonialism or colonial hang-over among African states. For example, Britain used an indirect rule system. Indirect rule policy concerns the ruling of the people through their own people or traditional institutions with close supervision from the British government. Also, the French, another significant colonizing European power in Africa, used the policy of assimilation and associations. Assimilation concerns the total integration of the French colonial colonies into the central French government in Paris. Association policy came at a later stage due to the problems the French government encountered from their initial application of the policy of assimilation. The French assimilated the "assimilable" and associated it with the "unassimilable." In Belgium-Congo, the policy differed from that of the British and French, which also applied to other European colonizing powers in Africa (Ocheni & Nwankwo, 2012, p. 51).

The fifth effect of colonialism on the development of digital technology in Africa is the legacy of colonial education. Colonial education was not rooted in African culture and could not foster meaningful development within the African environment because it had no organic linkage. Furthermore, colonial education was essentially literary; it had no technological base and was, therefore, antithetical to natural or industrial development (Ocheni & Nwankwo, 2012, p. 51). Colonial education was essentially aimed at training clerks, interpreters, produce inspectors, artisans, etc., which would help them exploit Africa's rich resources. Colonial education did not aim at the industrialization of African territories or stimulating technological development within the African environment.

Colonial education has created mental slavery and brought about distortion and disarticulation in the African indigenous pattern of education, which was rooted in African technology (Ocheni & Nwankwo, 2012, p. 51). Before fully embracing colonial education, Africans were good technologists, advancing at their rates with the resources within their environment. For example, Africans were good sculptors, carvers, cloth weavers, miners, blacksmiths, etc. They could provide and satisfy the technological needs of the various African societies. The introduction of colonial education made Africans abandon their indigenous technological skills and education in preference to one that mainly emphasizes reading and writing (Ocheni & Nwankwo, 2012, p. 51).

Colonialism brought about the emergence and institutionalization of classes and class struggle in the socio-economic and political life of the African people. This hindered the possibility of digital technology advancement in Africa. Colonialism aided a clear emergence and development of classes in Africa (Ocheni & Nwankwo, 2012, p. 51). These classes include the comprador bourgeoisie, petty bourgeoisie, proletariat, and peasant. The African petty bourgeoisie served as the conveyor belt through which the colonialists exploited and siphoned the economy of African countries (Ocheni & Nwankwo, 2012, p. 51). Colonial legacies influence contemporary African economic performance. The economic underdevelopment status of Africa is rooted in European colonization (Nunn, 2003, p. 2). As Rodney (1976, p.146) clearly expressed, Europeans have taken away the young, healthy, and the most productive forces of the continent, leaving the children, aged, and unhealthy section of the population. Thus, it is easy to deduce that the European powers contributed a lot to the economic underdevelopment of contemporary African states.

The colonizers left African poor and oppressed states, which made it difficult for digital technology to take root. Hence, Africans started their economic performance at a zero level. Colonialism turned the continent to produce primary products, cash crops, and raw materials, which had no more market value. This, in turn, caused unequal trade transactions (Farah & Mazongo, 2011, p. 3). As a result of the exploitation of necessary resources, the continent has been forced to rely on the importation of foreign products (Thomson, 2010, p. 192). This compelled African states to import very expensive manufactured goods from outside of what they gained from the export of their cheaper primary products. This leads to unequal exchange of market products and trade deficit. Hence, Africa's integration into the world capitalist economy with its less valuable products is unequal and usually makes it a risk taker.

Colonialism was not just about economic subjugation but about the ability to wrest control of the local economy from African rulers. Improving the production methods or strengthening the economy was not important. For example, the "British opposed the general development of palm oil plantations in Nigeria, despite the fact that the collection method of production was wasteful and produced a poor quality of oil" (Bimberg & Resnick, 1975, p. 347). Colonial powers instituted trade controls that limited colonial imports to those from the colonizing power and restricted exports to that same market (Settles, 1996, pp. 2-9). This reduced the freedom of choice in marketing goods previously available to commodity producers. Taking this power away under the guise of colonial development made traditional rulers weaker because their power base was destroyed (Settles, 1996, pp. 2-9). Since Africans were not allowed to improve their methods or to market their goods freely, they were forced into the colonial system. Europeans undermined the existing economic power structure by insisting on developing certain crops and made Africa reliant upon Europe for their economic destiny (Settles, 1996, pp. 2-9).

Economic development and education were considered unimportant and were left to the private sector (Bimberg & Resnick, 1975, p. 347). Designed to enrich the mother country, colonies were expected to foot the bill for their internal development and administration. One measure of the viability of a colonial territory was its financial self-sufficiency (Adebayo, 1990). "Ironically, the bill for colonialism was covered in the same way that West African nations had always paid for government: through the taxation of trade (Adebayo, 1990).

The sixth effect of colonialism on Africa's contributions to the wave of digital technology is the legacy of colonial governance, also known as proxy governance, which is still in Africa. This kind of governance has resulted from the inherited political institutions from the colonialists in post-independent African states. Colonialism did not allow for the industrialization of Africa. It assigned Africa the primary role of production of goods or raw materials in the international division of labor. Colonialism encouraged and intensified class struggle, tribalism, and ethnicity within the African colonies (Ocheni & Nwankwo, 2012, p. 51). The colonialists introduced these strategies to perpetuate or prolong their rule and domination of African territories (Ocheni & Nwankwo, 2012, p. 51).

Colonialism has had a far-reaching influence on the political aspect of the continent. The present political system of the continent is a direct reflection of the colonial system. Colonialism greatly influenced the politics of the continent by replacing indigenous institutions with strange administration (Farah et al., 2011, p. 2). Africa had a democratic culture in the pre-colonial era, which was later dismantled through the influence of slavery, colonialism, and neo-colonialism (Farah *et al.*, 2011, p. 3). Wyk (2007:3) interestingly noted that the "contemporary state in Africa is a remnant of a colonially imposed system". African states adopted the colonialists' centralized state system, which produced ethnic and authoritarian political culture (Wyk, 2007, p. 7).

Colonialists were neither interested in a flourishing democratic system in Africa nor prepared African states to administer themselves effectively in the post-independence era; instead, they used and threw them. Besides, it is noted that "the colonial state in Africa was an authoritarian bureaucratic apparatus of control and not intended to be a school of democracy"(Copson, 2001, p. 12). This shows the fact that state was used as instrument of exploitation throughout the colonial era. During the colonial era, there was no such concept of African popular participation (Copson, 2001, p. 12). The post-independence African leaders inherited this malpractice. It is due to this fact that Africa has suffered from, inter alia, evils of corruption and authoritarianism since the time of independence.

Hence, what post-African states inherited from colonialists is the spirit and attitude of mistrust among themselves and colonial institutions that entrench corruption and bring about undemocratic, dictatorial, and authoritarian political leadership. This is because the main purpose of colonialists was to exploit the resources of Africa; hence, they used an undemocratic system of administration, which has remained active in today's African political leaders. Putting it differently, African leaders have failed to consider the interests and realities of their people. Thus, when in power, they behave in the same way colonialists behaved. This brought about complex and challenging problems in Africa. The cruel administration used by colonialists for the exploitation of the continent has greatly impacted post-independence African leaders, leading to the autocratic political administration we see today in Africa. This is because this brutal colonialists' system of administration was inculcated in the minds of anti-colonial leaders of the time, who later became leaders of the independent African states.

Moreover, the notion of ethnicity left behind by the colonial powers has posed adverse problems in the overall political system of African states. The ethnic division, multiplied by the colonial system, left persistent rivalry and conflict in the continent, resulting in exclusion and marginalization in African political societies (Alemazung, 2010, p. 79). Kidane (2011, pp.15-16) also asserted that the fragmentation of ethnic identities into several states and the uneven socio-economic development among several ethnic groups become an obstacle to managing diversity. Instead of being rectified, this unholy colonial legacy has been exacerbated in the postcolonial period (Alemazung, 2010, p. 79). It is common practice to favor one ethnic group over the other and blurred inter-ethnic relations in the post-independence era. By doing so, post-independence leaders exacerbated inter-ethnic conflicts.

Colonialism has also brought about discrimination and segregation that have become a norm in the present Africa's political party system and decision-making processes. The colonial administrators and political elites made decisions on behalf of the indigenous population without considering their ideas or consulting them (Alemazung, 2010, p. 79). Inheriting from this monopolized system, African political parties become a top-down organizational structure and tend to be autocratic, causing more problems and making them intractable. Salih (2007, p. 41) argued that "current African political parties originated in the non-democratic setting of colonial rule, which was neither democratic nor legitimate." This shows that the African party system has inherited the undemocratic colonial tradition and culture.

Even though a multi-party system was introduced in Africa in the early 1990s, it is not genuinely exercised. Though parties are allowed to emerge, the environment has not been conducive to competing and holding the power of the government. Many opposition political parties in Africa are expected to be banned, criminalized for their activities, or undermined for their contribution to national policy. Most post-independent governments perverted into one-party states or military dictatorships characterized by corruption and inefficiency (Gordon, 2013, p. 4; Mazrui, 1994, p. 61). Any form of opposition to the state faces ruthless oppression and imprisonment (Gordon, 2013, p. 4). Hence, colonial legacies have compromised democratic governance, making it difficult and almost impossible for democracy to lead to good governance that would address African problems.

The other effect of colonialism that hinders Africa's contributions to the wave of digital technology is corruption. As Africa has not yet found time to address corruption that was left by the trans-Atlantic slave trade, colonialism cemented corruption on the continent. The African leaders in post-independent African states are known to be highly corrupt. This corruption can be traced back to the legacy of colonialism. Colonial powers' evil socio-political culture produced the habit of corruption in public service of contemporary Africa (Njoku, 2005, p. 99). From this, Njoku tells us that the existing corrupt behavior of contemporary leaders and Africans has caused tremendous problems on the continent due to the continuation of colonialism and its policies. Corruption is one of the most attributable factors to poverty, which leads to corruption, creating a vicious cycle. This situation is actual in Africa, and it is attributed to the far-reaching effect of colonialism. In many African countries, leaders enjoy political power to accumulate wealth (Wyk, 2007, p. 5) through corrupt means and ways. Thus, the greedy, selfish, and rent-seeking nature of contemporary African leaders is the result of colonialists' deeds. This has led to the state having its instrumental value for the wealth of the few political leaders. Ayittey, quoting Mazrui, wrote that "almost everything wrong that has gone in Africa is the fault of colonialism and imperialism" (Ayittey, 2002, p. 2).

The other effect that emerged from colonialism that affected the development of digital technology is that colonialists divided the African continent into smaller states, which brought about smaller and weak sovereignties. One of the significant colonial legacies that served as a source of many African problems is colonial boundaries inherited from colonialism by African leaders. The borders of African states were made in the Berlin Conference of peaceful partitioning of Africa by those who were strange to the continent based on their interest without considering African people's very interests and realities (Afisi, 2009, p. 64; Thomson, 2010, p. 14). They simply divided the continent to avoid intra-European war without analyzing the culture, language, ethnicity, and nature of indigenous black Africans. Thus, they arbitrarily divided the boundary of states across the continent. As a result, different heterogeneous groups with different languages, ethnicity, and culture were merged together, and at the same time, people with a common language, ethnicity, and culture were also disintegrated.

More significantly, Leakey (2002, p. 94) explained that some traditionally hostile groups of people were assigned to the same colony. At the same time, some other groups having longstanding peaceful coexistence were separated by the porous boundary of the colonizers. This unjust carving of the continent by the colonialists remained the root cause of African problems, particularly Xenophobia and conflicts in Africa (Kidane, 2011, p. 19 & Afisi, 2009:65). This is because, in post-independent Africa, the states have remained the creation of external colonial powers regardless of their realities.

Consequently, various conflicts across the continent have become a common practice. Accordingly, conflicts are identified as intra and inter-state conflicts that emanate from boundary-related problems and subsequent fragmentation and amalgamation of ethnic groups. Colonial borders have led to both interstate and intrastate conflicts in Africa, making Africa ungovernable and vulnerable to poverty, banditry, and lawlessness. The arbitrary division of ethnic communities and the subsequent occurrence of ethnic violence are the legacies of colonialism that ignored cultural differences during the creation of artificial state borders (Alemazung, 2010, p. 65). In this connection, Nguendi (2012, p. 1) noted that "many intra-state conflicts in Africa have been sparked by the forceful fusion of incompatible national groups into one state by the imposition of artificial boundaries by colonial powers."

The legacy of colonialism is not limited to intra-state conflicts across the continent. It also contributed to the outbreak of inter-state conflicts due to the fragmentation of the same ethnic groups into different countries. Arbitrary boundaries and subsequent fragmentation of ethnic groups also caused inter-state conflicts across several states of the continent. One notable example is the conflicts between Kenya and Somalia, where the segment of ethnic groups of the latter unfortunately merged into the former. This is the reason for the long tension between the two states. Thomson (2010, p. 25) noted that north-eastern Kenya has a large ethnic Somali population, many of whom identify more with their ethnic origins than the Kenyan state, and the Somali certainly wished to make this part of Kenya become Somali sovereign territory. The Somali ethnic groups are not fused with only Kenya but also other neighboring countries, which created Somali irredentism aiming at controlling those Somali-inhabited parts of the neighboring countries (Richards & Bekele, 2011).

This further resulted in different guerrilla wars in those provinces, complicating African problems. Therefore, the imperial boundary and its fragmentation of ethnic groups resulted in such persisting violence. As Thomson (2010, p. 25) noted, many Somalis living in Kenya still owe their political loyalties more to their relatives in Somalia than to the Kenyan government. Hence, the legacies of colonialism still have the potential to create future conflicts in Africa.

CONCLUSION

The chapter presented that the trans-Atlantic slave trade left the continent underdeveloped, disorganized, and vulnerable to the next phase of external forces. It undermined and uprooted political leadership that had served in Africa for a long time before the trade. The trans-Atlantic slave trade was a tragic saga as it led to Africa losing manual labor and force because it targeted young, energetic, vibrant people aged 15- 40 years old. It also led to population loss and retarded technological development that slowed economic growth and human development in current Africa. This is because the trans-Atlantic slave trade re-allocated the resources of African economies towards raiding, constrained the economic development of African states, encouraged ethnic and social division, provoked a culture of political violence and disregard for human life, and created widespread mistrust among Africans and brought inferiority complexes and contempt in Africans mindsets and attitudes. Social relations were restructured, and traditional values were subverted. The slave trade resulted in the development of predatory regimes, as well as stagnation or regression.

The chapter also found that colonialism did not allow industrialization to flourish in Africa. It encouraged and intensified class struggle, tribalism, and ethnicity within the African colonies. It perpetuated and prolonged autocratic rule and domination in Africa. Colonialism shaped both the economic and political structure of African colonies to be in line with the needs of the metropolis. It ensured that African economic and political structures, both in form and content, served the interest of their home government (European powers). This poisonous attitude and domination has continued in contemporary Africa. Instead of African leaders and elites serving the interests of their people and states, they continue to serve the interests of their colonialists. Being colonial collaborators, they have not yet cut off ties that link and unite them to colonizers. Colonialism, therefore, in all intents and purposes, was a disservice to Africa.

Before the partition of Africa, the continent had become increasingly integrated economically, with trade occurring north-south and east-west. The policies of the governing powers redirected all African trade to the international export market. Thus, today, there is little in the way of inter-African trade, and the pattern of economic dependence on colonizers continues, which has created more problems in Africa. Colonialism encouraged this development in some areas but, in many others, severely retarded the natal progress of the continent. In some decentralized African societies, to be shielded from slave traders, people evolved new styles of leadership that led to more rigid and hierarchical structures. In addition, European powers intervened in the political process to prevent the rise of the African centralized states that would have hampered their operations. Had the trans-Atlantic slave trade and colonialism never been imposed on Africa, its development would have been significantly different, and many of the problems that are currently overwhelming the continent would not exist. The trans-Atlantic slave trade and colonialism left the continent underdeveloped, disorganized, and vulnerable.

REFERENCES

Adebayo, A. G. (1990). The Production & Export of Hides and Skins in Colonial Northern Nigeria, 1900-1945. The Journal of African History, 33(2), 273 - 300.

Afisi. (2009). Tracing contemporary Africa's conflict situation to colonialism: A breakdown of communication among natives. *Philosophical Papers and Reviews, 1*(4), 59-66.

Alemazung, J. A. (2010). Post-Colonial Colonialism: An Analysis of International Factors and Actors Marring African Socio-Economic and Political Development. *The Journal of Pan African Studies, 3*(10).

Allain, J. (2008). Nineteenth Century Law of the Sea and the British Abolition of the Slave Trade. *British Yearbook of International Law, 78*, 342–88.

Allain, J. (2017). Contemporary slavery and its definition in law. In A. Bunting & J. Quirk (Eds.), *Contemporary Slavery: Popular Rhetoric and Political Practice* (1st ed., pp. 36–66). University of British Columbia Press. doi:10.59962/9780774832458-004

Arunachallam, S. (2007). *Global Village: A View from the Periphery.* Background paper commissioned by International Development Research Network (IDRC). Available at: www.idrc.ca/uploads/user-S/10583649160Arun_paper.pdf

Ayittey, G. B. N. (2002). *Sustainable Development: Promoting Progress or Perpetuating Poverty?* Profile Books.

Basil, D. (1961). *The African Slave Trade: Pre-Colonial History, 1450- 1850.* Atlantic Monthly Press.

Ben-Jochannan, Y. (1997). *Africa: Mother of Western Civilization. African-American Heritage Series*. Black Classic Press.

Bimberg & Resnick. (1977, June). Colonial Development' An Econometric Study. *The Journal of Economic History*, *37*(2), 489–491.

Boahen, K.A. (2012). The Impact of European Presence on Slavery in the Sixteenth to Eighteenth Century Gold Cost. *Transactions of the Historical Society of Ghana, 14*.

Bühnen. (1993). Ethnic Origins of Peruvian Slaves (1548-1650): Figures for Upper Guinea. *Paiduema*, 39.

Burnard, T. (2008). Review of Flight to Freedom: African Runaways and Maroons in the Americas. *The American Historical Review*, *113*(1).

Christopher, D. R. (2001). *An Archaeology of El-Mina: Africana and Europeans on the Gold-Cost, 1400-1900*. Smithsonian Institution Press.

Slavery Convention, 25 September 1926, 60 LNTS 253.

Copson, R. W. (2001). *Africa Backgrounder: History, U.S. Policy, Principal Congressional Actions.* CRS Report for Congress.

Curtin, P. D. (1967). *Africa Remembered: Narratives of West Africans from the Era of the Slave Trade*. University of Wisconsin Press.

Curtin, P. D. (1971). The Atlantic Slave Trade, 1600-1800. In History of West Africa (Vol. 1). Longman.

Dodge, M. A. (2013). The Search for Resistance: A Layperson's Reflection on the Historiography of Slavery in the African Atlantic. *The History Teacher*, *47*(1).

Doubleday & Co. Inc. (1976). *The overseers also forced the slaves to watch as they whipped the headless body of the rebellious slave*. Author.

Fadle, H.Y. (2005). *Studies in Sudanese History.* Khartoum: SUDATek Limited.

Farah, I., Kiamba, S., & Mazongo, K. (2011). *Major challenges facing Africa in the 21st century: A few provocative remarks At the International Symposium on Cultural Diplomacy in Africa - Strategies to confront the Challenges of the 21st Century: Does Africa have what is required?* International Symposium on Cultural Diplomacy in Africa 14th – 17th July 2011, Berlin.

Gordon, A. A., & Gordon, D. L. (Eds.). (2013). *Understanding Contemporary Africa* (5th ed.). Lynne Rienner Publishers.

Higginbotham, Jr. & Jacobs. (1992). The Law Only as an Enemy: The Legitimization of Racial Powerlessness through the Colonial and Antebellum Criminal Laws of Virginia. *N.C. L. Rev. 70*(969), 1063.

Honoré. (2012). Antony 2012. “Slavery: From Ancient to Modern. In J. Allain (Ed.), *The Legal Understanding of Slavery: From the Historical to the Contemporary* (pp. 9–39). Oxford University Press.

Horvath, R. J. (1972). A definition of colonialism. *Current Anthropology*, *13*(1), 45–57. doi:10.1086/201248

Inikori, J. E. (2003). *The Struggle against the Trans-Atlantic Slave Trade," in Fighting the Slave Trade: West African Strategies* (A. Diouf, Ed.). Ohio University Press.

Kusimba, C. M. (2004). Archaeology of Slavery in East Africa. *African Archaeological Review*, *21*(2), 59–88. doi:10.1023/B:AARR.0000030785.72144.4a

Leakey, R. E. (2002). *Exploration of Africa: The Emerging Nations: Central and East Africa. 1880 to the Present: From Colonialism to the Civil War*. Chelsea House Publishers.

Lovejoy, P. E., Mahadi, A., & Mukhtar, M. I. (1993). C.L. Temple's Notes on the History of Kano (1909): A Lost Chronicle on Political Office. Sudanic Africa, 4(40).

Maiese, M. (2003). *Dehumanization. In Beyond Intractability*. University of Colorado.

Manning, P. (2006). Slavery and Slave Trade in West Africa, 1430- 1930. In E. K. Akyeampong (Ed.), *Themes in West Africa's History*. Ohio University Press. doi:10.2307/j.ctv136c1xk.11

MartinaO. 2015. The Trans-Atlantic Slave Trade and the long-term effects on African Economies. Case study: Angola and Namibia. DOI: doi:10.13140/RG.2.1.3798.4808

Mazrui, A. A. (1994). *The bondage and boundaries*. IBRU Boundary and Security Bulletin.

Mengisteab, K. (2011). Critical Factors in the Horn of Africa's Raging Conflicts. In *Discussion papers 67*. Nordiska Afrikainstitutet.

Muhammad, E., & Patricia, M. (2003). The Trans-Atlantic Slave Trade: A Forgotten Crime Against Humanity as Defined by International Law. *American University International Law Review*, *19*(4), 883–947.

Nasidi N.A. (2022). Slavery And Its Characteristics. *A Journal of Contemporary Research 2022, 19*(2), 135-146.

Nathan, N., & Leonard, W. (2011, December). The Slave Trade and the Origins of Mistrust in Africa. *The American Economic Review*, *101*(7), 3221–3252. doi:10.1257/aer.101.7.3221

Nguendi, F. (2012). Africa's international borders as potential sources of conflict and future threats to peace and security. *Institute for Security Studies Paper*, *233*, 1–14.

Ninnes, P. (2000). Representations of indigenous knowledges in secondary school science textbooks in Australia and Canada. *International Journal of Science Education*, *22*(6), 603–617. doi:10.1080/095006900289697

Niyitunga, E. B. (2017). *Exploring the Challenges of the African Union (Au)-Led Mediation in Peace Processes: The Cases of Burundi, Libya, and Zimbabwe*. Thesis Submitted to the University of Johannesburg, Auckland Park.

Niyitunga, E. B. (2021). Colonial Legacies: Africa's International Relations and Agenda 2063. In Indigenous Systems and Africa's Development. Africa Institute of South Africa (AISA). The National Institute for the Humanities and Social Science Pretoria.

Njoku, U. J. (2005). Colonial Political Re-Engineering and the Genesis of Modern Corruption in African Public Service: The Issue of the Warrant Chiefs of South Eastern Nigeria as a Case in Point. *Nordic Journal of African Studies*, *14*(1), 99–116.

Nnoli, O. (2003). *Introduction to Politics* (Revised 2nd Edition). PACREP.

Nunn, N. (2008). The long-term effects of Africa's slave trades. *The Quarterly Journal of Economics*, *123*(1), 139–176. doi:10.1162/qjec.2008.123.1.139

Oba, N. F., & Onyije, E. L. (2011). Colonialism and education. *Proceedings of the International Conference on Teaching, Learning and Change, International Association for Teaching and Learning (IATEL)*. Accessed 28 September 2019, https://hrmars.com/admin/pics/170.pdf

Ocheni, S. B., & Nwankwo, B. C. (2012). Analysis of colonialism and its impact in Africa. *Cross-Cultural Communication*, *8*(3), 46–54.

Oliver, R., & Fage, J. D. (1969). Streams of Trade in Guinea States. In Africa from Early Times to 1800. Oxford University Press.

Olukayode, F. A. (2014). Africa and international relations theory: Acquiescence and responses. *Globalistics and Globalization Studies,* 154–163. Accessed 14 May 2019, https://www.sociostudies.org/almanac/articles/africa_and_international_relations_theory-_acquiescence_and_responses/

Pheko, M. (2001). *How Africans can Regain their Lost Power*. Tokloho Development Association.

Richards, S., & Gezu, B. (2011). *Conflict in the Somali Region of Ethiopia: Can Education Promote Peace-Building?* Feinstein International Center.

Rodney, W. (1972). How Europe underdeveloped Africa. *Beyond borders: Thinking critically about global issues*, 107-125.

Salih, M. A. M. (2007). Political Parties in Africa: Challenges for Sustained Multiparty Democracy. In *Africa Regional Report Based on research and dialogue with political parties*. International Institute for Democracy and Electoral Assistance.

Sanderson, S. K. (2005). World-systems analysis after thirty years. *International Journal of Comparative Sociology, 46*(3), 179–213. doi:10.1177/0020715205058606

Schumpeter, J. A. (1912). *The Theory of Economic Development*. Oxford University Press.

Settles, J. D. (1996). *The Impact of Colonialism on African Economic Development*. Chancellor's Honors Program Projects.

Snively, G., & Corsiglia, J. (2001). Discovering indigenous science: Implications for science education. *Science Education, 85*(1), 6–34. doi:10.1002/1098-237X(200101)85:1<6::AID-SCE3>3.0.CO;2-R

Stan, E., & Kenneth, S. (2000). Institutions, factor endowments, and paths of development in the new world. *The Journal of Economic Perspectives, 14*(3), 217–232. doi:10.1257/jep.14.3.217

Stilwell, S. (1999). *The Kano Mamluks: Royal Slavery in the Sokoto Caliphate, 1807-1909* [PhD Thesis]. York University.

Stilwell, S., Hamza, I., & Lovejoy, P. E. (2001). The Oral History of Royal Slavery in the Sokoto Caliphate: An Interview with Sallama Dako. *History in Africa, 28*, 273–291. doi:10.2307/3172218 PMID:17657904

Tabellini, G. (2008). The Scope of Cooperation: Values and Incentives. *The Quarterly Journal of Economics, 123*(3), 905–950. doi:10.1162/qjec.2008.123.3.905

Thomas, H. (1997). *The Slave Trade: The Story of the Atlantic Slave Trade: 1440-1870*. Simon & Schuster.

Thomson, A. (2010). *An Introduction to African Politics* (3rd ed.). Routledge. doi:10.4324/9780203857946

United Nations Development Programme & Third World Network of Scientific Organisations. (1999), *Sharing Innovative Experiences: Examples of Successful Initiatives in Science and Technology in the South.* Author.

United Nations Environmental Programme (UNEP). (2004). *Capacity Building Through Partnership andInformation and Communication Technology for Using Indigenous Knowledge for Nature Conservation in Africa.* Available at: www.un.org/esa/devaccount/thematic%20overview%20environment.html

van Wyk, J.-A. (2007). *Political Leaders in Africa: Presidents.* Patrons or Profiteers.

Warren, D. M. (1991). *Using Indigenous Knowledge in Agricultural Development: World Bank Discussion Paper No. 127.* World Bank.

Weitzman, M. L. (1998). Recombinant growth. *The Quarterly Journal of Economics, 113*(2), 331–360. doi:10.1162/003355398555595

Chapter 4
Revisiting the Effects of Colonialism on Africa's Indigenous Knowledge

Thivhavhudzi Muriel Badugela
https://orcid.org/0000-0001-9122-5915
University of Venda, South Africa

ABSTRACT

In Africa's history, colonialism deeply impacted indigenous knowledge, influencing cultural identity, knowledge transmission, and societal dynamics. Examining disruptions in education, governance, and environmental practices, the review synthesises insights from scholarly works to elucidate the marginalisation and erosion of traditional systems. It places particular emphasis on cultural identity and knowledge transmission. Simultaneously, the review explores the transformative potential of the fourth industrial revolution (4IR) in addressing global challenges, utilising a qualitative lens to extract nuanced perspectives. Advocating for a holistic approach, the review underscores the qualitative integration of indigenous knowledge into 4IR development, emphasising technology's ethical and culturally sensitive application. This qualitative research perspective aims to contribute valuable depth to our understanding of the intricate interplay between historical legacies, indigenous knowledge, and contemporary technological advancements for a more inclusive and sustainable global future.

INTRODUCTION

Colonialism, a historical epoch spanning several centuries, left an indelible mark on the continents it touched, notably Africa. The impact of colonial rule extended far

DOI: 10.4018/978-1-6684-7851-6.ch004

Copyright © 2024, IGI Global. Copying or distributing in print or electronic forms without written permission of IGI Global is prohibited.

beyond political boundaries and economic systems; it also profoundly influenced the rich tapestry of Indigenous knowledge woven into African societies' cultural, social, and spiritual fabric for generations (Omonov & Karimov, 2020).

This complex interplay between colonial forces and Africa's Indigenous knowledge systems has shaped the continent's trajectory in ways that continue to resonate today (Sellar & Battisti, 2023).

Colonial powers, driven by imperialistic ambitions, sought to exploit Africa's vast resources, reshape its social structures, and impose their cultural frameworks. Domínguez de Olazábal, (2022). In the process, they confronted and often undermined the wealth of traditional knowledge that had sustained African communities for centuries. The book chapter delves into the multifaceted effects of colonialism on Africa's Indigenous knowledge, exploring how the imposition of foreign ideologies, economic systems, and administrative structures disrupted and, in some cases, eroded the intricate networks of knowledge that were intrinsic to African societies (Fletcher et al.,2021).

The erosion of Indigenous knowledge occurred through various mechanisms, including the imposition of Western education, the marginalisation of traditional practices, and the disruption of intergenerational transmission. Rothenberg (2020) Western education, introduced to serve the interests of the colonisers, often marginalised local languages and traditional knowledge systems. Ngũgĩ (1986). This led to the loss of linguistic diversity and undermined the transmission of traditional wisdom from elders to the younger generations. Ngũgĩ emphasises the importance of language as a tool for preserving and transmitting traditional knowledge. He contends that African languages carry within them the collective wisdom, histories, and values of their respective cultures. However, the dominance of European languages in education, administration, and literature marginalised Indigenous languages, relegating them to the status of inferiority and diminishing their use in formal settings. (Ngũgĩ wa Thiong'o 1986).

Furthermore, the economic exploitation of Africa's resources, driven by colonial powers, disrupted traditional land management practices and agricultural systems. The imposition of cash-crop economies often supplanted sustainable farming practices, leading to environmental degradation and the loss of Indigenous agricultural knowledge. Additionally, the imposition of Western legal and administrative structures marginalised traditional governance systems, further weakening the role of Indigenous knowledge in decision-making processes. Ngungi (1986). Unravelling the effects of colonialism on Africa's Indigenous knowledge makes it evident that the legacy of this historical period continues to shape contemporary challenges. Recognising and revitalising Indigenous knowledge systems is essential for fostering sustainable development and addressing the ongoing repercussions of colonialism. The book chapter explores the dimensions in detail, shedding light on the intricate interplay between colonial forces and Africa's rich reservoir of Indigenous knowledge (Gómez-Baggethun, 2022).

THEORETICAL FRAMEWORK

In post-colonial contexts, the dynamics of power, resistance, and ethical considerations intersect to shape the complexities of societies grappling with the legacies of colonial rule. These dynamics are characterised by the persistence of power imbalances inherited from colonialism, the resilience of resistance movements challenging dominant narratives and structures, and the ethical dilemmas surrounding accountability and justice. This multifaceted understanding draws upon critical insights from scholars such as Edward Said, whose seminal work "Orientalism" (1978) illuminates the power dynamics inherent in colonial representations and the discursive construction of the Other, providing a foundational framework for analysing post-colonial contexts and their enduring challenges (Said,1978).

Post-Colonial Theory and Indigenous Knowledge

Critical Reflections: Post-colonial theory serves as a critical lens to examine the enduring impact of colonialism on Indigenous knowledge. It emphasises the power structures, cultural hegemony, and epistemic ide perpetuated by colonial forces, elucidating how these dynamics continue to shape the marginalisation of Indigenous knowledge in the post-colonial era ((Chilisa & Phatshwane, 2022).

Resistance and Resilience: Within the post-colonial framework, Indigenous communities are viewed not merely as passive recipients of colonial legacies but as agents of resistance and resilience. The theory underscores how communities draw on their cultural capital to resist cultural assimilation (Chilisa & Phatshwane, 2022).

Ethical Frameworks for Integration:

Ethics of Integration: The discussion navigates ethical frameworks for integrating Indigenous knowledge and 4IR technologies. This involves acknowledging cultural sensitivity, obtaining informed consent, and ensuring equitable benefits for Indigenous communities. Ethical considerations become pivotal in shaping a more inclusive and respectful approach to technological integration and reclaiming agency in their destinies. It calls for a change in basic assumptions that recognise the agency of Indigenous communities, values traditional wisdom, and harness the potential of 4IR technologies in ways that foster sustainable, inclusive, and culturally sensitive development (Kop, 2021).

LITERATURE REVIEW

Historical Context of Colonialism in Africa

The colonial era in Africa spanned the late 19th to mid-20th century, characterised by the scramble for Africa among European powers. The Berlin Conference of 1884-1885 formalised the continent's division, establishing artificial borders that often-ignored existing ethnic, cultural, and linguistic boundaries. European powers, including Britain, France, Germany, Belgium, Portugal, and others, sought to exploit Africa's vast resources and establish imperial dominance. This period significantly departed from Africa's pre-colonial history, introducing foreign ideologies, institutions, and practices (Çınar, 2020).

Motivations and Objectives of Colonial Powers: The motivations driving European colonial powers were multifaceted. Economic interests were paramount, as Africa was viewed as a source of valuable raw materials, minerals, and potential markets for finished goods—the desire for geopolitical dominance and strategic control of trade routes further fuelled colonial ambitions. Additionally, the European powers justified their colonial endeavours through notions of racial superiority and the "civilising mission," asserting that they brought progress and development to supposedly 'backward' societies (Franklin,2022).

Impacts of Colonialism on Political, Economic, and Cultural Landscapes:

Political Impact: Colonial powers implemented administrative structures that served their interests, often disregarding pre-existing governance systems. This led to the imposition of artificial borders, creating nations that did not align with the historical and cultural divisions of Indigenous communities. Traditional leadership structures were marginalised or replaced, contributing to the fragmentation and destabilisation of African societies (Ndlovu-Gatsheni, 2020).

Economic Impact: Colonial economic policies were designed to extract wealth from Africa, leading to the exploitation of natural resources and the establishment of cash-crop economies. This resulted in the disruption of traditional subsistence practices and the introduction of exploitative labour systems. The economic legacy of colonialism continues to shape the global economic disparities faced by African nations today.

Cultural Impact: Colonialism had a profound impact on African cultures. Indigenous languages and belief systems were often suppressed or replaced by European languages and religions. Western education, introduced to serve colonial objectives, marginalised traditional knowledge systems, contributing to the erosion of cultural practices, values, and ways of life (Ndlovu-Gatsheni, 2020).

Indigenous Knowledge in Africa

Definition and Scope of Indigenous Knowledge: Indigenous knowledge refers to the cumulative body of traditional skills, practices, wisdom, and innovations developed over generations within specific communities. This knowledge encompasses various domains, including agriculture, medicine, spirituality, resource management, and technology. It is deeply rooted in the local context, shaped by each community's unique experiences, environments, and cultural perspectives. The scope of Indigenous knowledge is broad, covering tangible practices and intangible aspects such as storytelling, folklore, and traditional rituals.

Significance of Indigenous Knowledge in African Societies: Indigenous knowledge holds immense significance in African societies, serving as the bedrock of community resilience, identity, and sustainable development. One key aspect is its adaptability to local environments and challenges. Indigenous knowledge systems are finely tuned to the specific needs of communities, offering practical solutions for agricultural practices, healthcare, and natural resource management. Integrating this knowledge fosters a harmonious relationship between communities and their surroundings (Rawat & Mishra,2021).

Furthermore, Indigenous knowledge is intertwined with cultural identity, providing a sense of continuity and connection to ancestral heritage. It serves as a repository of traditional values, social norms, and ethical principles, shaping how individuals and communities interact with one another and their surroundings. This cultural dimension contributes to social cohesion, preserving a rich, diverse tapestry of African identities (Rawat & Mishra,2021).

Traditional Systems of Education and Transmission: Indigenous knowledge is not acquired through formal education systems but through informal and experiential methods embedded in everyday life. Elders and community leaders are crucial in passing knowledge to younger generations through oral traditions, storytelling, apprenticeships, and direct experiences. This intergenerational transmission ensures the continuity of cultural practices and the preservation of wisdom accumulated over centuries (Olaopa & Ayodele,2022).

Traditional education systems emphasise an integrated approach, integrating practical skills with cultural values and ethical principles. Learning is embedded in the fabric of daily life, allowing individuals to absorb knowledge while participating in community activities seamlessly. This contrasts with the compartmentalised nature of formal education introduced during the colonial era, which often marginalised Indigenous knowledge (Olaopa & Ayodele,2022).

Colonialism's Impact on Education and Language

Introduction of Western Education in Africa: The introduction of Western education in Africa during the colonial era marked a significant departure from traditional learning modes. Colonial powers viewed education to instil their values, ideologies, and language, often to create a docile and assimilated workforce. Missionary schools and colonial educational systems were established, emphasising European languages, cultures, and curricula. This shift had profound implications for Africa's educational landscape, shaping future generations' trajectories.

Marginalisation of Local Languages and Traditional Knowledge: One of the enduring legacies of colonial education is the marginalisation of local languages and traditional knowledge. European languages, such as English, French, and Portuguese, were prioritised in schools, and Indigenous languages were often denigrated or ignored. This linguistic bias not only created a divide between the language of instruction and students' native languages but also contributed to the erosion of linguistic diversity. Indigenous languages, which encapsulated centuries of cultural and contextual knowledge, were pushed to the periphery, diminishing their status and functionality within communities (Ndiribe & Aboh, 2022).

The devaluation of traditional knowledge systems accompanied the marginalisation of local languages. The curriculum in colonial schools tended to exclude or downplay Indigenous knowledge, casting it as inferior or irrelevant. This dismissal of traditional wisdom not only undermined the self-esteem of Indigenous communities but also perpetuated a Eurocentric worldview that failed to appreciate the complexity and richness of local knowledge (Ndiribe & Aboh, 2022).

Disruption of Intergenerational Transmission of Indigenous Knowledge: Colonial education disrupted the organic intergenerational transmission of Indigenous knowledge, a cornerstone of African societies. Traditional systems of passing down knowledge from elders to the younger generation were marginalised or replaced by formal, Western-style education. As a result, skills and wisdom embedded in Indigenous practices faced the risk of being lost, creating a rupture in the continuity of cultural heritage.

The disruption of intergenerational transmission was exacerbated by the fact that the content of colonial education often did not align with the needs and realities of local communities. Younger generations were taught a curriculum that was detached from their lived experiences, further alienating them from their cultural roots. The consequences of this educational disjunction reverberate today, as communities grapple with the challenge of revitalizing and reconnecting with their Indigenous knowledge systems (Beyette, 2021).

Economic Exploitation and Agricultural Disruption

Exploitative Economic Practices During Colonial Rule: Colonial powers engaged in exploitative economic practices that sought to extract maximum resources from the African continent. This economic exploitation manifested in various forms, including forced labour, unequal trade relationships, and the establishment of cash-crop economies. The primary objective was to enrich the colonial powers at the expense of the local populations, leading to the systematic depletion of Africa's natural wealth (Naimah, 2022).

Disruption of Traditional Land Management and Agricultural Systems: One of the most profound impacts of colonial economic policies was the disruption of traditional land management and agricultural systems. Colonisers often implemented land tenure systems favouring European settlers and commercial interests, disregarding established Indigenous practices. Traditional communal land management, where communities deeply understood their local ecosystems and sustainable farming practices, was replaced by systems prioritising profit over sustainability.

Cash-crop agriculture became a dominant feature of colonial economies, with crops like cocoa, coffee, rubber, and tea cultivated for export. This shift altered the agricultural landscape and undermined food security and self-sufficiency. Indigenous farming practices adapted to local environments, and climates gave way to monoculture and large-scale plantations that often-depleted soil fertility and contributed to environmental degradation (Naimah, 2022).

Consequences for Indigenous Agricultural Knowledge: Economic exploitation and agricultural disruption during colonial rule were particularly severe for Indigenous agricultural knowledge. Traditional farming methods, finely attuned to local ecosystems, crop diversity, and seasonal patterns, were marginalised or replaced by Western practices prioritising profit and export-oriented production. This led to biodiversity loss, soil degradation, and reduced resilience to environmental changes.

Moreover, the commodification of agriculture under colonial rule contributed to the erosion of Indigenous knowledge related to food security and sustainable land use. The focus on cash crops meant that local communities were increasingly disconnected from cultivating staple foods essential for their subsistence. This shift not only jeopardised the nutritional well-being of communities but also weakened the transmission of knowledge related to diverse agricultural practices.

The disruption of Indigenous agricultural knowledge also had social and cultural ramifications. Traditional farming practices were practical and often embedded in cultural rituals, ceremonies, and communal cooperation. The shift towards exploitative economic practices severed these cultural ties, contributing to the breakdown of community cohesion and resilience.

Legal and Administrative Changes

Imposition of Western Legal and Administrative Structures: During the colonial era, one of the significant mechanisms of control was the imposition of Western legal and administrative structures on African societies. European colonial powers introduced legal systems based on their own norms and values, often disregarding, or undermining existing Indigenous legal traditions. The aim was to establish a legal framework that facilitated the extraction of resources, maintained social order from the colonial perspective, and consolidated their authority over the colonised territories.

Marginalisation of Traditional Governance Systems: The imposition of Western legal systems often went hand in hand with the marginalisation of traditional governance structures. Indigenous governance systems, which were deeply rooted in local customs, consensus-building, and communal decision-making, were supplanted by hierarchical and centralised administrative models. Traditional leaders and councils were often relegated to ceremonial roles, with decision-making powers transferred to European-appointed officials.

This marginalisation had far-reaching consequences, as traditional governance systems had been integral to the social fabric of African communities. Traditional leaders played crucial roles in conflict resolution, resource management, and community development. The erosion of their authority disrupted established power dynamics and led to a vacuum in leadership that, in many cases, fuelled tensions and conflicts.

Impact on Decision-Making Processes and Community Cohesion: The imposition of Western legal and administrative structures profoundly impacted decision-making processes within African communities. Decisions that were traditionally made through consensus-building and community involvement were now often centralised and dictated by external authorities. This alienated communities from the decision-making processes and led to policies that did not align with the local context, needs, or aspirations.

Furthermore, the disruption of traditional governance systems had implications for community cohesion. Traditional leadership structures fostered a sense of identity, belonging, and shared responsibility. The shift towards Western administrative models often created divisions within communities, as individuals and groups competed for limited positions of authority or sought favour with the colonial administration. This fragmentation weakened the social fabric that had sustained African societies for generations.

The impact on decision-making processes and community cohesion persists in contemporary Africa as post-colonial nations grapple with the legacies of imposed administrative structures. Efforts to reclaim and revitalise traditional governance systems are underway, recognising their importance in fostering community resilience, cultural identity, and sustainable development.

Environmental Consequences and Loss of Knowledge

Environmental Degradation Resulting from Colonial Practices: Colonial powers, driven by economic interests, engaged in exploitative practices that wreaked havoc on Africa's environment. Deforestation, a consequence of timber extraction and large-scale commercial agriculture, led to soil erosion and disrupted local water cycles. Overgrazing and mineral extraction further intensified environmental degradation. The immediate consequences were severe, with detrimental effects on both the natural landscape and the community is dependent on these ecosystems for their livelihoods.

Deforestation, for instance, not only altered local climates but also contributed to soil erosion, leading to decreased agricultural productivity. The extraction of minerals and other resources disrupted the delicate ecological balance, causing lasting harm to biodiversity. Large-scale commercial agriculture, often monocultural and unsustainable, depleted the fertility of the soil and contributed to the loss of traditional farming practices (Beyette, 2021).

Erosion of Indigenous Environmental Knowledge: Indigenous communities in Africa possessed a profound understanding of their local environments, incorporating sustainable land-use practices, resource management, and ecological balance into their ways of life. However, the imposition of exploitative economic systems and the disregard for traditional land management practices eroded this rich reservoir of knowledge (Beyette, 2021).

The marginalisation of Indigenous environmental knowledge during colonial rule was not only a loss of practical wisdom but also a rupture in the cultural and spiritual connection that communities had with their land. Traditional ecological knowledge, which had been passed down through generations, was dismissed as primitive or irrelevant by colonial authorities. This dismissal not only undermined the confidence of Indigenous communities in their own practices but also disrupted the transmission of invaluable environmental wisdom (Beyette, 2021).

Long-term Consequences for Sustainability: The long-term consequences of environmental degradation and the loss of Indigenous knowledge is palpable in contemporary Africa. Erosion of topsoil, deforestation, and disruptions to local ecosystems have left lasting scars on the landscape, making regions more vulnerable to the impacts of climate change. The depletion of biodiversity has diminished the resilience of ecosystems, affecting their ability to adapt to changing environmental conditions.

The historical loss of Indigenous environmental knowledge hinders efforts to address contemporary environmental challenges in post-colonial Africa. Traditional practices that once promoted resilience, such as agroforestry, water conservation, and crop diversification, were often replaced by unsustainable methods introduced by colonial powers. Rebuilding and integrating Indigenous knowledge into current conservation and sustainability practices are essential for fostering resilience and ensuring the long-term sustainability of Africa's ecosystems.

Contemporary Challenges and the Legacy of Colonialism

Introduction: The legacy of colonialism continues to reverberate in contemporary Africa, shaping the challenges faced by Indigenous communities and their traditional knowledge systems. This essay explores the ongoing repercussions of colonialism on Indigenous knowledge, emphasises the importance of recognizing and revitalizing this knowledge, and discusses how addressing current challenges requires a nuanced understanding of the historical context.

Ongoing Repercussions of Colonialism on Indigenous Knowledge: The impact of colonialism on Indigenous knowledge persists in various forms, from the erosion of traditional practices to the challenges of adapting to a rapidly changing world. The marginalisation of Indigenous knowledge during the colonial era has left communities grappling with the loss of cultural identity, weakened social cohesion, and a diminished capacity to address contemporary issues in a way that aligns with their own values and experiences.

The imposition of Western education, economic systems, and legal structures has entrenched a perception that devalues Indigenous knowledge. In the face of globalisation and modernisation, traditional practices often struggle to find relevance, leading to a decline in the transmission of Indigenous knowledge across generations. The ongoing repercussions manifest in health disparities, environmental degradation, and cultural disconnection within Indigenous communities.

Recognition of the Importance of Revitalizing Indigenous Knowledge: Acknowledging the significance of revitalising Indigenous knowledge is a crucial step towards addressing African societies' contemporary challenges. Indigenous knowledge holds intrinsic value, offering unique insights into sustainable practices, environmental stewardship, and community resilience. Recognising and revitalising this knowledge is a matter of cultural preservation and practical necessity for fostering sustainable development.

Addressing Current Challenges Through a Lens of Historical Understanding: Addressing contemporary challenges requires a nuanced understanding of the historical context shaped by colonial legacies. It involves dismantling structures that perpetuate inequality, marginalisation, and the devaluation of Indigenous knowledge. Policies and initiatives aimed at sustainable development, environmental conservation, and social cohesion must be informed by a historical awareness of the impact of colonialism on the continent.

Reclaiming and integrating Indigenous knowledge into decision-making processes is essential for fostering a more inclusive and holistic approach to development. Collaborative efforts between Indigenous communities, governments, and international organisations can promote policies that recognise and value traditional practices. Moreover, education programs that incorporate Indigenous perspectives contribute to breaking down the existing dichotomy between Western and traditional knowledge systems.

Revitalising Indigenous Knowledge for Sustainable Development

Revitalising Indigenous knowledge is crucial for sustainable development in contemporary Africa. This essay explores the importance of acknowledging and preserving Indigenous knowledge, examines initiatives and strategies for revitalising traditional practices, and highlights Indigenous knowledge's pivotal role in fostering sustainable development (Kohsaka & Rogel,2021).

Importance of Acknowledging and Preserving Indigenous Knowledge: Indigenous knowledge represents a repository of centuries-old wisdom, finely attuned to local environments, cultures, and sustainable practices. Acknowledging and preserving this knowledge is essential for several reasons. Firstly, it preserves cultural identity, ensuring that unique traditions, languages, and ways of life are passed down through generations. Secondly, it is a source of resilience, offering practical solutions to environmental challenges, agricultural practices, and healthcare. Thirdly, recognising Indigenous knowledge contributes to a more inclusive and equitable approach to development, valuing diverse perspectives and contributions (Kohsaka & Rogel,2021).

Preserving Indigenous knowledge also fosters a sense of pride and empowerment within Indigenous communities. When traditional practices are acknowledged and integrated into contemporary contexts, it empowers communities to take ownership of their development, reinforcing a connection to their cultural heritage (Kohsaka & Rogel,2021).

Initiatives and Strategies for Revitalizing Traditional Practices: Revitalizing Indigenous knowledge requires deliberate initiatives and strategies that empower communities to reclaim and transmit their traditional wisdom. One key approach is integrating Indigenous knowledge into formal education systems. This involves developing curricula that incorporate traditional practices, oral histories, and local environmental knowledge. By doing so, future generations can benefit from both

Western and Indigenous knowledge systems, fostering a more holistic and culturally sensitive approach to education.

Community-led initiatives play a crucial role in revitalizing traditional practices. Projects that focus on sustainable agriculture, natural resource management, and cultural preservation empower communities to reclaim their traditional practices. These initiatives should involve collaboration between Indigenous communities, government bodies, non-governmental organizations, and researchers to ensure a comprehensive and inclusive approach.

Furthermore, the use of modern technologies can be harnessed to document and transmit Indigenous knowledge. Digital platforms, community radio, and mobile applications can serve as tools for preserving and disseminating traditional wisdom, ensuring that it is accessible to a broader audience.

Role of Indigenous Knowledge in Contemporary Sustainable Development: Indigenous knowledge is not just a relic of the past but a dynamic and relevant resource for addressing contemporary challenges. In the context of sustainable development, Indigenous knowledge offers insights into environmentally friendly agricultural practices, conservation methods, and resilience-building strategies. For instance, traditional agroecological practices, such as crop rotation and agroforestry, are inherently sustainable and can contribute to food security while maintaining environmental balance.

In healthcare, Indigenous knowledge often includes holistic approaches that consider physical and spiritual well-being. Integrating traditional healing practices with modern healthcare systems can enhance the effectiveness of healthcare delivery, particularly in regions where access to Western medicine is limited.

The role of Indigenous knowledge extends beyond specific sectors; it contributes to communities' overall resilience and adaptability in the face of global challenges such as climate change. With their deep understanding of local ecosystems, Indigenous communities can offer valuable insights into sustainable resource management and climate adaptation strategies.

RESEARCH METHOD

Phenomenological Synthesis: A phenomenological synthesis approach to examine existing literature on African communities' lived experiences and perceptions in the context of colonialism's impact on Indigenous knowledge acknowledged. This framework facilitates an in-depth exploration of historical narratives, cultural shifts, and the erosion of traditional practices. (Cuthbertson et al., Blair, 2020).

Systematic Review: A systematic review of scholarly articles, books, reports, and other relevant literature on the effects of colonialism on Indigenous knowledge in Africa and comprehensive search strategies to ensure the inclusion of diverse perspectives and insights. (van Dinter et al., 2021).

DATA COLLECTION

Documentary Analysis: Engaging in documentary analysis of historical records, ethnographic studies, and oral traditions allows for extracting qualitative data from diverse sources, contributing to a comprehensive understanding of the historical context and the intricate interplay between colonial legacies and Indigenous knowledge (Morgan, 2022).

Thematic Coding: Utilisation of thematic coding to categorise and organise findings from the literature identifies recurring themes, patterns, and key concepts related to the impact of colonialism and the potential role of the Fourth Industrial Revolution is of importance.

DATA ANALYSIS

Content Analysis: Employ techniques to systematically explore selected literature's content. This method facilitates the identification of emerging trends, theoretical frameworks, and gaps in understanding Indigenous knowledge's historical and contemporary dynamics.

Comparative Analysis: Conduct a comparative analysis to discern variations and consistencies across different historical periods, regions, and cultural contexts. This approach enhances the synthesis of diverse perspectives within the literature.

Ethical Considerations

Critical Appraisal: Apply critical appraisal techniques to assess the selected literature's quality, reliability, and biases. Transparently acknowledge limitations and potential sources of bias in the existing body of work.

Integrative Synthesis: Employ an integrative synthesis approach to merge and synthesise findings from diverse sources. This ensures that the final analysis captures the complexity of the relationship between colonialism, Indigenous knowledge, and the evolving landscape of the Fourth Industrial Revolution (Du et al., 2020).

Trustworthiness and Rigor

Peer Review: Subject the review to peer review by experts in the field to ensure methodological rigour and enhance the credibility of the synthesis. Incorporate feedback to refine the analytical framework and interpretation of findings.

Reflexivity: Engage in reflexivity by acknowledging the researcher's positionality and potential biases. Transparency in the research process contributes to the overall trustworthiness of the synthesised findings.

CONTRIBUTIONS OF AFRICA'S INDIGENOUS KNOWLEDGE TO THE FOURTH INDUSTRIAL REVOLUTION

Indigenous Agricultural Practices: Traditional farming methods, deeply rooted in Indigenous knowledge, emphasise sustainable and environmentally friendly approaches such as agroforestry, crop rotation, and organic farming. These practices can contribute to sustainable agriculture, ensuring food security in the face of a growing population and changing climate. For example, crop rotation involves alternating different crops in the same field over sequential seasons. This practice helps break pest and disease cycles, replenish soil nutrients, and maintain soil structure. In regions where maize is a staple crop, farmers often rotate maize with legumes such as beans or peas. The legumes fix nitrogen in the soil, reducing the need for synthetic fertilisers and improving overall soil health (Rosati, et al, 2021).

Intercropping, another traditional farming practice, involves growing two or more crops simultaneously in the same field. For example, planting maize alongside beans or squash allows for efficient use of space and resources. The beans provide natural nitrogen fixation, while the squash acts as a ground cover, reducing weed growth and soil erosion.

Agroforestry integrates trees and shrubs into agricultural landscapes, providing multiple benefits such as shade, windbreaks, and additional sources of food and income. In sub-Saharan Africa, farmers often intercrop maize with nitrogen-fixing trees like Faidherbia albida. The trees enrich the soil with nitrogen through leaf litter decomposition, improving soil fertility and crop yields (Rosati., et al, 2021).

Traditional Medicine: Indigenous knowledge in Africa includes many traditional healing practices using local herbs and plants. Integrating these traditional medicine practices with modern healthcare can enhance healthcare delivery, especially in rural areas where access to Western medicine may be limited. One notable example of integrating traditional medicine with modern healthcare in Africa is found in the work of organizations like the Traditional Medicine Programme of the World Health Organization (WHO). This program collaborates with African governments and local communities to incorporate Indigenous healing practices into primary healthcare systems. For instance, in countries like South Africa and Kenya, traditional healers are being trained and integrated into the formal healthcare system to provide culturally relevant care alongside Western medical practitioners. Additionally, research initiatives such as those conducted by the African Centre for Traditional Medicine

(ACTM) in Zimbabwe focus on scientifically validating the efficacy and safety of traditional remedies, leading to their inclusion in national healthcare policies and programs. Such integration not only expands access to healthcare, particularly in rural areas where traditional medicine is often the primary source of healing, but also fosters mutual respect for diverse healing traditions and enhances healthcare delivery by combining the strengths of both traditional and modern approaches (Rosati, et al, 2021).

Local Environmental Knowledge: Indigenous communities deeply understand their ecosystems, biodiversity, and sustainable resource management. Leveraging this knowledge in conservation efforts can contribute to community-led initiatives for environmental protection, helping address climate change and biodiversity loss challenges.

Traditional Energy Sources: Indigenous communities have historically utilised renewable energy sources such as solar, wind, and hydroelectric power. Integrating these traditional practices with modern technologies can contribute to developing sustainable and decentralised energy solutions, especially in remote areas.

Cultural Heritage Preservation: Indigenous knowledge includes rich cultural practices, traditional arts, and craftsmanship needed to be leveraged in developing cultural tourism and promoting economic opportunities while preserving and celebrating cultural diversity.

Local Water Management Techniques: Indigenous communities often have effective water harvesting, storage, and conservation methods. These traditional water management practices can address water scarcity issues and ensure sustainable water use.

Oral Tradition and Storytelling: Indigenous knowledge is often transmitted through oral traditions and storytelling. Integrating these methods into modern educational approaches can enhance community-based education, fostering a more inclusive and culturally sensitive learning environment.

Technology for Knowledge Transmission: Utilizing digital platforms and technology to document, preserve, and transmit Indigenous knowledge can bridge the gap between traditional practices and the digital era. This approach ensures that traditional wisdom remains accessible and relevant in the context of the Fourth Industrial Revolution.

Collective Decision-Making: Indigenous governance structures emphasise collective decision-making and community involvement. Integrating these principles into social innovation initiatives can enhance community empowerment and ensure that technological advancements align with local needs and values. Africa's Indigenous knowledge can contribute significantly to the challenges and opportunities of the Fourth Industrial Revolution. By recognising, preserving, and integrating traditional practices into contemporary solutions, Africa can harness the strengths of its rich cultural heritage to foster sustainable development in the digital age.

(Mantzou., et al, 2023).

FINDINGS

Educational Erosion: The literature underscores how colonial powers disrupted traditional education systems, marginalising Indigenous knowledge in favour of Western-centric curricula. This led to a significant loss of cultural transmission and diminished the prominence of Indigenous wisdom in subsequent generations (Rawat& Mishra,2021).

Cultural Identity Shifts: Colonialism contributed to profound shifts in cultural identity. The imposition of Western values and norms eroded the foundations of Indigenous cultures, leading to a loss of linguistic diversity, traditional practices, and spiritual connections to the land.

Economic Exploitation: Exploitative economic practices during the colonial era disrupted traditional land management and agricultural systems. This had immediate economic consequences and resulted in the erosion of Indigenous agricultural knowledge and practices (Ndiribe &Aboh,2022)

Challenges and Resilience in the Post-Colonial Era

Challenges in Knowledge Transmission: The literature highlights ongoing challenges in transmitting Indigenous knowledge post-colonialism. Westernisation and globalisation continue to threaten the survival of traditional practices, with younger generations often facing challenges in understanding and adopting their heritage.

Resilience in Cultural Practices: Despite the challenges, the literature also reveals resilience within Indigenous communities. Efforts to preserve cultural practices, languages, and traditional ecological knowledge persist, highlighting the resilience of Indigenous communities in the face of historical disruptions (Ndlovu-Gatsheni,2020).

The Fourth Industrial Revolution as a Double-Edged Sword

Technological Opportunities: The Fourth Industrial Revolution presents opportunities for addressing global challenges. Advanced technologies, such as artificial intelligence, data analytics, and connectivity, can potentially enhance healthcare, education, and economic development.

Potential for Cultural Appropriation: However, the literature warns against the potential pitfalls of the Fourth Industrial Revolution, emphasising the risk of cultural appropriation. Rapid technological advancements may inadvertently overlook the significance of Indigenous knowledge or perpetuate Western-centric solutions, further marginalising traditional practices.

Call for Ethical Integration: Scholars advocate for the ethical integration of the Fourth Industrial Revolution. The literature suggests that meaningful collaboration between Indigenous communities and technologists is crucial to ensure that technological advancements align with cultural values, address local challenges, and contribute to sustainable development. Scholars emphasise the imperative of ethically integrating the Fourth Industrial Revolution (4IR) with Indigenous Knowledge Systems (IKS), advocating for the recognition and elevation of IKS in shaping technological advancements. For instance, in environmental conservation, initiatives like the Rainforest Connection leverage innovative technology and Indigenous knowledge to protect biodiversity and combat illegal logging. By repurposing old smartphones into solar-powered listening devices, this project enables Indigenous communities to detect and respond to sounds of deforestation in real time, thus empowering them as stewards of their ancestral lands. Similarly, in healthcare, collaborations between biomedical researchers and Indigenous healers have yielded promising results in the development of culturally appropriate treatments for ailments prevalent in Indigenous communities. Through the incorporation of traditional healing practices and plant-based medicines, these initiatives not only enhance healthcare accessibility but also uphold the dignity and autonomy of Indigenous peoples in managing their health. Such examples underscore the transformative potential of ethical integration, wherein 4IR technologies not only coexist with but also amplify the wisdom and resilience of Indigenous knowledge, fostering sustainable development rooted in cultural values and community empowerment.

Emerging Paradigms and Collaborative Solutions:

Paradigm Shifts in Development: The literature points to emerging paradigms in development that prioritise integrating Indigenous knowledge. Recognising the value of traditional practices in areas such as sustainable agriculture, environmental conservation, and community governance is seen as essential for holistic and culturally sensitive development.

Community-Led Innovation: Collaborative solutions are highlighted as a way forward. Literature advocates community-led innovation that draws on Indigenous knowledge and Fourth Industrial Revolution technologies. Initiatives focusing on digital storytelling, knowledge documentation, and culturally sensitive technological solutions emerge as promising avenues.

Cultural Preservation and Sustainable Development:

Revitalising Indigenous Knowledge: There is a growing consensus in the literature about the importance of revitalising Indigenous knowledge. Efforts to preserve and promote traditional practices, oral histories, and cultural heritage are integral to cultural preservation and sustainable development.

Holistic Development Approaches: Scholars propose holistic development approaches integrating Indigenous knowledge into policies, educational systems, and technological advancements. This approach aims to create a more inclusive and culturally sensitive path toward sustainable development.

RECOMMENDATIONS FOR HARNESSING AFRICA'S INDIGENOUS KNOWLEDGE IN THE FOURTH INDUSTRIAL REVOLUTION

Incorporate Indigenous Knowledge in Curricula: Education systems should integrate Indigenous knowledge into formal curricula at all levels. This ensures that future generations benefit from a comprehensive understanding of Western and traditional knowledge systems.

Establish Innovation Hubs: Create community-led innovation hubs that blend modern technology with Indigenous knowledge. These hubs can serve as centres for research, development, and collaboration, fostering a dynamic space for the co-creation of solutions.

Support Indigenous Knowledge Research: Invest in research initiatives documenting and studying Indigenous knowledge. This includes creating databases, archives, and digital repositories to preserve and share traditional practices, ensuring they are not lost with time.

Facilitate Collaboration: Encourage collaboration between Indigenous communities, research institutions, governments, and the private sector. Partnerships can facilitate the co-design and implementation of projects that leverage Indigenous knowledge for sustainable development.

Promote Cultural Heritage Preservation: Implement policies that safeguard cultural heritage, recognizing the importance of Indigenous knowledge in shaping cultural identity. Cultural preservation initiatives can include the protection of traditional languages, rituals, and arts.

Integrate Indigenous Knowledge in Policies: Governments should develop policies and legislation recognising, protecting, and promoting Indigenous knowledge. This includes incorporating traditional practices in agriculture, healthcare, and environmental conservation sectors.

Empower Indigenous Communities: Develop capacity-building programs that empower Indigenous communities to actively participate in decision-making processes related to technology adoption and development. This ensures that the benefits of the Fourth Industrial Revolution are inclusive and equitable.

Integrate Traditional Ecological Knowledge in Conservation: Environmental conservation efforts should integrate traditional ecological knowledge. This includes involving Indigenous communities in designing and implementing conservation strategies to ensure their sustainability and effectiveness.

Promote Technology Access: Ensure that Indigenous communities have access to digital technologies. This involves initiatives to bridge the digital divide, provide digital literacy training, and make relevant technologies accessible to remote areas.

Establish Ethical Guidelines: Develop ethical guidelines for using Indigenous knowledge in research and development. This includes respecting intellectual property rights, obtaining informed consent, and ensuring that communities benefit equitably from any commercialization of their knowledge.

Encourage Cross-Cultural Collaboration: Promote collaboration between Indigenous communities and technology experts. This fosters a shared learning environment where traditional wisdom and modern expertise intersect to create innovative and culturally sensitive solutions.

Support Community-Led Projects: Allocate resources to support community-driven development projects that leverage Indigenous knowledge. This empowers communities to address their unique challenges using traditional practices and modern technologies. Efforts to revitalise Indigenous knowledge involve empowering local communities to reclaim and transmit their traditional wisdom. This includes initiatives to integrate Indigenous knowledge into formal education systems, promote sustainable agricultural practices, and support community-led conservation efforts. Recognising the relevance of traditional healing practices, local governance structures, and environmental management strategies contributes to building more resilient and culturally rich communities.

CONCLUSION

The exploration of Africa's Indigenous knowledge and its role in the context of historical legacies and the Fourth Industrial Revolution reveals a profound tapestry of wisdom that can significantly contribute to sustainable development. The legacy of colonialism has left indelible marks on Indigenous knowledge systems, disrupting traditional practices, and eroding the rich cultural heritage of African communities. However, there is a growing recognition of the importance of revitalising and integrating Indigenous knowledge into contemporary solutions. The Fourth Industrial

Revolution, characterised by rapid technological advancements, presents challenges and opportunities for Africa. Harnessing the strengths of Indigenous knowledge is essential for ensuring that these advancements align with the continent's cultural values, address local challenges, and contribute to sustainable development.

The recommendations provided underscore the need for a comprehensive and collaborative approach. Integrating Indigenous knowledge into education, establishing innovation hubs, supporting research, and fostering partnerships are key strategies to bridge the gap between tradition and modernity. Recognising the importance of cultural preservation, ethical use of Indigenous knowledge, and empowering communities are critical components of this journey. Africa's Indigenous knowledge holds the potential to inform sustainable agriculture, eco-friendly technologies, healthcare solutions, and community-led conservation efforts. The emphasis on cultural heritage preservation, capacity building, and cross-cultural collaboration creates a framework for inclusive and equitable development. Nurturing Africa's Indigenous knowledge is a means of preserving cultural identity and a pathway to resilience, sustainability, and innovation. As the continent navigates the challenges and opportunities of the Fourth Industrial Revolution, embracing and integrating Indigenous wisdom will be instrumental in shaping a future that honours the past while fostering a harmonious and sustainable coexistence between tradition and progress.

REFERENCES

Beyette, A. K. (2021). *Environmental Legacies of Colonialism: How Past Colonization Affects Current Social and Environmental Outcomes*. Academic Press.

Çınar, M. A. (2020). Historical Perspective on Epidemics and New Beginning after *COVID-19*. Karabuk University.

Cuthbertson, L. M., Robb, Y. A., & Blair, S. (2020). Theory and application of research principles and philosophical underpinning for a study utilising interpretative phenomenological analysis. *Radiography*, *26*(2), e94–e102. doi:10.1016/j.radi.2019.11.092 PMID:32052765

Domínguez de Olazábal, I. (2022). *On Indigenous Refusal against Externally Imposed Frameworks in Historic Palestine*. Academic Press.

Du, H., Bradbury, T. N., Lavner, J. A., Meltser, A. L., McNulty, J. K., Neff, L. A., & Karney, B. R. (2020). A comparison of Bayesian synthesis approaches for studies comparing two means: A tutorial. *Research Synthesis Methods*, *11*(1), 36–65. doi:10.1002/jrsm.1365 PMID:31782621

Fletcher, M. S., Hamilton, R., Dressler, W., & Palmer, L. (2021). Indigenous knowledge and the shackles of wilderness. *Proceedings of the National Academy of Sciences of the United States of America, 118*(40), e2022218118. doi:10.1073/pnas.2022218118 PMID:34580210

Franklin, S. (2022). *Embodied progress: A cultural account of assisted conception.* Taylor & Francis. doi:10.4324/9781003284499

Gómez-Baggethun, E. (2022). Is there a future for Indigenous and local knowledge? *The Journal of Peasant Studies*, *49*(6), 1139–1157. doi:10.1080/03066150.2021.1926994

Kohsaka, R., & Rogel, M. (2021). Traditional and local knowledge for sustainable development: Empowering the Indigenous and local communities of the world. In *Partnerships for the Goals*. Academic Press.

Mantzou, P., Bitsikas, X., & Floros, A. (2023). Enriching Cultural Heritage through the Integration of Art and Digital Technologies. *Social Sciences (Basel, Switzerland), 12*(11), 594. doi:10.3390/socsci12110594

Naimah, I. (2022). The Practice of Imperialism and Colonialism in Development and Their Impact on the Community Economy: An Overview. *Heritage, 3*(1), 39-48.

Ndiribe, M. O., & Aboh, S. C. (2022). Multilingualism and marginalisation: A Nigeria diversity approach. *International Journal of Multilingualism*, *19*(1), 1–15. doi:10.1080/14790718.2020.1818752

Ndlovu-Gatsheni, S. J. (2020). *Decolonization, development, and knowledge in Africa: Turning over a new leaf.* Routledge. doi:10.4324/9781003030423

Ngũgĩ wa Thiong'o. (1986). *Decolonising the Mind: The Politics of Language in African Literature*. James Currey Publishers.

Olaopa, O. R., & Ayodele, O. A. (2022). Building on the strengths of African Indigenous knowledge and innovation (AIK&I) for sustainable development in Africa. *African Journal of Science, Technology, Innovation and Development*, *14*(5), 1313–1326. doi:10.1080/20421338.2021.1950111

Omonov, Q., & Karimov, N. (2020). Importance Of Ancestral Heritage. *The American Journal of Social Science and Education Innovations*, *2*(09), 196–202. doi:10.37547/tajssei/Volume02Issue09-29

Rawat, G., & Mishra, S. (2021). Spirituality and environment: significance of Indigenous knowledge systems in Africa and challenges. *Literary Endeavour, 12*(1).

Rothenberg, W. A., Sternberg, A., Blake, A., Waddell, J., Chassin, L., & Hussong, A. (2020). Identifying adolescent protective factors that disrupt the intergenerational transmission of cannabis use and disorder. *Psychology of Addictive Behaviors, 34*(8), 864–876. doi:10.1037/adb0000511 PMID:31524418

Said, E. (1978). *Orientalism.* Vintage Books.

Sellar, C., & Battisti, G. (2023). Europe. The Many Reasons for an Epoch-Spanning Crisis. A Long-Term Geohistorical and Geoeconomics Analysis of the Obstacles to European Integration. In *Geopolitical Perspectives from the Italian Border: Introducing Gianfranco Battisti, Triestino Geographer* (pp. 89–102). Springer International Publishing. doi:10.1007/978-3-031-26044-5_9

Chapter 5
Coloniality Forces Within the Cold War and Neocolonialism:
A Hindrance to Africa's Contributions to Digital Technology

Eric Blanco Niyitunga
University of Johannesburg, South Africa

ABSTRACT

The troubling question that one needs to answer is whether the African continent's supply of raw materials, which depends on the functionality and advancement of digital technology, is not a significant contribution to the wave of digital technology the global world faces today. If that is reality, how is it that Africa's contributions to digital technology are not recognized, and who will make the world recognize them? This is a troubling question that needs to be answered. The aim of this chapter is to examine coloniality forces within the Cold War and Neocolonialism events and expose factors emerging from them that have hindered Africa's contributions to the digital technology.

INTRODUCTION

The troubling question that one needs to answer is whether the African continent's supply of raw materials, which depends on the functionality and advancement of digital technology, is not a significant contribution to the wave of digital technology the global world faces today. If that is reality, how is it that Africa's contributions to digital technology are not recognized, and who

DOI: 10.4018/978-1-6684-7851-6.ch005

Copyright © 2024, IGI Global. Copying or distributing in print or electronic forms without written permission of IGI Global is prohibited.

will make the world recognize them? This is a troubling question that needs to be answered. This chapter aims to examine coloniality forces within the Cold War and Neocolonialism events and expose factors emerging from them that have hindered Africa's contributions to digital technology. The contemporary lingering coloniality of power and knowledge in Africa that is still trapping African political leadership and education has led to colonial domination in the African's ways of governance and ways of knowing, seeing, and imagining that the continent contributes to the wave of digitalization are not recognized. Scholars and political leadership classes still praise colonialists and, in their systems of governance and writing, keep praising them instead of protecting natural resources and raw materials on which the wave of digitalization depends.

One can see black political power and regimes in Africa and naively say that colonialism administrations have been entirely eradicated, and most of the periphery is politically organized into independent states, forgetting that African people are still living under crude European exploitation and domination (Ndlovu-Gatseni, 2018). The old colonial hierarchies of whites versus Africans are still alive in Africa and are entangled with the 'international division of labor and accumulation of capital at a world scale' (Grosfoguel, 2007, p.211). This means that while the establishment of international law and human rights organizations collapsed physical colonialism in the form of European forces imposing foreign governments on Africa, colonial legacies, colonial conditions, and power relations have remained alive and intact. These conditions are worse than colonialism and have affected Africa's contributions to the wave of digital technology. Coloniality refers to a long-standing pattern of power that emerged from colonialism but defines the culture, labor, intersubjectivity relations, and knowledge production well beyond the strict limits of colonial administrations (Maldonado-Torres, 2007, p.243). Thus, coloniality survives colonialism. It is maintained alive in books, in the criteria for academic performance, in cultural patterns, in common sense, in the self-image of people, in aspirations of self, and so many other aspects of our modern experience (Ndlovu-Gatseni, 2018). As modern subjects, Africans still breathe coloniality all the time and every day (Maldonado-Torres, 2007, p.243).

This chapter assesses the effects of coloniality on Africa's contributions to the wave of digital technology. This analysis is done by assessing the effects and problems caused by the Cold War, which came about immediately after the abolition of physical colonialism and neocolonialism forces that are still harming Africans. This analysis is particularly on social, political, and economic governance and their effects on digital technology advancement in Africa. This is because when colonialism was abolished, and African states started gaining independence to address the problems caused by the slave trade and colonialism, the Cold War and Neo-colonialism events kicked in, thwarting any efforts advanced by the Africans to address those problems. Coloniality events are devastating and worsening Africa's efforts towards contributing to digital technology.

This has made Africa's contributions to the wave of digital technology more challenging and complex because those two assaults are adding enormous problems in Africa. For example, under coloniality, the Cold War brought about colonial governance and administration and increased African proxy leaders. Neocolonialism, on the other hand, has led to regime changes and the installation of leaders who are colonial collaborators. This is to say that every political instability or civil war in Africa is because of the effects of coloniality during the event of Cold War and Neo-colonialism, with the lingering effects of trans-Atlantic slavery and colonialism disempowering the ability of Africa's contributions to the wave of digital technology.

Therefore, the question the chapter seeks to answer is to what extent has coloniality under Cold War and Neocolonialism events created problems in Africa that have hindered its contributions to the wave of digital technology? The chapter is structured into five main sections to achieve the aim and answer the question. The first section explains what coloniality is about, its characteristics, and its effects on Africa. The second section explains neocolonialism as a tool for coloniality and highlights the effects of neocolonialism on African indigenous knowledge and technology. This section also explains the nature and characteristics of those effects and how they threaten governance for technology in Africa. The third section explains what the Cold War was about and how its effects served as coloniality tools to affect the growth of indigenous knowledge of technology in Africa. The section explains factors making those effects a threat to African political leadership through regime changes and proxy political leaders, thus inhibiting Africa's contributions to digital technology. The fourth section explains the role of political governance for African indigenous knowledge and the realization of African indigenous technology to have informed the digital technology we see today. The fifth section consists of concluding remarks of the chapter.

Coloniality in the Era of Digital Technology

Coloniality is, metaphorically, the unconscious of Western civilization or the darker side of Western modernity (Mignolo, 2011). The concept captures the hidden atrocities and sufferings that have walked side-by-side with the golden promises expressed in the rhetoric of modernity with massive numbers of African human beings they called slaves. The concept of coloniality was coined by Quijano (1992) and is understood as a colonial matrix of power (Mignolo and Walsh, 2018, p. 246). Like any power structure, the colonial matrix of power consists of the permanent co-presence of domination, exploitation, and conflict (Ndlovu-Gatseni, 2015).

The control over domination and exploitation comes from creating and managing knowledge, which is a machine of world-making and world-managing rather than representing an objectively existing reality (Mignolo and Walsh, 2018, pp. 144-169). Quijano (2000) defines the term "coloniality of power" as racial and epistemological hierarchies that are entangled within structural hierarchies such as global capitalism, which continues to be a factor after the period of colonization. Coloniality of power assists in investigating how the current "global-political" order has been constructed, constituted, and configured into a racially hierarchized, Euro-American-centric, Christian-centric, patriarchal, capitalist, hetero-normative, hegemonic, asymmetrical, and modern power structure.

Miranda Fricker (2007, p.13) argued that the notion of social power means "a socially situated capacity to control others' actions and activities. Social power can be used "to create or preserve a given social order." This social power is displayed in various forms of enablement, such as disbelief, misinterpretation, and silencing of what others are doing that benefits the world. This has repercussions on the African indigenous knowledge that has covered every action that indigenous Africans have done through indigenous technology, which constitutes contributions to modern technology. Fricker (2007) argued that this power is "identity power" and has caused an identity crisis in Africa (Kochalumchuvattil, 2010). Fricker (2007) argues that this power's main interest is causing an identity crisis that manifests through identity prejudices we hold about ourselves, our ability and capabilities, and digital technology. Institutions from primary to secondary to tertiary levels and churches have become the tools for coloniality reproduction and advancement in Africa (Ndlovu-Gatsheni, 2013). One of the effects of trans-Atlantic slavery was the creation of mistrust and mental slaves unable to speak for themselves and to defend or protect their physical rights. Through colonized education in the institutions of learning and colonized Christianity through church fellowships and gatherings, the reproduction and continuity of mental slavery are rampant. Mental slavery has deprived Africans, particularly leaders and scholars, of their rights to independently defend African indigenous knowledge and present how digital technology we see today has its origins in Africa.

This failure has led to three kinds of coloniality manifesting themselves in power, knowledge, and being. There have been calls to decolonize the minds and the mental universe (university programs, infrastructures, and sites of knowledge production). Still, because of the lack of independence from mental slavery, the calls have become fruitless and never-ending. Coloniality is more horrible and destructive than the trans-Atlantic slave trade and colonialism. This, therefore, requires the necessity to revisit, reimagine, and redefine spaces where indigenous and non-indigenous people experienced the effects of colonialism, Cold War/ neo-colonialism, imperialism, and underdevelopment (Ndlovu-Gatsheni, 2013). In this post-Cold War era, the forces

Figure 1.

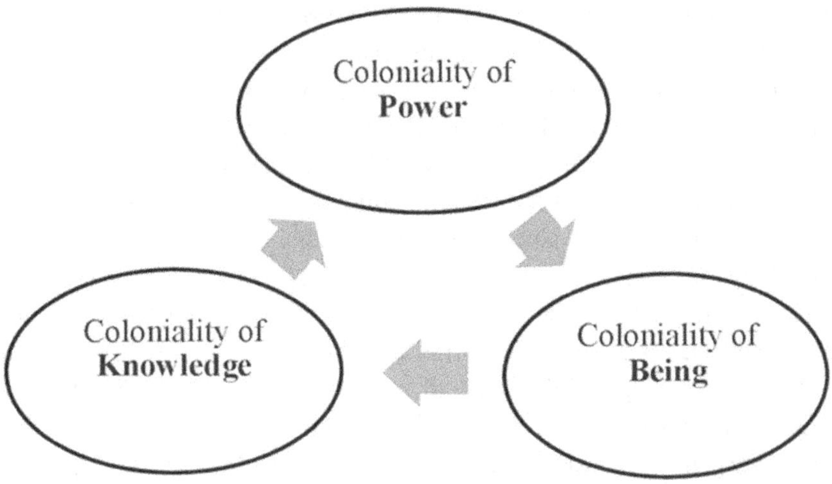

of coloniality under neocolonialism have colonized culture, intellect, language, aesthetics, religion, and indigenous people, thus preventing Africa's contributions to digital technology.

From the above diagram, one can see that three deadly coloniality forces have hindered any contributions of Africa to the world of digital technology. These coloniality forces must be overcome for total independence in academia, governance, and freedom of expression for Africa's contributions to the world of digital technology. The first neocolonialist force is the coloniality of power. Quijano (2000) conceptualized coloniality of power as racial and epistemological hierarchies entangled within structural hierarchies such as global capitalism. These hierarchies have become prevalent and are the key factor of neocolonialism. Coloniality of power assists in investigating how the current "global-political" order has been constructed, constituted, and configured into a racially hierarchized, Euro-American-centric, Christian-centric, patriarchal, capitalist, hetero-normative, hegemonic, asymmetrical, and modern power structure (Ndlovu-Gatsheni, 2013). This means that coloniality is centered on colonized Christianity and education that emanates from the Western world. Ndlovu-Gatsheni (2013) argued that the coloniality of power had used conversion to Christianity to colonize the consciousness and the minds of Africans to accept the Eurocentric hierarchization of power. This means that Christianity has colonized Africa's minds to the level of attribution of what Africa has contributed to digital technology to the Western world.

The second force of neocolonialism is the coloniality of Knowledge. The coloniality of knowledge has something to do with the impact of colonization on the different areas of knowledge production and application. Mbembe (2017) cautions that learning institutions cannot continue providing a Westernised kind of knowledge. Western education in Africa aims to promote Western indigenous technology and present the contributions of the West to digital technology while omitting those from Africa. The promotion of African indigenous knowledge and an Africa-oriented curriculum would erase colonized minds, thus enabling the achievement of a free mind that promotes Africanity to the world. Unfortunately, the continuation of colonial education has placed Africans as "things" or unproductive human beings who fear technology, mathematics, and physics. This is rhetoric because historical literature shows that Africa is the mother of civilizations, meaning the mother of engineering, mathematics, science, etc. It shows that the first University to offer a degree certificate was in Africa, thus making Africa the origin of education and academic knowledge. Coloniality of knowledge compels colonial scholars to understand how and why other knowledge systems, such as African indigenous knowledge, are primitive and useless (Grosfoguel 2013). Mpofu (2013, pp.109-110) enriches our debate in this article by saying:

The colonizer does not only distort the history of the colonized, slaughter their knowledge systems, and empty their heads of self-confidence and their hearts of the emotional stamina to live without colonial domination. But he goes ahead to manufacture accusations and labels against the colonized; among many of the accusations are laziness, drunkenness, backwardness, the propensity to violence, dirtiness, stupidity, ignorance, bad luck, and spiritual damnation, all of which require the colonizer to intervene and save the colonized from the abyss of many "lacks" and "deficits" that bedevil him and his lot.

When missionaries came to Africa, they consciously or unconsciously leveled damning accusations at indigenous people who were perceived as inadequate beings who deserved to be developed and civilized by inculcating Western knowledge in them. Different institutions of learning were used to promote Eurocentric knowledge. Foucault (2013) distinguishes between two kinds of knowledge that dominated these institutions: knowledge as organized in disciplines, scientific fields, or areas (reconnaissance) and general knowledge, which creates conditions for an object of knowledge to be taken up by a discipline (savoir). Fanon (1952, p.14) explains that it is imperative "to liberate the black man from the arsenal of complexes that germinated in the colonial situation." Fanon (1952) believes that indigenous people must overcome inferiority and superiority complexes hidden in Western knowledge systems. Biko (1978) noted that colonialism left indigenous people with an inferiority complex, which in some ways crippled them psychologically. Moreover, the coloniality of knowledge poses epistemological questions that are linked to:

- the politics of knowledge generation.
- who generates which knowledge and for what purpose.
- relevance and irrelevance of knowledge; and
- how some knowledge disempowered/empowered communities and peoples (Ndlovu-Gatsheni, 2013).

Coloniality of knowledge contributes to an understanding and exposure of epistemic silences, conspiracies, and epistemic violence that are hidden within Eurocentric epistemologies.

The third force of neo-colonialism is the coloniality of being. This concept of "coloniality of being" came into existence after scholars of colonialism/decolonialism concluded that colonial relations of power did not only leave indelible marks in authority, sexuality, knowledge, and the economy but also on the general understanding of being. It responded to the need to thematize the question of the effects of coloniality in lived experience and not only in mind (Mignolo, 2007, p.242). It penetrates issues of human ontology (Wynter, 2003, pp. 257–337), restoration of denied self-pride, and sovereign subjectivity (Blyden, 1967). Mignolo (2007, p. 256) summarises the coloniality of being as follows:

The coloniality of Being appears in historical projects and ideas of civilization, which advance colonial projects of various kinds that are inspired or legitimized by the idea of race. The coloniality of Being is, therefore, coextensive with the production of the color line in its different expressions and dimensions. It becomes concrete in the appearance of liminal subjects, which mark, as it were, the limit of Being, that is, the point at which Being distorts meaning and evidence to the point of dehumanization. The coloniality of Being produces the ontological colonial difference, deploying a series of fundamental existential characteristics and symbolic realities.

Ndlovu-Gatsheni (2013) argues that coloniality of being is important because it helps us to investigate how humanity for indigenous people was portrayed, which subsequently led to the "objectification"/ "thingification"/ "commodification" of these people.

The above three forces of neocolonialism have made Africa be excluded from the post-1648 Westphalian sovereign state system, and, which was physically partitioned after the Berlin Conference of 1884–5 was later incorporated into the post-1945 United Nations sovereignty state system (Mazrui, 1986; Ndlovu-Gatseni, 2013). One can add that the fragmented and weak African 'postcolonial' states were admitted into the lowest echelons of the Euro-North American-dominated state system of the world (Mazrui, 1986; Ndlovu-Gatseni, 2013). coloniality of power, knowledge, and being have made Africa be incorporated into a Euro-North

American-centric world culture and European languages, into a heavily Euro-North American-centric world of international law, and into the modern technological age including being 'swallowed by the global system of dissemination of information technology (Ndlovu-Gatseni, 2013). Mazrui concluded, therefore, that 'what Africa knows about itself, what different parts of Africa know about each other, have been profoundly influenced by the West (Ndlovu-Gatseni, 2013).

The end of colonialism stopped direct colonial administrations and oppression but did not give rise to total independence in Africa,' but rather coloniality. This coloniality comes under neocolonialism forces such as racially hierarchized, patriarchal, sexist, Christian-centric, hetero-normative, capitalist, military, colonial, imperial, and western education systems (Grosfoguel, 2011, p.15). The negative consequences of this modernity included the slave trade, imperialism, colonialism, apartheid, neo-colonialism, underdevelopment, and neo-liberalism, including the Washington Consensus, structural adjustment programs (Ndlovu-Gatsheni, 2013), and the Beijing Consensus. coloniality is a disruptive, 'decivilizing,' dehumanizing, exploitative, racist, violent, brutal, covetous, and 'thinking' system.

Coloniality is a helpful concept that names various colonial-like power relations existing today in those zones that experienced direct colonialism. Mignolo (2007) emphasized 'colonial difference' as a central leitmotif of coloniality. Maldonado-Torres (2007, p. 243) elaborated on the meaning of coloniality and its difference from colonialism in these revealing words:

"Coloniality is different from colonialism. Colonialism denotes a political and economic relation in which the sovereignty of a nation or a people rests on the power of another nation, which makes such a nation an empire. Coloniality, instead, refers to long-standing patterns of power that emerged as a result of colonialism but that define culture, labor, intersubjectivity relations, and knowledge production well beyond the strict limits of colonial administrations. Thus, coloniality survives colonialism. It is maintained alive in books, in the criteria for academic performance, in cultural patterns, in common sense, in the self-image of people, in aspirations of self, and so many other aspects of our modern experience. In a way, as modern subjects, we breathe coloniality all the time and every day."

Rodney (1973) understood coloniality in terms of 'underdevelopment' and 'dependency' perspectives. Ngugi wa Thiong'o detailed the workings of coloniality on the minds of its targets:

"The biggest weapon wielded and daily unleashed by imperialism against that collective defiance is the cultural bomb. The effect of a cultural bomb is to annihilate a people's belief in their names, in their languages, in their environment, in their

heritage of struggle, in their unity, in their capacities, and ultimately in themselves. It makes them see their past as one wasteland of non-achievement and want to distance themselves from that wasteland. It makes them want to identify with that which is furthest removed from themselves, for instance, with other people's languages rather than their own. It makes them identify with that which is decadent and reactionary, all those forces that would stop their springs of life. It even plants serious doubt about the moral rightness of struggle. Possibilities of triumph or victory are seen as remote, ridiculous dreams. The intended results are despair, despondency, and a collective death wish. Amidst this wasteland that it has created, imperialism presents itself as the cure and demands that the dependent sing hymns of praise with the constant refrain: 'Theft is holy.' Indeed, this refrain sums up the new creed of the neo-colonial bourgeoisie in many 'independent' African states" (Ngugi wa Thiong'o, 1986, p.12-13).

Coloniality is, therefore, an invisible power structure, an epochal condition, and an epistemological design that lies at the center of the present Euro-North American-centric modern world. At the center of coloniality is race as an organizing principle that not only hierarchizes human beings according to racial ontological densities but also sustains asymmetrical global power relations, thus affecting Africa's contributions to the wave of digital technology.

EFFECTS OF NEOCOLONIALISM ON THE AFRICAN INDIGENOUS TECHNOLOGY

The term 'neocolonialism' was first coined by Kwame Nkrumah. For Kwame Nkrumah, "neocolonialism represents the final stage of imperialism (Nkrumah, 1961). Neocolonialism is the worst form of Imperialism. While for those who believe it and who practice it, it means power without responsibility, those who suffer from it, it is exploitation without redress" (Northrop, 2012). Primarily, neocolonialism has no definite definition and has various versions of the definitions. In Encyclopedia Britannica, Sandra Halperin defined neocolonialism as the control of less-developed countries by developed countries through indirect means" (Northrop, 2012). It is the 'indirect means' as developed countries are indirectly dominating underdeveloped or developing countries by using colonial exploitation rules (Northrop, 2012). Colonialists first applied neocolonialism as a scheme to maintain control of African states under their domination (Northrop, 2012).

It has been argued that neocolonialism is a process by which a colonial mother country exposed exploiting rules and regulations to her newly independent underdeveloped countries or less developing countries for indirect dominating; the dominating can be economic political or cultural (Rahaman, Yeazdani and Mahmud, 2017). It has been noticed in economics, although this economic exploitation is cloaked by political and cultural exploitation (Rahaman et al., 2017). It is also seen in all forms of capitalism, which were spread in third-world countries in the aftermath of the Cold War. Nkrumah, who lived during the Cold War era, had seen when he equated neocolonialism to capitalism (Nkrumah, 1960). He further stated that neocolonialism is the worst form of imperialism and the highest stage of capitalism (Nkrumah, 1960: ix). Nkrumah further argued that the main goal of neocolonialism was to transform a newly independent country into an underdeveloped country and, at any cost, to remain at the mercy of the imperialist countries (Nkrumah, 1960: ix). Neocolonialism posed many threats to Africa, leading to immense problems. These problems are remarkable in all sectors of life, including social, economic, political, cultural, and literature—problems caused by neocolonialism on African social relations. The current neo-colonialism represents imperialism in its final and most dangerous stage.

The first effect of neocolonialism that hindered Africa's contributions to the wave of technology was the creation and support of regime change under the disguise of democracy. This led to proxy governance like colonial governance, the difference being a lack of oppression. The neo-colonial regime and regime change in the name of democracy have been established to serve the interests of the colonialists and imperialists. The essence of neocolonialism is that the State subject to it is, in theory, independent and has all the outward trappings of international sovereignty. The country facing neocolonialism means its economic system and political policy are directed from outside (Nkrumah, 1960: x). A venture into neocolonial experiences cannot be divorced from the goals of European imperialism and the study of the colonial state in Africa. This is because neocolonialism is prevalent in Africa, unlike other continents, and can be described as an advanced stage of colonialism. This is because it exists in the cultural, educational, industrial, and technological subjugation of a former colonial territory. It also exists in the former colonizer's economic domination of the independent state without an institutionalized political structure. Moreover, it is practiced in the absence of direct military presence for the physical control and direction of the state ideology and economy by the former colonizer over her former and liberated colonies.

Neocolonialism is perpetuated through the process of international diplomacy and protocols of state relations (Babatola, 2014). It is also perpetuated through an entrenched constitutional government inherited or imposed by the former colony that undermines the emergence of local political leadership in the newly independent states. The propagators of neocolonialism provide the former colonies a form of a market structure cum capital that controls the investment of the new state and the agenda for national development working in favor of the former colonizer (Babatola,

2014). This is enhanced through continuous tying to the former colonizer's apron strings in determining the former colony's current and future roles and relationships in world politics and economy (Babatola, 2014). It has been argued that neo-colonialism is evident from the stages of national development at the points where former colonialists continue to influence domestic affairs in the former colonies (Remmling and Campbell, 1970, pp.66-80).

Amechi (1986) asserted that the option left for the former colonizer, who gained independence from the colonizer, was to turn to the former colonialists for help and relations. This posed problems in Africa because it made inter-African relations scarce and limited. It has made Africans see each other as enemies and has led to increased African problems such as conflicts, war, and xenophobic cases. It has been argued that neocolonialism demolished African traditions and culture embedded in the society promoting friendly relationships. It also undermined African languages, calling them vernaculars, and promoted colonialist languages, which formed language barriers. Common African culture and language are undermined, and Africans struggle to learn foreign languages, which leads to skills shortages on the continent. This leads to a lack of expertise and technological transfer in existing industrial and trade complexes (Babatola, 2014). It also led to a lack of contending ideas to transform society and engage in the agenda for national development with the pressures of finding a path for national survival and growth guided by international forces and partners (Babatola, 2014).

Neocolonialism perpetuates an imbalance in the relationship between a former colonizer and her liberated colonies by providing a superstructure with irresistible economic relations (Amechi, 1986). It also perpetuates the potential for former colonizers to further exploit and extract the economic resources of the latter largely through the provision of capital and accumulation of profit (Amechi, 1986). It perpetuates the expansion of her home economy from the exploitative gains under the guise of running through a globally competitive market (Babatola, 2014). However, neo-colonialism was foisted on Africa but could not exist without existing transferred structures, internal collaborators, and partners whose responsibility is to enjoy the benefits and privileges of power. It perpetuates mistrust among Africans and makes African states depend on the former colonizer perpetually, thus being robbed of their capacity to develop and progress without the permission of the former colonizers (Babatola, 2014). This is a problem as it has incapacitated African states to the level of leaning on the West, now China, for survival. Lack of trust has been the source of every conflict, failed mediation, and conflict resolution efforts, and has been the trigger of Xenophobic violence against foreigners in South Africa.

The second effect of neocolonialism on the African continent that has prevented Africa from contributing to the wave of digital technology is poverty and economic stagnation, and retardation in Africa. The economy is the driving force for the development of digital technology in any country. That means any solutions to solve African problems must involve the economy. This is because there can be no peace without economy or development and vice versa. Moreover, without economic development, no state can achieve development and social order, which leads to stability. In the African context, imperialist countries had applied economic tools as an arm to dominate the African states (Rahaman, Yeazdani & Mahmud, 2017), causing more problems in Africa. it has been argued that "the features and development of neo-political economy in African states translated the scale of economic growth and development processes into substituted imports of intermediate and producer goods for imports of consumer goods" (Williams, 1981, p. 41). Williams (1981:41) argued that "this consolidates rather than undermines dependence on foreign suppliers, since production, as well as consumption, now depends on imports." It creates problems in Africa in the sense that the continent continues to suffer from exploitations that it went through during the trans-Atlantic slave trade and colonialism.

This is because those exploitations are not the only way of colonial economic exploitation, but colonial power continues to exploit Africa in several means, rendering Africans to remain and become beggars. The result is that Africa becomes incapacitated to the level of not being able to grow capital, technologies, experts, and management to develop itself and solve its problems. The agents of neocolonialism that continue to create problems in Africa include former imperialist states, international organizations such as the International Criminal Court (ICC), the World Bank (WB), the International Monetary Fund (IMF), and The International Finance Investment and Commerce Bank Limited (IFIC), Commonwealth among others, had taken this opportunity appropriately to bound Africa in the neo-colonial iron curtain (Rahaman et al., 2017). Nkrumah (1963) warned African states, newly independent, not only to join international organizations but also to fear them because they serve the interests of the colonialists and create unbearable problems in Africa. Nkrumah (1961) argued that it is with many tears and sadness to learn that our fellow African states have joined international organizations because they do not understand that they are joining neocolonialist forces that would undermine their states and plunge the continent into intractable problems.

Neocolonialism has led to economic drain and debt in Africa, affecting the growth and development of digital technology on the continent. From the colonial period, economic drain and debts became endless, creating many problems and challenges in Africa. Colonial economic drain and debts are responsible for Africa's underdevelopment, poverty, and political instability. After the decolonization, the African leaders tried to stop the drain, but the donor group gave clear-cut threats to the African state's authority (Rahaman et al., 2017). Delwar (2016) said that if the anti-drain economic acts were active, neocolonialism agents would be totally unable to invest any amount of money in Africa. For example, it has been opined that "a rough diamond mined in Africa costs about $40 per carat, and a diamond cut and polished in Europe increases to $400 per carat" (Rahaman et al., 2017, p. 12).

Furthermore, the same stone fetches around $900 per carat when it reaches the consumer (Rahaman et al., 2017). Further examples include Zimbabwe, known for producing the best quality tobacco in the world. In 2016, the state earned $650 million from selling raw tobacco and would have earned $6.5 billion had the state processed the crop into cigarettes rather than exporting it as a raw good (Rahaman et al., 2017). Examples include Nigeria, which sells its crude oil for a measly $9 a barrel, only to import refined gasoline, diesel, and kerosene made from its oil resources for hundreds of dollars per barrel (Chengu, 2015). This economic drain reveals the inequality and economic injustice that African states go through every day. They continue to create more problems in Africa, such as underdevelopment, poverty, and weak political institutions that create and intensify conflicts, which are given a platform to flourish.

In the contemporary world, neo-colonialist agents in Africa have included China and Japan, which share a history of colonialism with African states. For example, the Asian giant China has invested a huge amount of dollars in Africa with tough terms and conditions, draining un-statistical dollars from Africa (Rahaman et al., 2017). Debt is another form of coloniality and a pivotal component of neocolonialism, strongly affecting economic growth and development in Africa. It also affects human growth and stability in Africa. African failures to echo the continent's contributions to digital technology emanate from the massive amount of unpayable debts that have re-colonized Africans and deprived them of the power to be independent in political leadership and academia (knowledge) (Rahaman et al., 2017). However, the decrease in debt interest by the World Bank and IMF has not yet helped the continent escape the debt trap because the terms and conditions have remained difficult to overcome. This debt trap has created problems in Africa in that the African economy has failed to grow because it does not run and cannot cope in the competitive world.

The other effects of neocolonialism on Africa's ability to contribute to the wave of digital technology are deep-rooted poverty and underdevelopment. Neocolonialism has not only created income inequality but also poverty and underdevelopment situations in Africa. Rodney argued that neocolonialism in Africa is responsible for economic subjugation and underdevelopment and then asserted that:

African economies are integrated into the very structure of the developed capitalist economies, and they are integrated in a manner that is unfavorable to Africa and ensures that Africa is dependent on the big capitalist countries. Indeed, structural dependence is one of the characteristics of underdevelopment. Underdeveloped countries are dependent on metropolitan capitalist economies. It is also true that metro poles are dependent on the wealth of the exploited portions of the world.

This is a source of their strength and a potential weakness within the capitalist/ imperialist system...The capitalist countries are technologically more advanced and are therefore the sector of the imperialist system which determined the direction of change...It is for this reason that a formerly colonized nation has no hope of developing until it breaks effectively with the vicious circle of dependence and exploitation... (Rodney, 1972).

He further opined that the exploration of neocolonialism is a new type of deadly exploitation of African states and a major source of problems Africa faces today. He asserted that:

the economic backwardness of a given African nation...the symptoms of underdevelopment and the secondary factors that make for poverty...lies in seeking out the relationship between Africa and certain developed countries and recognizing that it is a relationship of exploitation...More far-reaching than just trade is the actual ownership of the means of production in one country by the citizens of another...So long as foreigners own land, mines, factories, banks, insurance companies, means of transportation, newspapers, power stations, etc, then for so long will the wealth of Africa flow outwards into the hands of those elements (Rodney, 1972).

The problem of economic injustice and subjugation by neocolonialism is clearly explained in the dependency theory. It has been argued that "former colonizers institutionalized structurally unequal economic relationship between itself and their former colonies through changing of guards to systematically coordinate the processes of extracting resources and raw materials in those territories for their economic benefits" (Amechi, 1986, pp. 296-320). Babatola (2014, p.13) asserted that this practice in itself is similar to "the interdependence of state system, wherever it perpetuates the process and capacity to appropriate concept of balance of power through domination and subjugation of weaker states in the international economic order". This is what is further described as "a phase when the systems and structures of exploitation have reached such a high state of perfection that they do not need the physical presence of and supervision by the colonizers" (Amechi, 1986, pp. 296-320).

Neocolonialism represents the capitalist domination of Africa in the post-colonial historical experience, a period generally characterized as the early stages of the political experience of the independent African states after decolonization. Furthermore, the advanced stages of imperialism manifest through the exploitation of unequal and dominant economic processes and structures to superimpose the former colonizers on their liberated colonies as dependent states (Amechi, 1986, pp. 296-320). It has been argued that "the tendency of neocolonial experiences and practices in Africa has been largely directed at international conspiracy and transnational trading domination by multinational and transnational corporations at the expense of local entrepreneurs" (Akeredolu-Ale, 1975, p. 23). Akeredolu-Ale (1975, p. 24) argued that "this was highlighted by the existence of 'expatriate domination of investment opportunities...their superior access to credit, supplies, and the technology and managerial skills necessary to industrial production, (which) inhibits the accumulation and reinvestment of capital by indigenous entrepreneurs who lack the resources necessary to compete with vertically integrated multinational corporations."

Greig et al. (2007) argued that dependency theory explains development and underdevelopment in the light of a "situation whereby some countries have their economy conditioned by the development and expansion of another". Relatively, when a country lack most of the factors mentioned in modernization theory, underdevelopment is imminent, as will be detected by the development theory. When underdevelopment strikes, that country must depend on other countries that possess these development factors, as the dependency theory detects. The creation and/or maintenance of unequal economic, cultural, and territorial relationships, usually between states based on domination and subordination (Johnston, 2000, p. 375), causes underdevelopment.

Hence, the stage of economic domination and subjugation of the African continent as a raw material exploration field and a finished products exploitative market causes many problems in Africa (Johnston, 2000, p. 375) that hold the continent in poverty and underdevelopment statuses. Corruption, being a neocolonialism tool, exists among African leaders is a menace in most African countries today, and has remained the most significant cause of underdevelopment. Contemporary African leaders misuse the political power they acquire from citizens and use it for their private gains. Every year, political African leaders steal public funds and keep them in former colonizers' banks. These funds should be used for development projects that stimulate economic growth. Corruption in African countries today has so far gone deep into people that it has almost become a cultural phenomenon and is a language of the day in Africa.

The other effect of neocolonialism on Africa's ability to contribute to the wave of digital technology is dependency syndrome. Dependency syndrome is another practice by neo-colonialists that has created many problems in Africa. Babatola (2014, pp.11-12) argued that although political and economic powers were handed over to modern African states, the continued interference from colonizers in Africa, such as France, Belgium, British, and now the USA and China pose problems in Africa as these foreign powers are interfering Africa's political, economic and social life. This has caused a lot of problems in Africa as it has turned the continent into a blood-letting zone for wars, civil wars, genocides, and political instability.

These problems arise from the evils of neocolonialism and the unequal structure of the African states that has left the continent perpetually under the control and intervention of the global coloniality powers (Babatola, 2014, pp. 11-12). For example, the political crisis in the French and British colonies is associated with the exploitation of economic resources and political imbalance. Babatola (2014, p. 12) argued that resolving political differences and disputes in French colonies (Mali, Great Lakes Region, Cameroun, Central Republic, etc.) has presented more harm and problems to the continent because of the direct interference of France, the neocolonialists. The same experience occurs in British colonies such as Uganda, Kenya, Zimbabwe, Sudan, etc, and Portugal colonies such as Angola and Mozambique, to mention a few others (Babatola, 2014, pp. 11-12).

To explore the effects of neocolonialism in Africa, it has been well stated that 'foreign faces have withdrawn after local faces (Offiong, 1980, p. 65) have been groomed and installed to perpetuate the continued influence of the former colonizers (Babatola, 2014, p. 14). This created a dependency syndrome that made the economic exploitation of Africa fail to change after independence. However, the corresponding political responsibility and international stigma had been removed from the necks of the former colonizers (Amechi, 1986, pp. 296-320). African political leaders continued depending on the former colonizers for direction and leadership. This problem continues today; every newly appointed leader must get direction and political guidance from the colonizers. Amechi (1986, pp. 296-320) opined that though there was political liberation that granted independence to African states, neocolonialism has come up with tactics that make former colonizers still operate coquettishly and covertly, exploiting nations, maneuvering men and governments by remote control. This can be called colonialism by remote control and is creating proxy political leaders or colonial collaborators who serve the interests of the colonizers at the cost of African people.

Babatola (2014, p.12) argued that "local entrepreneurship was developed and made to compete and create capital, many of them became compradors, that is, intermediaries between foreign interests and the indigenous polity and economy, which are transited to the former colony as a source of both capitals and contracts." Many of such "entrepreneurs were known to have established companies with foreign interest and partnership at various stages of their contract awards and execution as well as the transfer of funds to overseas countries resulting in capital flight" (Babatola, 2014, p. 12).

The sad reality is that Africa has no power to share its contributions to digital technology because of the trap of neocolonialism and the effects of the trans-Atlantic slave trade and colonialism, which are still alive and rampant. This is because African states are still experiencing slavery and colonialism; hence, they have no options to be free from the imperialist curtain. The World Bank is lending loans to African states, which has increased debts in Africa. As Niyitunga (2014) said, loans have plunged Africa and her people into horrible slavery with no ability to be autonomous in the decisions concerning their states. This is because a debtor is always a creditor's slave. Thus, African states become slaves to the creditors and are ready to serve their masters as they become helpless at the mercy of the creditors.

This means that being a slave in a situation of slavery becomes so impossible for one to raise up. Raising up requires revolution, and mostly revolutions lead to bloodshed. This is a major problem that the continent is facing and to get rid of it necessitates liberation war to be waged against neocolonialism, its forces, and agents. Freedom from neocolonialism forces would lead to freedom of mind, power, and knowledge thus one is become independent and fruitful. For example, the liberation of the coloniality of mind would lead to Africans be able to control economy, control their raw materials, natural resources, control of authority and independency in political governance for technology. It would enhance control over knowledge production, generation, and application.

The other effect of neocolonialism on Africa's ability to contribute to the wave of digital technology is introduction of class structures in Africa. Neocolonial experience in Africa has created class structures. It is widening class structures creating dominance and manipulation in perpetuation of power and access to capital and development opportunities (Babatola, 2014, p.15). From the perspective of political leadership, political administration and state bureaucracy inherited from the colonisers is the first pointer to the largesse of class divisions between the ruling elites, educated senior public servants and other cadres of state employees. Political institutions inherited from the colonisers created a wide gap leading to class differentiation, hero worship and the craze for money making at the expense of good governance and disciplined public life (Babatola, 2014, p.15). It is vital to note that the economic structure inherited from the colonisers was a class structure that created gaps between the senior cadre and junior workers, with the salaries and nature of benefits derived from government. This resulted in the existence of "highly inegalitarian colonial administrative, salary and tax structure, with its complex of fringe benefits (car and child allowances, health facilities and housing subsidies, available to the earners of high salaries)" (Babatola, 2014, p.15).

The matter was worsened by the regime of salary structure and allowances imposed on various categories of government employees. As it is across the continent, Awolowo (1970: 25) stated that:

One of the evil things we inherit from the British is a colonial salary structure. In the early stages of colonialism... all senior posts in Government and mercantile services were exclusively reserved for and held by Europeans. The salaries paid ...were fixed...in Britain in identical, equivalent, and comparable posts...By contrast, however, under the colonial administration, unquestioning obedience, cringing, and fawning on the part of Nigerian civil servants often counted more towards personal advancement than efficiency and initiative.

In the Nigerian experience, attempts by the 1971 Adebo Commission and the 1974 Udoji Commission to reverse the trends in salary and wages regime failed to change class differentiation. Instead, it widened the existing gaps regardless of the improvements introduced at the lower base. The same has remained in place over the years. The higher a salary earner becomes, the better the state of life, comfort, and even opportunities to make money (Babatola, 2014, p. 16). In the private sector, for example, the same pattern existed where the owners of means of production determine the cost, profitability, and access to the medium of production and services at the expense of the masses (Amechi, 1986). Babatola (2014, p.17) highlighted that it shows "the limited economic framework of the Nigerian state with deeper reflection on the morale of workers and citizens and the failure of the political institutions left behind by colonial administration to sustain the state and her human resources.

The implication of the above is that it has created problems in Africa, such that working-class people fail to meet their basic human needs because their earnings are less than the cost of life. Therefore, neocolonialism is increasing problems in Africa because it influences the control of African states by their former colonizers. This is remarked in the following:

- "Unstable political institutions foisted on the state without the consent of the citizens.
- Internationally guided behavioral patterns and attitudes fuelled by colonial experiences and past relationships where the former acts as the principal tutor and guardian of the interest of her former colonies at a price.
- Continuous economic relations arising from monopoly domination inherited from the colonial era and the perpetuation of economic exploitation and subjugations.
- Maximisation of exploitative mechanisms through patronage system, trade balancing, and determination of commodity pricing in the international market
- Destruction of the culture and education of the African States through continued advancement of foreign culture and language to supplant the African culture" (Babatola, 2014, p. 18).

The other effect of neocolonialism on Africa's ability to contribute to the wave of digital technology is the Structural adjustment programs (SAPs) through the coloniality of power. The existing economic conditions based on Western policy prescriptions also play a significant role in African problems because they foster the deprivation of African populations of the essential services that are vital to development (Hong, 2000, p. 14). It can be argued that it is not the responsibility of the "investors" to improve the living conditions of the people in the regions where they profit. However, the cosmetic expression of multinational companies' corporate responsibility policies, the global economy's integrated nature, and security concerns dictate that such factors cannot be ignored or overlooked. Hong (2000, p.14) argued that "the economic policies in the form of Structural Adjustment Programmes (SAPs) packages entailed sweeping economic and social changes designed to siphon the indebted country's resources and productive capacity into debt payments and to enhance international (TNCs) competition." Clarke (1995, p.301) argued that "massive deregulation, privatization, currency devaluation, social spending cuts, lower corporate taxes, export-driven strategies (i.e. export of agricultural products and natural resources) and removal of foreign investment restrictions" constitute factors that have created African problems.

These economic conditions in Africa severely impact essential services such as health, education and infrastructure. Clarke (1995, p.301) argued that "cuts in public expenditure under SAPs led to a drastic decline in control and prevention measures against diseases such as cholera, yellow fever, and malaria that were once under control while new diseases remain a big challenge." He further stated that "the absence of these fundamental services implies that generations and generations go through deprivation which impacts upon their potential to be productive members of their societies" (Clarke, 1995, p. 302). In the context of SAPs under which budgets for essential services like education were cut, it becomes difficult and impossible for African states that implemented these austerity measures to overcome the problem of illiteracy (Hong, 2000, p. 15). Hence, the problem might look African, but the cause is western based financial institutions whose interference through conditionality and economic policy prescriptions, contribute to the instability in Africa (Hong, 2000, p. 16).

As noted, the accumulative and exploitative tendencies, "along with demographic and multifarious other pressures, greatly increased government indebtedness and reinforced a spiral of decline in the delivery of public services and the economy generally" (Southall, 1998, p. 102). In countries where SAPs were implemented, "there were massive layouts and retrenchment in the civil service, in this case, many people lost their source of livelihood; hence increased levels of poverty and struggles for scarce state resources, which subsequently led to internal conflicts as the appropriation of state resources take an ethnic or nepotism dimension" (Hawley, 2000, pp. 7-8). Privatisation of public enterprises and downsizing of the civil services have "engendered the spread of corruption as multinationals supported by Western governments and their agencies continue to engage in corruption on a vast scale in North and South alike" (Hawley, 2000, p. 8).

These conditions, therefore, "reinforced the systematization of corruption and graft that is common in Africa, as ruling politicians use such opportunities of incumbency to loot the public domain" (Ajulu, 1997, p. 12). Such bad governance practices and loopholes for plunder of public funds demand an extensive regulatory system of contract rules and an adequate legal supervision of their observation. As long as these mechanisms are being flouted in Africa, the markets, even though touted as the solution to economic development problems, have still shown their failures due to misuse of power in relationships (Ajulu, 1997). This included corruption, rent-seeking behavior, and the existence of externalities and related problems of free riding (Ajulu, 1997).

The market outcome of the allocation of resources, especially in Africa, is socially unacceptable; they cause massive deprivation and inequalities, marginalization, and, in some cases, trigger interethnic conflicts as various groups struggle for scarce state resources (Hawley, 2000, p. 8). Moreover, this is more common in countries less endowed with natural resources. In contrast, regional conflagrations emerge in those cases with natural resources as different parts of the country compete or lay claim to such resources. Furthermore, external pressure combined with internal governance has created enormous problems in Africa that have severe implications for political stability. For instance, in Kenya, "the presidential centralization of power and the intensification of kleptocracy in the 1980s were backed by an increasing resort by the regime to greater repression and intimidation" (Southall, 1998, p. 102). These kinds of hostile political environments had consequently contributed to the exodus of many highly qualified persons, a further contributing factor to the problem of brain drain in Africa (Adepoju, 2008). It has also led to the exodus of the next of kin, or general refugee migration, as part of populations seeking safety in foreign lands, including Europe and North America (Adepoju, 2008).

The other effect of neocolonialism on Africa's ability to contribute to the wave of digital technology is creating a culture of political instability in Africa through the coloniality of mind. During the onset of multi-party democracy in the so-called third wave of democratization (Huntington, 1991), most African regimes did not fully embrace the changes accompanying the transition. Many autocratic regimes, for instance, accepted multi-party democracy out of Western donor pressure and agitation for change by local civil society groups (Adepoju, 2008). Consequently, the constitutional frameworks and state institutions have been tampered with to create an uneven playing field against the opposition (Adepoju, 2008). Some of these processes have seen sporadic violence during electioneering, leading to political instability. However, the degree of violence and how they are perpetrated vary from country to country, as the general elections and party formations in Africa reveal.

The reluctance of some incumbent regimes in Africa has been due to the fear of being held accountable for past misdeeds. In this context, a combination of external and internal factors, namely, "the impact of a global economy driven by the logic of the market, sectarian and particularistic tendencies, ethnic, religious, linguistics or cultural differences, or political and economic insecurities" (Mohamoud, 2006, p. 15). This combination plays a significant role in shaping the political processes in Africa and how the incumbent regimes respond to demands for change (Mohamoud, 2006, p. 15).

Mohamoud (2006, p.17) argued that "with many countries holding their third or fourth multi-party elections, internal power struggles have taken a violent dimension, thereby threatening the very existence of many countries." Since the onset of multi-party democracy in most African countries, political violence has accompanied electoral processes as part of democratization (Lindsey, 2004; Huntington, 1991). Most violence in such cases is often state-sponsored by external forces, either former colonizers or China, to the advantage of the incumbent, while various groups that struggle for state power also deploy hired violence in the form of informal groups, militias, and gangs. In the context of all the violence, it is the ordinary people who lose out in terms of lives and property. The tragedy is that the political conflict is not about alternative political programs that could address significant problems such as poverty, disease, and illiteracy but merely a fight over who has access to state resources.

The African elites bent on hanging to power at all costs and for primitive accumulation, have perfected the art of political expediency even when these acts threaten the stability of their countries (Lindsey, 2004; Ajulu, 1997; Murunga, 2004; Oyugi et al., 2004). Similar cases are found in numerous African states. The electoral process presents how colonizers' interests play a significant role in exacerbating electoral conflicts in Africa. Democracy has not brought about a culture of good democratic governance in Africa. Electoral voting processes remain touted for Western colonization over African states to remain intact (Mohamoud, 2005, p. 20). This means that the so-called democracy exported to Africa is colonized thus leading to the continuation of colonial governance in Africa. For instance, during general elections, many external observers have violated electoral laws to promote massive rigging in favor of colonizers. This implies that colonizers have used undemocratic means in the name of democracy to tilt the electoral outcomes to allow the installation of leaders of their own choice or incumbents serving their interests to continue (Mohamoud, 2005, p. 20). This amounts to regime change and the installation of puppet leaders who serve the interests of the colonizers under the disguise of democracy.

This is the case in democratic and political elections in African states. Introducing a multiparty system that brought about democracy has not yet solved the African problem. It has instead worsened the African problems. Murunga (2004) stated that the exporters of democracy in Africa, who often emphasize democracy as the way forward for Africa, are often the same forces that provide legitimacy to flawed processes. Murunga (2005) further argued that these exporters of democracy are the same forces that support and sustain regime change through democratic coup d'états. A democratic coup d'état is a situation whereby the former colonizers export democracy in the former colony and suggest it as a way forward, and at the same time, the exporter participates in the rigging to favor their own candidate. This is what the British and USA are doing in Africa, particularly Rwanda, Uganda, and Nigeria. A recent example is working with the current Paul Kagame in Rwanda to destabilize the Democratic Republic of Congo (DRC), and the blind eye they have shown when Kagame violates internal laws by invading DRC and the blind of an eye they have shown when the same president is vying for the presidency for a fourth time, a move that contradicts the values of true democracy. Moreover, the EU and other countries have taken a different position based on the Rwanda invasion of DRC, and the AU and its regional economic communities have gone mute while democracy is at stake. This confirms that democracy is a tool for coloniality and is part of the forces that have hindered Africa's contributions to the wave of digital technologies.

The other effect of neocolonialism on Africa's ability to contribute to the wave of digital technology is creating a brain drain problem in Africa through the coloniality of mind and knowledge. The issue of political instability in Africa is also directly related to the question of migration and development. Africa's life-threatening political and economic conditions contribute enormously to the massive exodus of highly qualified and lowly educated populations. These include civil conflicts, bad governance, and poor economic conditions (Mohamoud, 2005, p. 20). Many people are on the move as asylum seekers or refugees in several countries. They are escaping from civil wars and oppressive regimes working in cohorts with external special interests, especially in countries endowed with natural resources. The result is forced and wilful migration, and the final destination is the north.

On the other hand, the sending countries struggle to address the challenges posed by brain drain (Mohamoud, 2006). The problem of 'brain drain' has affected Africa more severely than any other region. The problem is more serious for higher education and research institutions in Africa as they face severe shortages of qualified academics and researchers. According to the International Organisation for Migration (IOM) report shows that approximately 100,000 skilled Africans left the continent to working in the EU and North America (IOM, 2003). The World Bank report shows that the number of skilled Africans leaving the continent increases yearly (World Bank, 2002). This causes brain drain in this context to hinder potential economic growth in Africa, thus handicapping the growth of industrialization in Africa (Mohamoud,

2006: 8). It is through the growth of industrialization that Africa's contributions to the wave of digitalization would be made visible.

The other effect of neocolonialism on Africa's ability to contribute to the wave of digital technology is weak and colonized political governance in Africa through coloniality of mind and being. African states achieved independence in the 1960s, but former colonial powers still hold their bobbin. As follows, African and ant-colonialist scholars said that political liberty is in vain without economic liberty. Besides, economic unity makes political unity. It has been argued that ''economic unity to be effective must be accompanied by political unity. The two are inseparable, each necessary for the future greatness of our continent, and the full development of our resources" (Nkrumah, 1965; Economic Development in Africa Debt Sustainability, 2004:5). Nkrumah argued that (1965), the colonial powers balkanized Africa, making the continent ungovernable. He further argued that a group of small and large countries was created through balkanization, which made Africa weak and reliant on former imperialist countries (Nkrumah, 1965, p. 173). For Africa to have good governance that would enhance Africa's contributions to a wave of digitalization, the necessity of having continental sovereignty that eliminates the divide and rules, one of the colonial policies, is significant. The divide-and-rule policy has weakened the AU and has made it become a colonial tool that recolonizes Africans. Through continental sovereignty, Africa's contributions to the wave of digital technology would be visible, and the continent would become one of the most significant contributors to digitalization.

Political governance and political systems in most African states have remained colonial in nature. Paternalism, clientelism, dictatorship, corruption, and other such pejorative labels are often used to describe the type of politics prevalent in most of Africa today (Osaghae, 2006; Adebanwi & Obadare, 2013; Falola, 2016), and yet these characteristics justify the continuation of colonial governance in Africa. Existing economic and political institutions have excluded the majority of the citizens from playing meaningful roles in their respective states' economic and political affairs and have brought poverty and destitution to the general population. Similarly, several studies have explained Africa's perverse political arrangements on the characters and personalities of the leaders who have ruled the continent since the 1960s.

Wallechinsky (2006) states that Africa is the continent with the most dictatorial political leaders. Scholars have argued that destructive legacies are the main foundations of the perverse political culture that defines contemporary Africa (Acemoglu et al., 2001; Acemoglu, Robinson 2010). Hence, coloniality resulting from colonialism legacies has created extractive institutions in Africa. African states are still designed as organs of the colonial state as instruments of exploitation because the entire political system remained the same after colonial rule (Acemoglu et al., 2001; Acemoglu, Robinson, 2010). The result of retaining the colonial political structures has been a continuation of bad governance records that create poverty and instability in most of the continent.

The other effect of neocolonialism on Africa's ability to contribute to the wave of digital technology is the falsification and destruction of African history and literature through the coloniality of knowledge. After a long period of oppressive rule, for the collective struggle of the Africans, when the Europeans left Africa, they left their ill spirits behind in Africa. Ushie (2008) argued that immediately after the independence, those African leaders who attained power were the colonial collaborators and agents of their colonizers. These leaders still exist today in Africa and have hindered Africa's contributions to the wave of digital technology. This is because they have become "tax agents and court messengers, and, armoured by the new laws and authority, they treat the community without the usual African fellows with no mercy or love. They also used their political powers and positions to oppress their fellow Africans because their power had come from elsewhere outside the African society (Ushie, 2008). African leaders, whether in the public sector or private sector, are influenced by outsiders and guided by the Eurocentric philosophy. The West had sponsored this Eurocentric philosophy for political, economic, and cultural gain. Olatunji (2010, pp. 8-9) argued that colonialists deformed and bastardized African history and culture.

EFFECTS OF THE COLD WAR ON THE AFRICAN INDIGENOUS TECHNOLOGY

While the Cold War was an ideological war between capitalism and communism and was cold in the Northern Hemisphere, Africa became its hot theatre. The Cold War was a political, ideological, and military confrontation, sometimes indirect, that occurred after the Second World War in 1945-1991. It was an ideological confrontation between the USA, the Soviet Union, and their respective allies (Calhoun, 2002). The term Cold War means a conflict in which one of the belligerents does not formally declare war on the other and has led both sides to war over the other using the media "propaganda," art, and secret means such as secret agents and spies (Bell, 2001). Thus, the Cold War has been defined as an ideological conflict (Calhoun, 2002) in which the USA and the Soviet Union tried to dominate the world with their ideology (Bell, 2001). It is called Cold because it was not an actual shooting war between the US and the USSR (Calhoun, 2002; Westad, 2007).

The effects of the Cold War on Africa have been the perennial economic and political struggles, as poverty and hardship. Scholars have said that the causes of African problems have been institutional explanations that implicate a unique political and leadership culture that has promoted anti-development in Africa (Acemoglu et al., 2001; Kalu, 2017). The foreign policies of both the US and the Soviet Union actors of the Cold War towards Africa had essential impacts on the political culture that evolved in Africa. It has been said that through financial and military support provided to domestic actors in African states, the prominent Cold War actors helped to create and sustain instability (Kalu, 2018). These insecurities ultimately became a permanent feature of Africa's politics (Kalu, 2018).

The first effect of the Cold War on African indigenous technology is the dependency syndrome and foreign aid. This was broadly explained in the previous section on neocolonialism. One can say that the Cold War deepened it and cemented it in the minds of the Africans through the coloniality of mind, power, and being. Cold War also created a more complicated dependency syndrome and increased the attitude and mindset of the Africans to wait to be dictated by the Western world on what to do and how to do it. This is because during the Cold War, many African states "remained at the receiving end in terms of policy directions from advanced countries who are often providers of loans and development aid, buyers of Africa's commodities, sources of supply for industrial goods, and providers of technical/ policy assistance to the continent" (Schmidt, 2013, p. 21). In effect, "while one cannot discount the contributions and sacrifices of domestic actors in the fight for independence, political and economic development in African states is defined by a constellation of colonizers" (Kalu, 2020, p. 12).

Scholars have argued that the Cold War brought about a similar master-servant mentality that existed during the period of the slave trade, and this has uprooted the principles of transparency and accountability from the worldviews and minds of many political leaders in contemporary Africa (Wallechinsky, 2006; Adebanwi, Obadare, 2013). It is important to note that in contemporary Africa, citizens are most times treated as slaves or servants by the political leaders who control the instruments of both political and economic power. While many changes have been taking place across the continent that promote democratic accountability as many African states, such as South Africa, have embraced regular elections, the frequency of democratic elections has not yet produced absolute transparency in the political process (van der Walle, 2003). This has created problems in Africa, leading to human rights violations that have promoted intractable conflict in Africa. It has led to the failure of institutionalization of democratic elections across many African states not to produce democratic governance and liberal democracies that would make peace and security reign in Africa.

The other effect of the Cold War on African indigenous technology that prevented Africa's contributions to digital technology is the change of political culture in Africa through coloniality of mind and power. Almond and Verba (1965, p.13) defined political culture as "the distribution of patterns of orientation toward political objects among the nation's members." Based on this definition, political culture relates to the citizens' feelings, responses, and attitudes to the politics of their society. As citizens hold different views and orientations towards political objects, different political cultures exist in different societies (Connelly, 2000). Oppression and force were used as it did not matter whether the majority of African citizens perceptibly preferred communist ideas, but what was more important was what USA wanted for Africa (Chakrabarty, 2000) instead of what Africans wanted for themselves. Capitalism is equated with deadly neocolonialism; hence, embracing it either by force or freely is embracing neocolonialism (Kalu, 2020, p. 15). While many African states benefited from accessing aid to finance and military support of one or all the opposing Cold War actors, the continent suffered long-term defects due to a convoluted political development process (Kalu, 2020, p. 15). The activities of the prominent Cold War actors laid the foundation for violent political culture and institutions in African states (Kalu, 2018). By providing military support to opposing forces in the region, the Cold War actors in some way institutionalized the use of force, oppression, and violence in Africa's political process (Kalu, 2018).

The financial and military support provided to domestic forces from either side of the divide created tensions, led to more conflicts and civil wars, and eventually laid the foundations for a violent political culture in many African countries (Kalu, 2020, p. 15). It caused proxy leaders in Africa, regime changes, and the installation of Africa's worst dictators (Kalu, 2018). By creating an environment for violent political culture, the Cold War actors made Africa's decolonization and political independence less meaningful (Chakrabarty, 2000).

The other effect of the Cold War on African indigenous technology is the creation of divisions among African leaders through the coloniality of mind and being. During the Cold War, newly independent African countries had unprecedented opportunities to adopt new development policies and institutions and form new international alliances (Mazrui & Tidy, 1984; Cooper, 2002). The excitement, promise, and uncertainty of these new opportunities and the sheer number of critical evaluative decisions created conditions ripe for reliance on ideology as a guide (Jost, 2006; Zaller, 1992). Young (1982) persuasively argued that ideology was not simply rhetoric and affected the development strategies of African countries in significant ways. Furthermore, Gorman (1984) documented that the Cold War strengthened divisions among African states as they were aligned according to which ideology they embraced. Scholars have shown that due to the path-dependent nature of development policies, even initial or temporary ideological alignment could have differential economic consequences outlasting the Cold War (Acemoglu & Robinson, 2012; Pierson, 2000).

Nunn (2008:139) stated that although many African nations supported the non-aligned movement, the continent could not avoid the pressures of the Cold War". African leaders who were strongly anti-communist received Western aid and support, and those who embraced communism were considered suspicious by the

West (Kalu, 2020, p. 19). This created more divisions and corruption among them, which later brought about conflicts and violence in Africa. The rival superpowers were drawn into African conflicts and problems (Kalu, 2020, p. 19). For example, the US and Soviet Union-supported forces took part in the liberation struggles in southern Africa. In the Horn of Africa, along the Red Sea, the superpowers became involved in a long, drawn-out war between Ethiopia and Somalia (Kalu, 2020, p. 20). Elsewhere, the superpowers supplied arms to governments they favored. In the process, they helped increase the power of the military, thus adding to an already unstable situation that hindered Africa's contributions to the wave of digitalization.

The other effect of the Cold War on the African indigenous knowledge that hindered Africa's contributions to digital technology was the creation of proxy wars through the coloniality of power. During the Cold War, "proxy war refers to the superpowers' use of allied factions or states to pursue their global rivalry outside the strictures of Northern-Hemisphere nuclear deterrence" (Kalu, 2020, p. 21). Great-power hostility expressed through client states. The phrase may be rooted in a proxy fight, an attempt to get control of corporate management through a contest for stockholders' proxy votes (William, 1995; Schlesinger, 2004). Proxy war has also been taken to mean both "localized conflict" with outside sponsors and "brush-fire war" (a war likely to spread quickly unless put out) (William, 1995; Schlesinger, 2004). The uniform character of proxy warfare during the Cold War was "the provision of military aid and advisors, by one or both superpower blocs, in support of local political entrepreneurs like Somalia's Siad Barre or Angola's Jonas Savimbi" (Minter, 2008, p. 13).

The dominance of these patterns of superpower behavior to cause proxy wars in Africa was so complete (Kalu, 2020, p. 21) that South Africa's African National Congress (ANC) enjoyed finance and support from the Soviet Union while the US authoritatively sponsored rebel groups in Angola and Mozambique. Minter (2008) went so far as to propose using the term "contra warfare" for all proxy wars. Somalia, Rwanda, Liberia, Sierra Leone, the DRC, the Congo, Sudan, Uganda, and the Central African Republic, to name just a few examples, represent this wave of conflict that confronted Africa and still faces the legacies of such today. Those mostly troubled areas included Central Africa, Southern Africa, West Africa, and the Horn of Africa (extended in western Sudan and eastern Chad). Each of these regions, especially the first two, is still experiencing new kinds of conflict as internal conflicts burst around access to resources but, at the same time, are overflowing with transnational instabilities due to population and the resultant contagion phenomenon (Young & Kent, 2004).

CONCLUSION

The chapter examined the role of neocolonialism and the Cold War in creating problems in Africa and making them intractable, which means they are hard to resolve. The chapter found that after the end of colonialism, Africa suffered from neocolonialism and Cold War events that prevented the African indigenous knowledge and technology from developing and growing. The chapter found that neocolonialism and the legacies of the Cold War have made colonial forces weaken the continent and prevent it from participating in digital technology. The end of colonialism, imperialism attained neocolonialism, which became a new phase that polarised the newly independent states in Africa. It undermined any efforts that were advanced to solve the legacies of colonialism in Africa. African leaders and their people found themselves at the mercy of the imperialists because they had failed till today to free themselves from the cruel clutches of coloniality. Even if time has changed, and side by side, the color of imperialism also changed. However, the nature of imperialism within coloniality has remained unchanged, thus hindering Africa's contributions to the wave of digital technology. The African continent has remained the victim of coloniality and its agents. To overcome this victimization and set up paths for total liberation from the coloniality of power, being, and mind, there is a need for the unity of Africans. Africa should embrace its original ideology of communalism and reject capitalism, which means embracing coloniality. It is through communalism that African indigenous knowledge and technology are given their true meaning in African society, and their contributions to digitalization are made visible to the global world.

REFERENCES

Acemoglu, D., Johnson, S., & Robinson, J. (2001). The Colonial Origins of Comparative Development: An Empirical Investigation. *The American Economic Review, 91*(5), 1369–1401. doi:10.1257/aer.91.5.1369

Acemoglu, D., & Robinson, J. (2010). Why is Africa Poor. *Economic History of Developing Regions, 25*(1), 21–50. doi:10.1080/20780389.2010.505010

Acemoglu, D., & Robinson, J. (2012). *Why Nations Fail: The Origins of Power, Prosperity and Poverty*. Crown.

Adebanwi, W., & Obadare, E. (2013). *Democracy and Prebendalism in Nigeria: Critical Perspectives*. Palgrave Macmillan. doi:10.1057/9781137280770

Adepoju, A. (2008). *Migration in sub-Saharan Africa*. Nordiska Afrikainstitutet.

Ajulu, R. (1997). *Kenya's 1992 Election and its implications for Democratisation in Sub-Saharan Africa.* Foundation for Global Dialogue Occasional Paper No.9., Johannesburg.

Akeredolu-Ale, E. O. (1975). *The Underdevelopment of Indigenous Entrepreneurship in Nigeria.*

Almond, G. A., & Verba, S. (1965). *The Civic Culture — Political Attitudes and Democracy in Five Nations*. Little, Brown and Company.

Awolowo, O. (1970). *The Strategy and Tactics of the People's Republic of NIGERIA*. Macmillan.

Babatola, J. T. (2014). *Neo-Colonialism In Africa: A Perpetuation Of Western Interest And Subjugation Of Africa.* Available at: https://www.researchgate.net/publication/313666367_NEO-COLONIALISM_IN_AFRICA_A_PERPETUATION_OF_WESTERN_INTEREST_AND_SUBJUGATION_OF_AFRICA

Bell, P. M. H. (2001). *The world since 1945, and international history*. Oxford University Press Inc.

Biko, S. (1978). *I Write What I Like: Selected Writings*. University of Chicago Press.

Blyden, E. W. (1967). *Christianity, Islam and the Negro Race*. Edinburgh University Press.

Calhoun, C (2002). Cold War. *Dictionary of the Social Sciences.*

Chakrabarty, D. (2000). *Provincialising Europe: Postcolonial Thought and Historical Difference*. Princeton University Press.

Chengu, G. (2015). *Africa's "Second Liberation" against Today's Neo-Colonialism.* https://www.globalresearch.ca/africas-second-liberation-against-todays-neo-colonialism/5451647

Clarke, R. V. (1995). Situational Crime Prevention. In M. Tonry & D. Farrington (Eds.), *Building a Safer Society: Strategic Approaches to Crime Prevention* (pp. 91–150). University of Chicago Press.

Connelly, M. (2000). Taking Off the Cold War Lens: Visions of North — South Conflict during the Algerian War for Independence. *The American Historical Review*, *105*(3), 739–769. doi:10.2307/2651808

Cooper, F. (1940). Africa since 1940: The past of the present. Cambridge University Press.

Delwar H. & Abu Md. (2016). *Africar Itihas* [History of Africa]. Dhaka: Bishobidiyaloy Prokashoni.

Economic Development in Africa Debt Sustainability. (2004). *Oasis or Mirage?* United Nations.

Falola, T. (2016). *Nigerian Political Modernity and Postcolonial Predicaments.* Pan-African University Press.

Fanon, F. (1952). *Black Skin, White Masks.* Grove Press.

Foucault, M. (2013). The right of death and power over life. In Biopolitics: A Reader. Duke University Press.

Fricker, M. (2007). *Epistemic Injustice: Power and the Ethics of Knowing*. doi:10.1093/acprof:oso/9780198237907.001.0001

Gorman, R. F. (1984). Soviet Perspectives on the Prospects for Socialist Development in Africa. *African Affairs, 83*(331), 163–187. doi:10.1093/oxfordjournals.afraf.a097604

Greig, A., Hulme, D., & Turner, M. (2007). *Challenging Global Inequality*. Palgrave Macmillan. doi:10.1007/978-0-230-20840-7

Grosfoguel, R. (2007). The Epistemic Decolonial Turn: Beyond Political Economy Paradigms 1. *Cultural Studies, 21*(2-3), 2–3. doi:10.1080/09502380601162514

Grosfoguel, R. (2011). Decolonizing Post-Colonial Studies and Paradigms of Political-Economy: Trans-modernity, Decolonial Thinking, and Global Coloniality. *Transmodernity: Journal of Peripheral Cultural Production of the Luso-Hispanic World, 1*(1). Retrieved from https://escholarship.org/uc/item/21k6t3fq doi:10.5070/T411000004

Hawley, S. (2000). *Exporting Corruption: Privatization, Multinationals and Bribery.* http://www.psrast.org/corrupt.htm

Hong, Y. X. (2000). *Research on the Transformation of Economic Growth Mode.* Nanjing University Press.

Huntington, S. P. (1991). *The Third Wave: Democratization in the Late Twentieth Century.* University of Oklahoma Press.

Johnston, R. J. (2000). The Dictionary of Human Geography (4th ed.). Wiley-Blackwell.

Jost, J. T. (2006). The end of the end of ideology. *The American Psychologist, 61*(7), 651–670. doi:10.1037/0003-066X.61.7.651 PMID:17032067

Kalu, K. (2017). State–Society Relations, Institutional Transformation and Economic Development in Sub-Saharan Africa. *Development Policy Review, 35*(S2), O234–O245. doi:10.1111/dpr.12320

Kalu, K. (2018). Africa and the Cold War. In M. S. Shanguhyia & T. Falola (Eds.), *The Palgrave Handbook of African Colonial and Postcolonial History* (pp. 661–680). Palgrave Macmillan. doi:10.1057/978-1-137-59426-6_27

Kalu, K. (2020). The Cold War and Africa's Political Culture. Vestnik RUDN. *International Relations, 20*(1), 11–21. doi:10.22363/2313-0660-2020-20-1-11-21

Kochalumchuvattil, T. (2010). The Crisis of Identity in Africa: A Call for Subjectivity. Academic Press.

Lindsey, L. (2004). *Democratization in the Third Wave: Political Violence as a Transition Tactic.* Centre for the Study of Democracy Department of Political Science University of California, Irvine; Presentation at the Midwestern Political Science Association Annual Meeting Chicago, IL.

Maldonado-Torres, N. (2007). On The Coloniality Of Being. *Cultural Studies, 21*(2-3), 2–3, 240–270. doi:10.1080/09502380601162548

Maldonado-Torres, N. (2007). On the coloniality of being: Contributions to developing a concept. *Cultural Studies, 21*(2-3), 2–3. doi:10.1080/09502380601162548

Mazrui A. (1986). *The Africans: A Triple Heritage.* Academic Press.

Mazrui, A., & Tidy, M. (1984). *Nationalism and New States in Africa.* Heinemann Educational Books Inc.

Mbembe, A. (2017). *Critique of Black Reason.* Duke University Press.

Mignolo, W. D. (2011). *Local Histories/Global Designs; Mignolo, The Darker Side of Renaissance.* Mignolo, The Darker Side of Western Modernity.

Mignolo, W. D., & Walsh, C. E. (2018). *On Decoloniality: Concepts, Analytics, Praxis.* Duke University Press.

Minter W. (2008). *Apartheid's Contras: An Inquiry into the Roots of War in Angola and Mozambique.* Academic Press.

Mpofu, P. (2013). *Multilingualism, localism and the nation: identity politics in the Zimbabwe Broadcasting Corporation*. University of South Africa. http://hdl.handle.net/10500/18663

Mpofu, W. J. (2013). Coloniality in the Scramble for African Knowledge: A Decolonial Political Perspective. *Africanus: Journal of Development Studies*, *43*(2), 105–117. doi:10.25159/0304-615X/2305

Murunga Godwin, R. (2004). The State, Its Reform and the Question of Legitimacy in Kenya, Identity. *Cultural Politics*, *5*(1 & 2), 179–206.

Ndlovu-Gatsheni, S. J. (2013). *Empire, Global Coloniality and African Subjectivity*. Berghahn Books.

Ndlovu-Gatsheni, S. J. (2015). Decoloniality as the Future of Africa. *History Compass*, *13*(10), 485–496. doi:10.1111/hic3.12264

Ndlovu-Gatsheni, S. J. (2018). *Epistemic Freedom in Africa: Deprovincialization and Decolonization*. Routledge. doi:10.4324/9780429492204

Niyitunga, E.B. (2014). Who owns mediation at the African Union (AU)? Understanding the causes of the AU mediation ineffectiveness in the Libya crisis in 2011. *African Insight*, *43*(4).

Nkrumah, K. (1963). *Africa Must Unite*. Frederick Praeger. A. Publisher.

Nkrumah, K. (1965). *Neo-colonialism: the Last Stage of Imperialism*. International Publishers.

Northrop, D. (2012). *A Companion to World History*. Willi Blackwell. doi:10.1002/9781118305492

Offiong, D. (1980). *Imperialism and dependency*. Fourth Dimension.

Okolo, A. (1986). Dependency: The Highest Stage of Capitalist Domination in Africa. In The Future of Africa and the New International Economic Order. Macmillan.

Okoth, A. (2006). *A history of Africa: African nationalism and the de-colonisation process*. East African Educational Publishers.

Olatunji, S.A. (2010). Thematic changes in postcolonial African literature: From colonialism to neocolonialism. *SinoUS English Teaching, 7*(10).

Osaghae, E. (2006). *Ethnicity and the State in Africa*. Afrasian Center for Peace and Development Studies.

Pierson, P. (2000). Increasing Returns to Scale, Path Dependence and the Study of Politics. *The American Political Science Review*, *94*(2), 251–267. doi:10.2307/2586011

Quijano, A. (1992). Colonialidad y modernidad/racionalidad. *Perú Indígena*, *13*(29), 11–20.

Quijano, A. (2000a). Coloniality of Power, Eurocentrism, and Latin America. *International Sociology*, *1*(3), 533–580. doi:10.1177/0268580900015002005

Rahaman, S., Yeazdani, R., & Rashed, M. (2017). The Untold History of Neocolonialism in Africa (1960-2011). *Historical Research*, *5*(1), 9–16. doi:10.11648/j.history.20170501.12

Remmling, G. W., & Campbell, R. C. (1970). *Basic Sociology: An Introduction to the Study of Society*. Littlefield Adams & Co.

Rodney W. (1972). *How Europe Underdeveloped Africa*. Bogle-L'Ouverture Publications.

Schlesinger Arthur, M., Jr. (2004). The Imperial Presidency. Houghton Mifflin.

Schmidt, E. (2013). *Foreign Interventions in Africa: from the Cold War to the War on Terror*. Cambridge University Press. doi:10.1017/CBO9781139021371

Southall, R. (1998). Moi's flawed mandate: The crisis continues in Kenya. *Review of African Political Economy*, *25*(75), 101–111. doi:10.1080/03056249808704295

Ushie, J. (2008). *Two Africans in One: Neo-colonialism and the African Writer*. http://www.africaresearch.org/ncol.htm

Van der Walle, N. (2003). Presidentialism and Clientelism in Africa's Emerging Party Systems. *The Journal of Modern African Studies*, *42*(2), 297–321. doi:10.1017/S0022278X03004269

Wa Thiong'o, N. (1993). *Moving the Centre: The Struggle for Cultural Freedoms*. James Currey.

Wa Thiong'o Ngugi. (1986). *Decolonizing the mind: The politics of language in African literature*. James Currey.

Wallechinsky, D. (2006). *Tyrants: the World's 20 Worst Living Dictators*. Harper Collins Publishers.

Westad, O. (2007). *The Global Cold War*. Cambridge University Press.

Williams, G. (1981). Nigeria: The Neo-colonial Political Economy. In Political Economy of Africa (Selected Readings). Longman.

Wynter, S. (2003). Unsettling the coloniality of being/power/truth/freedom: Towards the human, after man, its overrepresentation: An argument. *CR (East Lansing, Mich.)*, *3*(3), 257–337. doi:10.1353/ncr.2004.0015

Young, C. (1982). *Ideology and Development in Africa*. Yale University Press.

Zaller, J. (1992). *The nature and origins of mass opinion*. Cambridge University Press. doi:10.1017/CBO9780511818691

Chapter 6
Green Energy Solutions in Africa:
Rediscovering Indigenous Knowledge for the Fourth Industrial Revolution

Joshua Mawere
https://orcid.org/0000-0001-9492-9846
University of Venda, South Africa

ABSTRACT

This study robustly explores green energy solutions in Africa, underscoring the imperative rediscovery of indigenous knowledge amid the fourth industrial revolution (4IR). It aims to underscore the manifold advantages of incorporating indigenous knowledge for developing sustainable energy solutions across the continent. Addressing a central problem—the energy deficit exacerbated by Eurocentric systems, overlooking Africa's unique needs—the study uses a secondary data analysis methodology, drawing insights from diverse sources including academic articles, policy documents, and reports from international organisations. Rooted in Afrocentricity theory, the research adopts a firm decolonial perspective, unveiling cultural and historical dimensions crucial for seamlessly integrating indigenous knowledge into green energy practices. The investigation explores the transformative potential of minerals and digital technology in shaping innovative, contextually relevant solutions.

1. INTRODUCTION

In the pursuit of sustainable development in Africa, the imperative of embracing green energy solutions cannot be overstated. Endowed with abundant renewable

DOI: 10.4018/978-1-6684-7851-6.ch006

Copyright © 2024, IGI Global. Copying or distributing in print or electronic forms without written permission of IGI Global is prohibited.

energy resources, the continent stands poised to meet escalating energy demands while preserving its environment. Despite formidable challenges, Africa has made considerable progress in advocating for green energy solutions. However, the untapped potential of African indigenous knowledge (AIK) remains a critical frontier, holding the promise of culturally sensitive, affordable, and universally accessible sustainable energy solutions.

Taking a decolonial perspective, this chapter delves into the latent capabilities of AIK in propelling the development of green energy solutions. The integration of traditional knowledge systems into sustainable development has garnered increasing recognition, with AIK systems emerging as reservoirs of invaluable wisdom that can harmonise with modern science and technology. An essential acknowledgment is that Africa possesses indigenous knowledge (IK) capable of contributing significantly to the Fourth Industrial Revolution (4IR), independently of Eurocentric ideologies.

The overarching aim of this chapter is to furnish an extensive overview of the current state of green energy in Africa while delineating the potential role of IK in forging sustainable energy solutions. Drawing on insights from eminent African Studies scholars, this chapter seeks to underscore the value of AIK systems in sustainable development and elucidate how these traditional systems can inform and augment modern sustainable energy solutions in Africa.

As posited by Olukoshi and Odhiambo (2016: 37), the integration of indigenous knowledge systems (IKS) and practices has the potential to yield innovative solutions that resonate with the diverse cultural identities of Africa's populations. Echoing this sentiment, Simelane and Amutabi (2019: 3) argue that the utilisation of IK in sustainable development is inherently transformative, ethically grounded, and conducive to decolonisation.

Amidst a global shift towards sustainable energy to combat climate change, Africa finds itself at the forefront of an opportunity to lead in green energy solutions, leveraging its vast renewable energy sources. Nevertheless, the full exploitation of the potential of IK in developing and implementing green energy solutions on the continent remains unrealised. This book chapter not only explores the integration of Africa's IK into green energy solutions through illustrative case studies from diverse African countries but also adopts a decolonial lens to underscore the imperative of decolonising the energy sector, recognising the pivotal role of Africa's minerals in the development of digital technology.

2. THE IMPORTANCE OF DECOLONISING THE ENERGY SECTOR

In the pre-colonial era, Africa's rich tapestry of IK systems in energy production thrived, providing sustainable solutions that catered to the diverse needs of local communities (Showers, 2014). Traditional practices, often rooted in harmony with nature, allowed for the generation of energy sufficient for the continent's requirements (Showers, 2014). However, the exploitative nature of colonialism disrupted and dismantled these indigenous systems, replacing them with Eurocentric models that prioritised the interests of European powers and neglected the unique needs of African societies (Alessandrini, 2017).

The imposition of centralised energy grids during colonial rule exemplified this disregard for Africa's IK. These grids failed to consider the intricacies of local communities, resulting in a widespread lack of access to electricity across the continent (Touray & Subramanian, 2019). Furthermore, the suppression of IK during this period contributed to the marginalisation of local communities in energy decision-making processes (Omoju & Faloye, 2018). Therefore, the imperative to decolonise and indigenise Africa's energy sector is underscored by the need to revive and harness the wealth of traditional knowledge that was disrupted by colonial forces. Kinyua et al. (2020) argue that decolonisation involves dismantling colonial power structures, recognising the inherent value of IK, and indigenisation promotes local ownership and control of energy resources. This dual approach aims to rectify the historical injustices imposed by colonialism and pave the way for a more sustainable and equitable energy future for Africa.

The revitalisation of IK systems can serve as a guiding force for developing locally tailored and sustainable energy systems. By acknowledging the value of traditional wisdom, Africa has the potential to empower local communities to actively participate in shaping their energy landscape. This shift towards decentralisation, embracing mini-grids and off-grid systems, is vital in ensuring that local communities gain reliable access to energy, thereby fostering sustainable development at the grassroots level (Touray & Subramanian, 2019). African perspectives on energy and sustainable development, rooted in a decolonial perspective, emphasise the need for a balanced approach that considers economic development, environmental sustainability, and social inclusion (Osha, 2020). The African Union's Agenda 2063 stands as a testament to this commitment, prioritising renewable energy development and rural electrification with the ambitious goal of achieving universal access to electricity by 2025 (African Union, 2015). This continental vision not only addresses historical disparities but also serves as a blueprint for aligning energy development with broader sustainable goals.

Overall, Africa possessed robust IK systems for energy production that were disrupted and suppressed during colonialism. The scars of this historical injustice persist in the form of widespread energy poverty. Adopting a decolonial perspective is crucial for rectifying these historical injustices and establishing a basis for Africa to rely on its IK to contribute significantly to green energy production. This transformation not only benefits Africa but also holds the potential to contribute positively to global sustainability efforts. The ongoing efforts, as reflected in the African Union's Agenda 2063, signal a promising shift towards a more equitable and sustainable energy landscape for the continent and the world at large.

3. THEORETICAL FRAMEWORK

The Fourth Industrial Revolution (4IR) presents both opportunities and challenges for Africa, notably the exacerbation of the energy deficit due to Eurocentric systems primarily serving Europe's needs. This section delves into the theoretical underpinnings by examining the role of Afrocentricity theory as a framework fostering a decolonial perspective, facilitating Africa's use of IK for sustainable energy solutions.

Afrocentricity theory posits that individuals of African descent should draw upon their distinctive cultural heritage, experiences, and knowledge systems to shape their social and political realities (Asante, 2013). Rooted in a decolonial perspective, the theory challenges the historical dominance of Eurocentric worldviews that have historically subjugated Africa. Asante (2013) emphasises that Afrocentricity theory encourages a sense of pride in African cultural heritage, facilitates critical examination of African experiences, and promotes the rediscovery of IK systems. It advocates for grounding African solutions in cultural and social contexts unique to the continent.

Africa possesses a wealth of IKS that can be explored for the development of green energy solutions. For example, principles from traditional African wind sails, a technology used by the Bantu people for centuries, can potentially inform the design of wind turbines (Chowdhury, 2019). Similarly, traditional African practices, such as using livestock manure with a high methane content, a method employed by the Fulani people in West Africa, could be considered for biogas production (Owusu et al., 2017). The embrace of IKS presents an avenue for Africa to pioneer green energy solutions attuned to its social and cultural needs.

A decolonial perspective regarding green energy solutions in Africa necessitates the acknowledgment and rectification of the adverse impacts stemming from imposed Eurocentric systems. These systems have inadequately addressed Africa's energy requirements, resulting in an energy deficit and an undue reliance on non-renewable sources. Imperative to this perspective is the redirection of energy policy and investments towards solutions aligned with African cultural and social contexts. This shift empowers Africa to develop sustainable energy solutions that resonate with its unique social and cultural dynamics.

Afrocentricity theory serves as a theoretical foundation for adopting a decolonial perspective in the exploration of green energy solutions in Africa. By harnessing IKS, Africa has the potential to formulate sustainable and culturally resonant green energy solutions. This transformative theoretical shift necessitates a reorientation of energy policies and investments towards solutions that prioritise African cultural and social contexts.

In summary, Afrocentricity theory provides a comprehensive theoretical framework for exploring and understanding the implementation of green energy solutions in Africa. The use of IKS is fundamental to the development of sustainable solutions that align with Africa's unique cultural and social contexts. A decolonial perspective, guided by Afrocentricity theory, emerges as a key theoretical approach for addressing the energy challenges imposed by Eurocentric systems and lays the groundwork for a sustainable and culturally responsive energy future for Africa.

4. RESEARCH METHODOLOGY

This chapter employs a secondary data analysis approach to examine case studies pertaining to green energy solutions that leverage indigenous knowledge in the African context. The data sources encompass a spectrum of scholarly articles, policy documents, and reports sourced from reputable international organisations. This analytical framework is designed to provide a comprehensive examination of the intersection between indigenous knowledge and green energy initiatives in Africa, drawing on established academic literature, policy insights, and the expertise of international organisations.

5. CASE STUDIES IN GREEN ENERGY SOLUTIONS USING INDIGENOUS KNOWLEDGE

Examining case studies in green energy solutions within Africa reveals the transformative impact of incorporating IKS. These case studies underscore the pragmatic application of traditional wisdom in addressing contemporary energy challenges.

5.1 Solar Energy Projects in Mali Using Indigenous Knowledge

The Malian Government, in its pursuit of universal electricity access by 2030, has strategically harnessed IK in solar energy projects. Incorporating local building techniques and materials, such as mud bricks and traditional water channels, has proven instrumental. The use of indigenous practices not only reduces carbon footprint and project costs but also enhances cooling systems around solar panels, increasing efficiency (Ecofin Agency, 2021; Jordan, 2018).

This approach extends beyond technological efficiency; it addresses socioeconomic aspects. Local labor engagement in the construction of solar plants has curbed unemployment, while knowledge transfer enhances the capacity of communities for long-term project maintenance (USAID, 2021). The case of Mali exemplifies Africa's potential to leverage IKS for sustainable and cost-effective energy solutions. The Afrocentric framework underscores the cultural significance and social capital embedded in indigenous practices, preserving identity and fostering community development.

A decolonial perspective becomes imperative when recognising the historical obstacles imposed by colonialism on IK. The hierarchical preference for Western scientific epistemologies perpetuates cultural inequality. By challenging this hierarchy, the decolonial perspective advocates for the co-creation and acknowledgment of diverse knowledge, empowering marginalised groups (Gallagher et al., 2019; Olashore, 2021).

This paradigm shift aligns with studies emphasising the importance of integrating IK into development practices across various contexts (Hossain & Raihan, 2020; Mengesha & Hassen, 2019; Scoones, 2017; Tengeh, 2017). By weaving together these sources, the interconnected discussion accentuates Africa's potential to address its energy challenges sustainably and highlights the critical role of Afrocentric and decolonial frameworks in valuing and integrating IKS into modern development.

5.2 Mini-Hydro Power Plants in Rwanda Using Traditional Water Management Practices

Mini-hydro power plants play a crucial role in addressing Rwanda's energy needs, especially in rural areas, where affordability and reliability are paramount (Mwirigi, 2020). However, their development has encountered challenges, rooted in the historical legacy of colonialism that imposed Eurocentric models on water management and energy production in Africa (Rawat et al., 2013). The adverse impacts of this Eurocentric approach are evident in its disregard for IKS, negatively affecting the environment and the socio-economic well-being of local populations (Mazimpaka et al., 2018). The recognition of the potential of IK to contribute to sustainable development and energy production marks a pivotal shift (UNESCO, 2019).

In Rwanda, some mini-hydro power plants exemplify the integration of traditional water management practices, such as diversion weirs and canal systems, effectively controlling water flow and reducing erosion (Mwirigi, 2020; Cook et al., 2017). Decolonising the systems employed in the development and management of mini-hydro power plants is not only a necessary step but also an efficient one. Incorporating IK enables the tailoring of energy production methods to local needs while ensuring environmental sustainability. Beyond these immediate benefits, embracing IKS in Rwanda's energy sector holds multifaceted advantages. It promotes access to affordable energy, fosters employment opportunities, drives economic growth, and contributes to overall infrastructure development (Olaopa & Ayodele, 2022).

As a result of adopting a decolonial perspective and leveraging IK, Rwanda and other African nations have the potential to showcase a more sustainable and locally attuned approach to energy production. This approach not only addresses immediate energy needs but also positions Africa as a key player in the global pursuit of the Fourth Industrial Revolution (4IR) and broader energy sustainability. Africa's ability to use its own resources and knowledge systems stands as a testament to its potential to lead in shaping a more sustainable and equitable energy future.

5.3 Bioenergy Projects in Tanzania Using Locally Sourced and Managed Biomass

In recent years, Tanzania has strategically embraced bioenergy initiatives using locally sourced and managed biomass, offering a sustainable alternative to conventional energy sources. Recognising the potential of such projects extends beyond energy production; scholars like Chove et al. (2015) contend that they hold the key to enhancing the socio-economic life of local communities. This shift away from Eurocentric systems, critiqued for marginalising African communities, finds support from Nyssens and Lensink (2015), who advocate for IKS as more effective in addressing local challenges.

Illustrating the success of this transition is the Biogas Program for the Agroforestry Development Project (BFADP) in Babati District, where local communities manage the project. Through using cow dung and organic waste to generate biogas for cooking and lighting, the initiative not only reduces dependence on traditional biomass but, as revealed by Moshi et al. (2019), also improves community health by mitigating the harmful effects of indoor air pollution. Another noteworthy endeavor is the Jatropha bioenergy project in the Tanga region. Here, local farmers spearhead the production of biodiesel from the Jatropha oilseed plant, thriving in marginal lands. Supported by non-governmental organisations like Sustainable Agriculture

Tanzania (SAT) (Nyssens & Lensink, 2015), this community-led project not only provides a sustainable energy source but also generates employment opportunities for local farmers.

These examples underscore the efficacy of transitioning away from Eurocentric models towards IKS that prioritise community involvement and sustainability. As emphasised by Chove et al. (2015), the active participation of local communities in the management and operation of bioenergy projects is pivotal to their success. This involvement ensures not only the sustainability of these projects but also their alignment with the nuanced needs of the local communities. In conclusion, the use of locally sourced and managed biomass emerges as a potent avenue for uplifting the socio-economic wellbeing of African communities, concurrently fostering sustainability and diminishing reliance on traditional energy sources.

5.4 Wind Energy Projects in Ethiopia Using Traditional Knowledge of Wind Patterns and Turbine Design

In Ethiopia, the development of wind energy projects showcases a departure from conventional energy solutions by incorporating traditional knowledge of wind patterns and turbine design, thereby challenging Eurocentric systems that historically marginalised African communities both economically and in energy provision. Bekele and Arnqvist's (2015) study underscores the significance of IK in the success of wind energy projects in Ethiopia. Indigenous communities rely on natural indicators such as tree movements, dust patterns, and cloud and bird behavior to predict wind patterns (Bekele & Arnqvist, 2015). This knowledge is seamlessly integrated with modern wind measurement instrumentation for accurate assessment of wind energy potential. The study also highlights the adaptation of traditional turbine designs to local conditions, using locally available materials to create low-cost and durable turbines.

This approach signifies a departure from western-centric energy solutions that often disregard IKS, recognising the pivotal role of local wisdom in energy use. Consequently, it has led to more sustainable energy practices and improved socio-economic outcomes for local communities. Moreover, aligning with global efforts to reduce carbon emissions and transition to a greener economy, the use of IKS in renewable energy production corresponds with the African Union's recognition of renewable energy as a key pillar in its Agenda 2063 development plan. This agenda aims for sustainable development and prosperity in Africa, using IK to identify areas with high potential for renewable energy production. This ensures that development aligns with local communities, contributing to their economic growth.

This alternative approach to energy provision stands in contrast to donor-driven development, which often perpetuates external dependence. Instead, it emphasises the potential for African countries to emerge as leaders in renewable energy production, contributing significantly to global climate action on their own terms.

5.5 Geothermal Projects in Kenya Using Indigenous Knowledge of Hot Springs and Geysers

Kenya, strategically positioned in East Africa, boasts abundant geothermal resources, owing to its location on the East African Rift System. Traditionally, communities in the region have relied on IK of hot springs and geysers for medicinal and domestic purposes (Ogola, Davidsdottir & Fridleifsson, 2012). Over the years, there has been a transformative shift towards harnessing this indigenous wisdom for sustainable energy solutions, challenging Eurocentric systems that have historically marginalised African communities (Ogola, Davidsdottir & Fridleifsson, 2012). IK has played a pivotal role in the evolution of geothermal energy in Kenya. Local communities residing near geothermal areas have developed a profound understanding of this natural phenomenon, using it for their well-being. For example, the Maasai community has traditionally employed geothermal water for medicinal uses, passed down through generations (Hughes & Rogei, 2020). Similarly, the Borana community in northern Kenya ingeniously directs hot spring water into their homes through small channels to warm their living spaces during cold seasons (Chocian, 2017).

Recognising the potential of geothermal energy, the Kenyan government, particularly through the Kenya Electricity Generating Company (KenGen), has made substantial investments. The Olkaria geothermal field, located in Naivasha, stands as the largest geothermal power plant in Africa, generating electricity for the national grid (Miller, 2016). With a capacity of 520 MW, it was slated to double in the near future (Miller, 2016). Decolonising Eurocentric systems that marginalise African communities is essential for fostering a sustainable future. The integration of IK into the development of geothermal energy in Kenya has enhanced local community participation and spurred economic empowerment. It is paramount to view IK not as a substitute for scientific knowledge but as a complementary and valuable asset in the pursuit of sustainable development. This perspective challenges the misconception that IK is outdated or unsophisticated, emphasising its unique contribution to holistic and contextually relevant solutions.

Through acknowledging the complementary nature of IK and scientific expertise, societies can harness the full spectrum of available wisdom, propelling sustainable development initiatives toward greater success. Drawing on the work of Mawere and Awuor (2019), the incorporation of IKS is deemed vital for sustainable development in Africa. Traditional ecological knowledge, according to the authors, can contribute to the development of culturally relevant and environmentally sound strategies. In the case of Kenya's geothermal energy development, IK has complemented scientific expertise, resulting in successful harnessing. This underscores the importance of preserving traditional knowledge to foster economic, social, and environmental sustainability in Africa. Africa can emerge as a key player in global green energy development by embracing IKS.

In the next section, the chapter proceeds to examine the role of Africa's minerals in digital technology.

6. AFRICA'S NATURAL RESOURCES AND THE DIGITAL REVOLUTION

This section of the chapter explores the complex nexus between Africa's abundant natural resources and the ongoing digital revolution, guided by three key perspectives: the interplay between minerals and the digital revolution, the assessment of environmental and socio-economic impacts of mineral extraction, and the advocacy for sustainable mining practices in Africa. Emphasising the imperative to rediscover IK for green energy solutions, the analysis acknowledges potential challenges associated with resource utilisation, such as environmental degradation, poverty, and diseases resulting from Eurocentric mining methodologies. The chapter aims to proactively anticipate and mitigate these challenges by providing well-informed recommendations to safeguard the well-being of local communities, aligning with IK principles. It positions itself as a proactive guide, offering concise and robust suggestions to promote sustainable mining practices that balance economic development with environmental preservation and community welfare.

6.1 The Connection Between Africa's Mineral Wealth and the Digital Revolution

The intersection of Africa's mineral wealth and the digital revolution stands as a focal point within academic discourse, particularly in the realm of African studies. According to Moyo and Yeros (2019), Africa's economic growth has been significantly propelled by its mineral wealth, encompassing resources such as gold, copper, cobalt, and uranium. Concurrently, the digital revolution has ushered in transformative changes across various sectors, reshaping the way people live and work, including domains like education, health, and commerce. Despite the potential synergy between Africa's mineral wealth and the digital revolution, challenges persist. Maru and Masu (2020) highlight that the exploitation of mineral resources often leads to environmental degradation, social disparities, and the displacement of indigenous communities. Furthermore, critiques of the digital revolution argue that it can perpetuate inequalities and reinforce global dominance, particularly by players from developed countries.

In response to these challenges, scholars advocate for the incorporation of IKS in Africa. Hountondji (2019) asserts that African communities possess rich knowledge, especially in agriculture, medicine, and technology, which can foster sustainable development. By departing from Eurocentric systems that marginalise African communities, Hountondji contends that Africa can achieve greater socio-economic development and offer solutions to global challenges. Chihambakwe's (2021) work underscores the efficacy of IKS in addressing contemporary challenges, exemplified by the innovative solutions developed by local communities during the COVID-19 pandemic. Relying on their own knowledge and resources, African communities showcased adaptability and resilience, mitigating the pandemic's impact.

This discussion sets the stage for examining the impact of mining on the environment and communities, emphasising the shift towards green energy solutions and the role of indigenous knowledge in shaping sustainable practices.

6.2 The Impact of Mineral Extraction on the Environment and Communities

The extraction of natural resources in Africa has engendered substantial environmental and societal repercussions. Numerous studies attest to the deleterious effects of resource extraction, including deforestation, soil degradation, and water pollution, contributing to biodiversity loss and undermining the livelihoods of local communities (Adedeji & George, 2016). The historical context reveals that Eurocentric models, both during colonial and postcolonial eras, have perpetuated a paradigm of extractive economies, prioritising natural resource exploitation with scant regard for environmental and local community welfare. This exploitative approach persisted beyond colonialism, as multinational corporations continued resource exploitation, resulting in land loss and displacement of indigenous populations without adequate compensation (Friedman & Schipper, 2011).

Despite the extensive resource extraction, Africa grapples with energy poverty, with over 600 million people in sub-Saharan Africa lacking access to electricity, significantly impacting health, education, and socio-economic development (IEA, 2019). Addressing this energy deficit requires a departure from conventional practices towards alternative, sustainable approaches. Recognising the inherent value of IKS emerges as a crucial aspect in this paradigm shift. Integrating traditional practices and beliefs with modern technologies presents an avenue for the development of green energy solutions that concurrently benefit the environment and local communities. Noteworthy examples, such as community-based renewable energy projects in South Africa, underscore the potential of this integrated approach, fostering job creation, improved electricity access, and sustainable development (Sánchez & Botelho, 2019).

In summary, the profound impact of natural resource extraction on Africa's environment and communities necessitates a departure from conventional practices. The historical exploitation of resources, primarily benefiting external entities, has perpetuated energy poverty in Africa. To chart a sustainable course, the integration of IKS is imperative for the development of green energy solutions. This holistic approach not only serves the immediate needs of African communities but also positions the continent to contribute significantly to global green energy initiatives. Therefore, this chapter advocates for the imperative of adopting sustainable mining practices in Africa.

6.3 The Need to Develop Sustainable Mining Practices in Africa

The imperative to develop sustainable mining practices in Africa extends beyond addressing local challenges within the mining industry; it plays a pivotal role in ensuring Africa's contribution to global green energy initiatives. As the mining sector stands as a significant driver of the African economy, its transition toward sustainable practices, particularly those aligned with green energy objectives, becomes paramount. The mining industry's influence on the African economy is undeniable, contributing significantly to economic growth and development. However, the conventional extractive approaches have resulted in environmental degradation, community displacement, and mineral depletion, hindering the continent's overall development. To align with global efforts toward sustainable and green energy, Africa must reevaluate its mining practices and integrate IKS into the sector.

Sustainable mining practices, as defined by Boadi et al. (2021), seek to maximise mineral resource utilisation while minimising negative impacts on the environment and local communities. The adoption of IKS, a repository of traditional knowledge, skills, and practices, offers a promising avenue to mitigate these negative impacts. Indigenous communities in Africa have historically relied on their knowledge systems to manage natural resources sustainably, making IKS a valuable tool for minimising environmental degradation associated with mining activities. The integration of IKS into the mining industry not only addresses environmental concerns but also holds the potential to contribute significantly to green energy initiatives. As Matimu et al. (2021) suggest, involving local communities in mining activities, guided by IKS, can foster socio-economic development by creating job opportunities and enhancing local livelihoods. Moreover, the application of IKS can provide innovative solutions for sustainable energy practices within the mining sector.

For instance, the use of IKS can contribute to the reduction of water pollution through the application of local plants to absorb pollutants from mining sites (Chikwata et al., 2020). Additionally, the restoration of vegetation cover, facilitated by IKS, can mitigate the environmental impact of mining operations, promoting a more sustainable and ecologically balanced landscape. Such environmentally conscious practices align with the goals of global green energy initiatives.

Overall, the integration of sustainable mining practices using IKS in Africa is not only imperative for local economic, social, and environmental sustainability but also positions the continent to contribute meaningfully to global green energy efforts. Through leveraging IKS to guide mining operations, Africa can address the challenges within its mining industry, aligning with the principles of sustainable development and making a substantial contribution to the global transition to green energy. It is essential for stakeholders, including governments, mining companies, and local communities, to collaborate in this endeavor to ensure a sustainable and environmentally responsible mining future for Africa.

7. RECOMMENDATIONS

The chapter advocates for the indigenisation of green energy in Africa through the following key recommendations:

7.1 Centering IK in Decolonised Green Energy Policies and Strategies

Prioritise the centering of IK in the development of decolonised energy policies and strategies for Africa. Collaborate extensively with local communities and IK experts to ensure that policies are rooted in the wisdom and practices of Indigenous communities. This shift will not only enhance the relevance and effectiveness of green energy solutions but also contribute to the decolonisation of energy governance.

7.2 Investment in Research and Development of Indigenous Green Energy Technologies

Redirect funding from governments and international organisations towards a decolonial approach in researching and developing indigenous green energy technologies. Empower local communities by investing in initiatives that prioritise the unique energy needs of each community, ensuring that technologies are not only efficient and sustainable but also aligned with local contexts, knowledge, and values.

7.3 Decolonial Awareness and Recognition of IK

Promote a decolonial awareness and recognition of the inherent value of IK in shaping sustainable energy solutions in Africa. Governments, policymakers, and stakeholders should actively engage in decolonial education and training programs, specifically targeting young people. Recognition of IK should be integral to decolonial policy-making processes, ensuring that the diversity of Indigenous perspectives is acknowledged and respected in the realm of green energy.

7.4 Fostering Collaboration and Partnership Between Industry and Indigenous Communities

Facilitate decolonial collaboration and partnership between industry and Indigenous communities to ensure the development of sustainable and culturally appropriate green energy solutions. Establish frameworks and policies that prioritise Indigenous perspectives, allowing industry to learn from local knowledge and innovations, while local communities benefit from technological advancements without compromising their cultural integrity.

7.5 Strengthening Institutional Capacity for IK Research

Prioritise the decolonial strengthening of institutional capacity for researching IK in the context of energy solutions in Africa. This involves dismantling colonial structures and fostering collaborative research initiatives that respect and uphold the diverse ways of knowing within Indigenous communities. Such an approach will contribute to the development of context-specific, decolonized green energy solutions.

7.6 Empowering Local Communities

Emphasise the decolonial empowerment of local communities by prioritising their active involvement and leadership in decision-making processes related to green energy projects. This engagement should be based on equitable partnerships that recognise and value Indigenous perspectives, fostering a sense of ownership and ensuring that solutions are culturally sensitive, decolonised, and aligned with the specific needs and aspirations of the communities.

8. CONCLUSION

The development of sustainable energy solutions in Africa depends on recognising and valuing IK. The fourth industrial revolution offers an opportunity to take advantage of this IK and provide solutions that are appropriate, efficient, and sustainable in Africa. It is imperative that governments, policy makers, and stakeholders invest in understanding IK and incorporating it into green energy solutions in Africa.

IK is essential in developing sustainable energy solutions because it is adapted to the local ecological conditions and contexts. Africa has had a long tradition of using natural resources to generate energy, and this knowledge should be recognised and incorporated into the development of green energy solutions. For instance, through innovation and creativity, communities in Africa have been able to develop technologies such as wind pumps, biogas, and solar cookers that are efficient and suitable for their needs. Such IK can be harnessed to develop more sustainable and affordable energy solutions for communities in Africa.

To indigenise green energy in Africa, a multi-disciplinary approach is needed that incorporates social, economic, and environmental factors. There is a need for education, training, and support for communities to integrate IK in green energy adoption, deployment, and operation. Furthermore, existing policies and standards should recognise and support the value of IK in developing appropriate green energy solutions in Africa.

Looking to the future, the potential of IK in developing sustainable energy solutions in Africa is limitless. The fourth industrial revolution provides an opportunity to rethink energy systems and solutions that are suited to Africa's unique needs and context. Empowering local communities and investing in IK will contribute to a greener energy future in Africa.

In summary, the potential of IK in developing sustainable energy solutions in Africa is immense, and it is essential that it is recognised and valued. Indigenising green energy in Africa requires multi-disciplinary approaches and investment in education, training, and policy development. With such approaches, Africa can take advantage of the fourth industrial revolution to develop efficient, sustainable, and appropriate green energy solutions that will benefit communities in Africa.

REFERENCES

Adadi, M., Hilson, G., & Garvin, T. (2020). Indigenous knowledge systems and non-indigenous actors in Ghana's artisanal and small-scale gold mining sector. *Journal of Cleaner Production*, *244*, 118849.

Adedeji, A. A., & George, G. P. (2016). The impact of mineral resources on the environment and community development: A critical analysis. *International Journal of Developmental Research*, *6*(11), 10326–10332.

African Union. (2015). *Agenda 2063: The Africa We Want*. Retrieved from https://au.int/sites/default/files/documents/36453-doc-africa_2063_eng.pdf

Alessandrini, M. (2017). Colonialism and energy policies in Africa. *Journal of Energy in Southern Africa*, *28*(2), 1–10.

Asante, M. K. (2013). *Afrocentricity: The theory of social change*. CRC Press.

Bekele, G. Z., & Arnqvist, P. (2015). Indigenous knowledge for renewable energy production: Insights from small-scale farmers in Ethiopia. *Journal of Cleaner Production*, *87*, 870–879.

Boadi, L. N., Markwei, C. M., & Osei-Bonsu, A. (2021). Rethinking the concept of sustainable mining practices in Ghana: The need for a 'mining development agenda'. *Journal of Cleaner Production*, *280*, 124485.

Chihambakwe, N. (2021). Covid-19, Africa and indigenous knowledge systems. *African Journal of Science, Technology, Innovation and Development*, *13*(1), 59–66.

Chikwata, C., Mangara, N., & Tsamba, A. (2020). The Use of Indigenous Plant Species for Mine Rehabilitation in the Copperbelt Province of Zambia. *Journal of African Earth Sciences*, *163*, 103739.

Chocian, E. (2017). *Wherever it rains, we go there: Afar nomadic architecture: tradition and challenge* (Doctoral dissertation).

Chove, B. E., Mwita, E., Mlyuka, H., & Mrema, G. E. (2015). Assessment of the potential of bioenergy for electricity generation in Tanzania. *International Journal of Renewable Energy Research*, *5*(4), 977–986.

Chowdhury, T. (2019). *Wind Energy in Africa: A Case Study*. Springer.

Cook, C., Koenigstorfer, J., & Biswas, A. (2017). Comparing hydro-power resource and needs of India, Nepal, Bhutan, and Rwanda. *Renewable & Sustainable Energy Reviews*, *77*, 12–21.

Ecofin Agency. (2021, January 14). *Mali inaugurates its first solar power plant in Kita*. Retrieved August 3, 2021, from https://www.ecofinagency.com/energy/1401-42941-mali-inaugurates-its-first-solar-power-plant-in-kita

Egbon, M. (2021). Impact of Mining on Sustainable Development in Africa: A Review. *Journal of Sustainable Mining*, *20*(1), 1–10.

Friedman, L., & Schipper, L. (2011). Mining colonial legacies: Extractive regimes and indigenous movements in the Andes. *Development and Change*, *42*(4), 1017–1035.

Gallagher, K. P., Matsuura, M., & Tang, X. (2019). Rediscovering African sustainability: Integrating indigenous knowledge and social capital into development. *Ecology and Society*, *24*(3), 21.

Hossain, M. M., & Raihan, A. (2020). An enquiry into knowledge management practices of indigenous communities: A case study of Bagdi tribe in Bangladesh. *African Journal of Science, Technology, Innovation and Development*, *12*(1), 1–11.

Hountondji, P. (2019). *African modes of thought: The challenges of indigenous knowledge*. Routledge.

Hughes, L., & Rogei, D. (2020). Feeling the heat: Responses to geothermal development in Kenya's Rift Valley. *Journal of Eastern African Studies : the Journal of the British Institute in Eastern Africa*, *14*(2), 165–184. doi:10.1080/17531055.2020.1716292

IEA. (2019). *Africa Energy Outlook 2019*. International Energy Agency.

Jordan, C. (2018, May 17). *Incorporating indigenous knowledge into 21st-century innovation*. Retrieved August 3, 2021, from https://www.un.org/africarenewal/magazine/may-2018/incorporating-indigenous-knowledge-21st-century-innovation

Kinyua, E. A., Maina, M., & Ngaruiya, G. M. (2020). Energy poverty in developing countries: A review of literature on the history, causes, and solutions of energy poverty. *International Journal of Energy Economics and Policy*, *10*(1), 468–481.

Maru, Y., & Masu, S. (2020). *Navigating Digital Disruptions and Opportunities for Inclusive Development: Proceedings of the 9th International Conference on ICT for Development, Education and Training*. Springer.

Matimu, V. F., Tawodzera, R., & Chikodzi, D. (2021). Sustainable mining, enhancing community livelihoods and socio-economic development through artisanal small-scale mining in Zimbabwe. *Journal of Cleaner Production*, *278*, 123938.

Mawere, M., & Awuor, F. O. (2019). Local knowledge systems and sustainable development in Africa: A discursive view. *African Journal of Science, Technology, Innovation and Development*, *11*(6), 699–707. doi:10.1080/20421338.2018.1560587

Mazimpaka, J. D., Nzeyimana, I., & Harelimana, F. (2018). Colonialism and economic development in Rwanda: Can history explain the post-colonial state of affairs? *Journal of International Development*, *30*(3), 497–518.

Mengesha, T., & Hassen, N. (2019). Bridging the gap between indigenous knowledge and scientific knowledge for improved land use management in Ethiopia. *African Journal of Science, Technology, Innovation and Development*, *11*(5), 601–610.

Miller, S. T. (2016). *Geothermal Electricity Compared To Electricity From Other Sources In Kenya* (Master's thesis).

Moroka, V. (2021). Sustainable mining practices in Africa with particular reference to Botswana. *Journal of Sustainable Mining*, *20*(1), 29–37.

Moshi, A. P., Msangi, E. S., Njau, K. N., & Khalid, R. (2019). The impact of biogas technologies on household air pollution in Babati District, Tanzania. *Energy for Sustainable Development*, *46*, 89–99.

Moyo, S., & Yeros, P. (2019). *Africa in the age of the fourth industrial revolution*. Pluto Press.

Mwirigi, J. N. (2020). Sustainability assessment of mini hydro power projects in Rwanda: A comparative analysis. *Energy Sources. Part B, Economics, Planning, and Policy*, *15*(8), 675–687.

Ndlela, L. L. (2019). Green energy solutions in Africa: Rediscovering indigenous knowledge for fourth industrial revolution. In F. Van Wyk (Ed.), *African Studies: Current and Future Challenges*. Harvard Citation Style.

Nyssens, M., & Lensink, R. (2015). Explaining the success of social enterprise in the renewable energy sector: The case of Tanzania. *Journal of Cleaner Production*, *98*, 75–83.

Ogola, P. F. A., Davidsdottir, B., & Fridleifsson, I. B. (2012). Potential contribution of geothermal energy to climate change adaptation: A case study of the arid and semi-arid eastern Baringo lowlands, Kenya. *Renewable & Sustainable Energy Reviews*, *16*(6), 4222–4246. doi:10.1016/j.rser.2012.01.081

Olaopa, O. R., & Ayodele, O. A. (2022). Building on the strengths of African indigenous knowledge and innovation (AIK&I) for sustainable development in Africa. *African Journal of Science, Technology, Innovation and Development*, *14*(5), 1313–1326. doi:10.1080/20421338.2021.1950111

Olashore, O. (2021). African-centered development: An Afrocentric critique of development theories and practices. In J. C. Okeibunor & M. M. Kone (Eds.), *African Development* (pp. 91–113). Springer.

Olukoshi, A., & Odhiambo, I. (2016). Indigenous Knowledge Systems (IKS) and Sustainable Development in Africa. In Indigenous Knowledge Systems and Development in Africa (pp. 33-48). Springer.

Omoju, O. E., & Faloye, F. A. (2018). The role of indigenous knowledge in renewable energy development in Africa: A review. *Renewable & Sustainable Energy Reviews*, *81*, 1856–1868.

Osha, S. (2020). African perspectives on energy transitions and sustainable development. *Energy Policy*, *140*, 111337.

Owusu, P. A. (2017). Renewable energy in Ghana: Status, prospects, and policies. *Renewable & Sustainable Energy Reviews*, *68*, 14–24.

Rawat, S., Kala, C. P., & Bhatt, D. (2013). Traditional water management system in Western Himalayas, India: Need for revival. *Journal of Mountain Science*, *10*(3), 399–413.

Sánchez, D. L., & Botelho, T. (2019). Renewable Energy Transition in Africa: The Role of Community-Based Energy Development. *World Development*, *119*, 42–52.

Scoones, I. (2017). Sustainable rural livelihoods and the links between agriculture and HIV/AIDS in Africa: A case study from Zimbabwe. *The Journal of Development Studies*, *53*(7), 923–936.

Showers, K. B. (2014). Europe's long history of extracting African renewable energy: Contexts for African scientists, technologists, innovators and policy-makers. *African Journal of Science, Technology, Innovation and Development*, *6*(4), 301–313. doi: 10.1080/20421338.2014.947200

Simelane, T., & Amutabi, M. N. (2019). Decolonizing Knowledge: The Use of Indigenous Knowledge in Sustainable Development. In Handbook of Research on Science Education and University Outreach as a Tool for Regional Development (pp. 1-18). IGI Global.

Tengeh, R. K. (2017). The role of indigenous knowledge and innovation in creating sustainable agri-food systems in Africa: Evidence from Cameroon. *Journal of Agriculture and Rural Development in the Tropics and Subtropics*, *118*(2), 219–232.

Touray, I. M., & Subramanian, V. (2019). Decolonizing renewable energy transitions: Perspectives from Africa. *Energy Research & Social Science*, *47*, 125–134.

UNESCO. (2019). *Indigenous knowledge for sustainable development*. UNESCO.

USAID. (2021). *Mali: Energy*. Retrieved August 3, 2021, from https://www.usaid.gov/mali/energy

Chapter 7
Technology for Digitalization in Africa: The Contribution of Africa's Raw Minerals

Sandile Moloi
University of Johannesburg, South Africa

Dimakatso John Manthosi
University of Pretoria, South Africa

ABSTRACT

Africa has been on the development trajectory since its liberation from colonialism. Scholars, policy makers, and the citizens of Africa unanimously agree that freedom means nothing without development, and development means nothing without freedom. Over the years, Africans have been in search of development modals and strategies that suits the conditions of the peoples of Africa, which will remedy the historical injustices. As the fourth industrialization (4IR) unravels, many questions are raised: (1) Is 4IR compatible to Africa's developmental agenda? (2) Can 4IR help Africa tackle poverty and inequality, enhance inclusion, and foster sustainable development?

INTRODUCTION

Africa has been on the development trajectory since its liberation from colonialism. Scholars, policy makers, and the citizens of Africa unanimously agree that freedom means nothing without development, and development means nothing without

DOI: 10.4018/978-1-6684-7851-6.ch007

Copyright © 2024, IGI Global. Copying or distributing in print or electronic forms without written permission of IGI Global is prohibited.

freedom (Bouzar, 2015). Over the years, Africans have been in search of development modals and strategies that suits the conditions of the peoples of Africa, which will remedy the historical injustices (Murithi, 2007). As the forth industrialisation (4IR) unravels, many questions are raised: (1) is 4IR compatible to Africa's developmental agenda? (2) Can 4IR help Africa tackle poverty and inequality, enhance inclusion, and foster sustainable development? 4IR also known as industry 4.0, is a term coined by Klaus Schwab, to describe a world "where individuals move between digital domains and offline reality with the use of connected technology to enable and manage their lives" (Schwab, 2016).

African scholars have had mixed reactions to 4IR, with scholars such as, Tshilidzi Marwala (2020), Njuguna Ndung'u and Landry Signé (2020), among others - urging Africans to join the cause, or risk missing out. On the other hand, scholars such as, Everisto Benyera (2020), have warned Africans of recolonisation through 4IR data collection. Other scholars have questioned Africa's readiness for 4IR, with scholars such as, Desmond Tutu Ayentimi and John Burgess (2019), going as far as to ask, "Is the fourth industrial revolution relevant to sub-Sahara Africa?" Despite diverse views on 4IR in/and Africa, a view that has gained traction among African scholars and policy makers alike, is that 4IR has the potential of transforming Africa's struggling economies and other sectors such as health and education, and foster sustainable development.

Scholars generally agree that 4IR represents the emerging technologies, which are blurring the traditional boundaries between the physical, digital, and biological worlds (Serumaga-Zake and Van der Poll, 2021:1). This fusion is driven by artificial intelligence (AI), robotics, the Internet of Things (IoT), genetic engineering, quantum computing, and more. Digital technologies are key components of 4IR (Serumaga-Zake and Van der Poll, 2021:1). As argued by Nyagadza, Pashapa, Chare, Mazuruse, and Hove (2022:4), digital technologies are as a result of the emergence of 4IR dynamic disruptive technologies. Digital technologies have the potential of unlocking sustainable industrial innovation in emerging economies, by facilitating the inter-linkages in the enhanced production process. According Nyagadza, Pashapa, Chare, Mazuruse, and Hove (2022:4), industrialised economies possess a reputation for innovation, many other developing economies do not or simply lag behnd. Technological advancement is frequently linked to the development of novel and useful products through ingenuity and invention (Akileswaran & Hutchinson, 2019; Schwab, 2016). However, in many developing economies, it is more prevalent to implement and scale technological solutions developed elsewhere (International Finance Corporation, 2017). Nevertheless, this is not to suggest that those in the developing economies are less capable and uninventive. It is worthwhile to state that the developing economies often act as an appendage for the developed economies by providing capable personals through brain drain, and mineral resources for the realisation of the technological innovations and advancements. Furthermore, provide the market for the consumption of the end products.

Beneath the Africa soil lies a wealth of mineral resources of enormous value. Africa is home to about 30 percent of the world's mineral reserves, 12 percent of the world's oil and 8 percent of the world's natural gas reserves according to the United Nations. In 2019, the continent produced almost 1 billion tonnes of minerals worth $406bn. These African mineral resources often end up in industrialised economies where they are engineered into various technologies which includes digital technologies.

The relationship between developed and developing economies has been explored through various models and theories. For the purpose of this chapter, the centre-periphery theory will be adopted to assess the role of Africa's minerals and raw materials in manufacturing the 4IR technology and infrastructure for digitalisation in Africa and the world. This chapter begins by tracing the origins of 4IR and assessing its goals. It moves on to discuss the characteristics of 4IR. Furthermore, this chapter assess Africa's role in the construction of the 4IR infrastructure and the making of 4IR technology.

Understanding the Fourth Industrial Revolution: Origins and Goals

The concept of 4IR is fairly new, yet it has attracted tremendous attention from scholars and policymakers alike. Like any new concept, it has been received with both optimism and scepticism (Benyera, 2020). As aforementioned, the term 4IR was coined by Klaus Schwa in 2016, and since then, it has been further explicated and refined by its advocates and critiques. As opined by Marwala (2020), 4IR builds on the previous industrial revolutions in particular the third industrial revolution. Thus, it is important to briefly deliberate on the evolution of the industrial revolutions with an intention of understanding how they differ from one another. The first industrial revolution changed the lives and economies of the Europeans from an agrarian and handicraft economy to one premised on industry and machine manufacturing (Uwizeyimana, 2022:10). It is unequivocal that Africa served as an appendage (and continues) to European and North American industrialisation endeavours by providing the necessary raw material and cheap labour to enable the first revolution. Oil and electricity brought mass production in the second industrial revolution (Uwizeyimana, 2022:11). In the third industrial revolution, information technology facilitated automate production (Uwizeyimana, 2022:11). Although each industrial revolution is often considered a separate event, together they can be better understood as a series of events building upon innovations of the previous revolution and leading to more advanced forms of production (Marwala, 2020).

Figure 1.

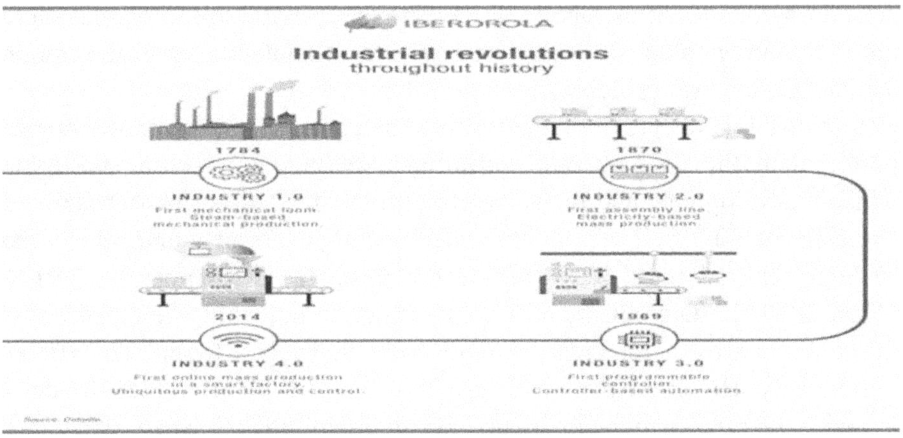

The concept of 4IR surfaced around 2014, with the appearance of smart factories and online production management. However, it was only in 2016 that Klaus Schwab, the founder of the World Economic Forum, used the term Forth Industrial Revolution, in his book bearing a similar title, to describe a world in which "virtual and physical systems of manufacturing cooperate with each other in a flexible way at the global level". While 4IR is widely known for its smart and connected machines and system, its scope is much broader. As noted by Williams and Otto (2022) other key features of 4IR include gene sequencing to nanotechnology, from renewable energies to quantum computing. It is the fusion of these technologies and their interaction across the physical, digital and biological domains that make the 4IR fundamentally different from previous revolutions.

Building the 4IR Technology and Infrastructure: Assessing Africa's Role

A bulk of African economies have been characterised as extractive economies. By their nature, extractive economies rely on harvesting, mining and exporting of raw resources with little or no processing (Wilk, 2004:286). A large part of Africa's contribution to the global economy is through its exportation of the raw mineral resources to the industrialised economies for processing, which leaves Africa as a consumer of finished goods as opposed to being innovators and manufacturers (Murombedzi, 2016:2). This has been a feature of Africa's economy since colonialism (Benyera, 2020). As the world prepares for digitalisation in various sectors, Africa too is gearing up for a digitalised world, but Africa is set to lose out if it does not invest in digital technology and infrastructure (Marwala, 2020). In this section, we look at various resources and raw materials that are used in manufacturing 4IR technologies and infrastructure. For example, we will look at mineral such as: titanium, lithium, and cobalt, which are key in the manufacturing of smartphones, computers and electric vehicles.

Before delving deeper into the main subject of this section, we begin by grappling with the concept of digilisation and its potential role on Africa's sustainable development and inclusive growth. Digitalisation has been defined as the "transformation of all sectors of our economy, government and society based on the large-scale adoption of existing and emerging digital technologies" (Vironen and Kah, 2019: 1). Digitalisation offers a chance to unlocking new pathways for rapid economic growth, job creation, innovation, and access to services efficiently and effectively. However, there is a growing concern about digital divide, and increased cyber risks, which require urgent and organised action to mitigate. While this is beyond the scope of this chapter, the proliferation of technologies has led to transformations in the role of the state. Across the globe, governments have responsibilities of coming up with new laws to safeguard against the abuse of digital technology and ensuring maximum benefit from this revolution.

Scholars, policy makers, and other non-state actors have been searching for initiatives on how and where digital technologies could tackle some of the world's most pressing environmental, economic and social challenges, to attain the 17 Sustainable Development Goals (SDGs). A 2020 report by the World Economic Forum (WEF) suggests that 70 per cent of the 169 targets underpinning the SDGs could be enabled by existing 4IR technology applications. For all the potential benefits that digital technologies offer, however, they could also put greater stress on the earth, its resources, and on society. Therefore, it is important to ensure that technologies are harnessed in proper manner, to fulfill their potential to revolutionise the world, change the lives of people, and unlock new pathways to prosperity - fast-tracking sustainable development globally.

In line with the SGDs, countries across the globe seek to transition away from fossil fuels to clean energy. Minerals like lithium and cobalt are key components in the batteries that could support the world's transition away from fossil fuels. The Democratic Republic of Congo (DRC), Africa's second-biggest country by land mass, is rich in both those elements. DRC is rich in natural resources and has 70 per cent of the world's cobalt. Cobalt is essential in lithium-ion batteries for phones and electric vehicles, or EVs. Cobalt, which is mainly produced as a byproduct of copper and nickel, this commodity was ignored for a long time in favor of those more in-demand minerals. The advent of 4IR technology, has ignited interest in this mineral resource.

According to the Organisation for Economic Co-operation and Development (2020), between 18% and 30% of the cobalt produced in the DRC is mined through artisanal mining. This makes it difficult to ascertain how -- and how much of -- this cobalt makes its way into the supply chain, and ultimately into the world's cellphones and EVs. Among other things, cobalt keeps the batteries, which power everything from cell phones to electric cars, from catching fire. As global demand for lithium-ion batteries has grown, so has the price of cobalt. The artisanal mining industry is mostly informal, and it is a cause of conflicts. Moreover, government policies have been ineffective in formalising and protecting these artisanal miners from exploitation and intra-conflict. Safety standards are also very low with no regulations, there are little to no personal protective equipment available to miners, and child labor is rampant in amongst artisanal miners. Furthermore, the rush in mining of these mineral resources put enormous strain on the environment with cities being in danger of being wiped out. The diagram below shows various minerals used in mobile phone.

DRC is also rich in other natural resources, this includes oil, timber, gas, gold and diamonds, as well as minerals critical to the digital technology transition, like cobalt and copper. It is not only the DRC in Africa that is rich in mineral resources that are critical to the manufacturing of the digital technologies. Other countries that have significant deposits of minerals critical to digital technology includes Mozambique, Zimbabwe, South Africa, and Angola, to mention but a few. Below we discuss various mineral resources found in Africa that are used in digital technologies.

Figure 2.
Source: Al Jazeera

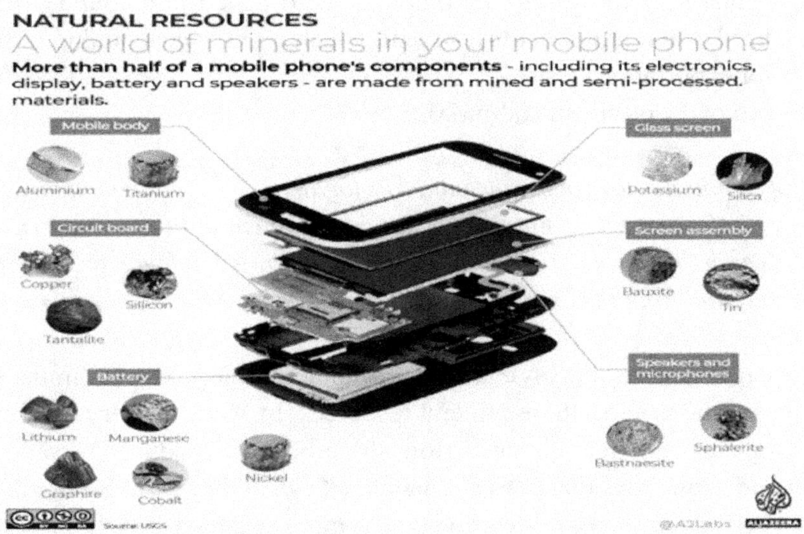

The Zimbabwean economy has been severely bruised from decades of sanctions from United States of America and its western allies. The country has since been a desperate situation due to widespread poverty and massive unemployment. Hyperinflation was a significant problem from about 2003 to April 2009, when the country suspended its own currency. Zimbabwe faced 231 million per cent peak hyperinflation in 2008 (. The dawning of 4IR has reignited hope for the people of Zimbabwe as Chinese companies are making multimillion-dollar acquisitions to secure lithium supplies amid worldwide race to go green. The Chinese are also joined by other interested buyers as many countries are scrambling to get their hands on key minerals – notably lithium, nickel, copper, cobalt, graphite and manganese – as demand for them is set to increase. According to the IEA, lithium demand could grow by more than 40 times by 2040, while graphite, cobalt and nickel demand could grow 20 to 25 times. Due to the high demand and strains on the supply chain, prices have been surging since 2021. All these minerals are essential to the manufacturing of the technologies.

According to data from the United States Geological Survey, Zimbabwe produced just over 1% of global lithium in 2021. While its production share is small, however, the country still ranks as the sixth-largest global producer of the mineral. The demand for this critical mineral in the manufacturing of digital technology has attracted attention of potential investors who are scrambling for this precious mineral resources. With 220,000 tonnes of reserves identified so far, Zimbabwe has the largest lithium reserves in Africa. Zimbabwe is also one of the largest exporters of nickel ore, which is also key in the manufacturing of technology gadgets. The Zimbabwean government has lamented less gains made from these precious mineral resources; hence it took a decision in December 2022 to ban the export of lithium to other countries. A move seen by others as detrimental to a country that is in desperate need of foreign direct investment, which often comes mining of mineral resources for exportation.

As argued by Isabeau van Halm (2023):

By forcing companies to process the minerals locally, the government hopes to grow its resource industry and eventually see the development of lithium batteries in the country. Keeping parts of the supply chain in Zimbabwe will mean the country will see more revenue from the extraction of its minerals.

A move that aims to address challenges of underdevelopment in Zimbabwe and benefit those who unearth these mineral resources. In Marx's theory of alienation, the working class is often alienated from the product it produces because of the exorbitant price one pays that does not match with what they earn. In case of export of lithium in Zimbabwe, the miners and government in general benefit far less when compared to the money generated from the finished goods that manufactured from these mineral resources.

Mozambique has one of the largest deposits of high-quality graphite in the world – somewhere between 20% and 40% of total global reserves. Graphite is a form of coal that is used in lithium-ion batteries, which power mobile phones, portable electronics and, increasingly, green vehicles. China is the largest exporter of graphite in the world, supplying up to 80% of the 900,000tn demand for the commodity in 2017. Like in the DRC, the precious mineral found Mozambique has been the cause of jihadist attacks in Cabo Delgado, one of the country's northern-most provinces where most of the graphite is. Before the outbreak of jihadism in the region, four Australian companies dominated the mining of graphite, these are: Triton Minerals, Mustang Resources, Battery Minerals and Syrah Resources.

South Africa is one of many African countries with a cobalt supply, with the Kruisrivier mine currently being worked commercially, with operations focused solely on cobalt (Van der Veen, 2022). In 2020, South Africa produced 1 800 metric tonnes of cobalt, which is only a fraction of the total annual global cobalt mine production of roughly 140 000 metric tonnes. Apart from DRC and South Africa, other countries that mine cobalt include: Madagascar, Morocco, Zambia and Zimbabwe.

Africa's Pursuit for a Digital World

Africa plays a significant role in providing mineral resources and cheap labour for the manufacturing of 4IR technologies and services. This exploitation of African resources and its peoples is now common in data annotation industry wherein Africans provide cheap labour, but the continent reaps little benefit. The likes of Google, Microsoft and Yahoo use tech labourers in Kenya through US firm Samasource (Marwala, 2020). While this creates employment, Kenya sees little of the technology its labourers are powering. When compared to other countries, Africa states still lags behind when it comes to digital infrastructure and technology. Below we assess Africa's digitalization challenges and opportunities. The table below illustrates various Information and Communications Technology (ICT) in Africa:

The above diagram shows how low Africa ranks in various ICT sectors when compared to developed and developing countries. While its ranking is deemed low, however, in recent years, the ICT sector in Africa has shown considerable growth, a trend which scholars say is likely to continue (Ndung'u and Signé, 2019:62). Thus, it is essential for Africa to maintain upward trajectory pace in order to keep up with the rest of the world and reap maximum benefits brought by digitilisation in various sectors. Below we deliberate on the challenges and opportunities of digital technology in Africa's developmental agenda.

Figure 3.
Source: Africa Growth Initiative

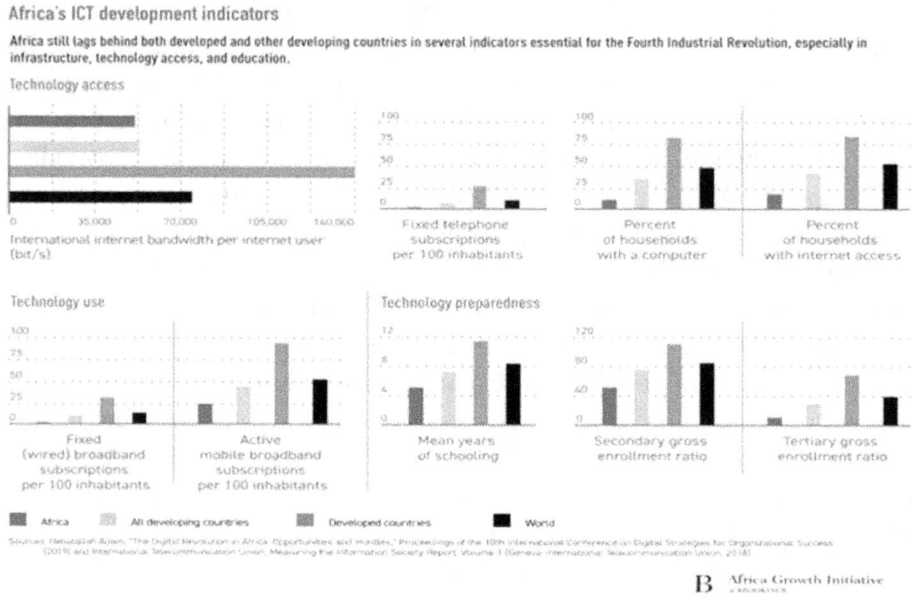

In May 2013, the heads of state and government of the African Union (AU) converged in Addis Ababa, Ethiopia, to celebrate the Golden Jubilee of the formation of the Organisation of African Unity (OAU), which was later transformed into the AU (African Union, 2013). As part of the celebrations, the AU member states reflected on the successes, failures, and challenges this continental organisation faced throughout the years of its existence. However, these heads of states and government went beyond reflecting on the past by considering their aspirations for the future of Africa. Transforming Africa into the global powerhouse of the future became the main ambition of the heads of states and government that met in May 2013 (African Union, 2013).

As an affirmation of their commitment to transforming Africa into the global powerhouse of the future, the AU member states signed the 50th Anniversary Solemn Declaration during the Golden Jubilee celebrations. This declaration paved a way for the adoption of Agenda 2063 as Africa's blueprint and master plan towards attaining this goal. Thus, Agenda 2063 is the concrete manifestation of how the continent intends to achieve its vision within a 50 year period from 2013 to 2063. Aspiration 1 of the Agenda 2063 is the attainment of "a prosperous Africa based on inclusive growth and sustainable development" (African Union, 2014). Digital technology and innovation have been considered as cornerstones for the attainment of this aspiration. Innovations and digitalisation are stimulating job creation and contributing to addressing poverty, reducing inequality, facilitating the delivery of goods and services, and contributing to the achievement of Agenda 2063 and the Sustainable Development Goals (African Union, 2014).

The AU did not just confine its aspirations to its Agenda 2063 blueprint; however, it took various initiative in an endeavour to conceptualise and put policies in place towards the attainment of a digitalised continent. Below we look at some of these initiatives by the AU. At the 32nd AU Assembly of Heads of State and Government, the AU Executive Council Decisions related to ICT, ECA Resolution (812-XXXI) on the African Information Society Initiative and the Smart Africa Board meeting. The AUC undertook to develop a comprehensive Digital Transformation Strategy for Africa in collaboration with the UN Economic Commission for Africa, Smart Africa, AUDA-NEPAD, Regional Economic Communities, African Development Bank, Africa Telecommunications Union, Africa Capacity Building Foundation, International Telecommunication Union and the World Bank.

The Digital Transformation Strategy for Africa(DTSA) addresses the objectives of the 4IR within the AfCFTA(Bharadwaj 2013; Chanias, Meyers and Hess 2019: 1-17). The DTSA seeks to convergence centripetal processes and systems within AfCFTA frameworks such as the Policy and Regulatory Initiative for Digital Africa (PRIDA), the Programme for Infrastructure Development in Africa (PIDA), the African Union Financial Institutions (AUFIs), the Single African Air Transport Market (SAATM); and the Free Movement Persons.

Against this background, this chapter underscores the importance of 4IR in Africa's pursuit for inclusive growth and sustainable development. It argues that Africa contributes substantially to the advancement of 4IR technology and infrastructure to attain a digital world, but it also notes that Africa's contribution is peripheral, for two reasons. First, Africa lags behind in 4IR innovation, which makes it the consumer of finished goods (technologies) as opposed to designers (innovators), manufacturers, and maintainers of the 4IR technology. Second, Africa does not have the financial capability to construct and maintain the 4IR infrastructure. Thus, we share a view advanced by Songwe (2020) that African countries should start to think of their natural resources not solely as export commodities but reservoirs for geological knowledge, as well as a basis for advancing material science and technology.

The pace, scope, scale and political nature of the 4IR and our normative aspirations that it engenders human-centred welfare gains makes an equitable global digital governance imperative. First, the indisputable realisation that there will be disruptive changes that are faster with global complex ramifications than ever before and therefore the designing of policies should be within the framework of the SDGs 2030. Secondly being objective and explicit about the political nature of technologies legitimises a mandate for an equitable global digital governance in the uneven and asymmetrical area of emerging technologies and their applications. In this regard identification of where and how values are formed in the development and application of technologies and determining harmonised political frameworks for integrating divergent values. Thirdly, positioning mandates and values that enhance public commons ownership and welfare gain as priorities for global digital governance can direct the development and use of emerging technologies legitimately within a transparent and accountable framework. The foregoing suggests that global digital governance must become a multistakeholder endeavour (Cockburn 2000; Cummins 2008).

CONCLUSION

The notion of the 4IR holds great promise for technological innovations that can grow the economy while improving people's lives. However, it has been received with mixed reactions especially in Africa. For some scholars and observers, 4IR presents Africa with an opportunity to leapfrog towards sustainable development and close the lingering gap between Africa and the developed western world. While for some, 4IR is a hoax used to trap Africa into neo-colonialism and technological underdevelopment through data colonisation. These diverse views were not the focus of this chapter but were worth noting. Our main focus in this chapter was to assess the role of Africa's raw materials in the manufacturing of 4IR technology and infrastructure to attain digitalisation in Africa and the world. The ability of African countries to develop, disperse and utilise digital innovations are shaped by a range of historical, social, cultural, political and economic factors. Therefore, any policy attempts at attaining a digitised Africa would require a systemic approach that takes into cognisant the country's specific socioeconomic needs. As opined by Mazibuko-Makena and Kraemer-Mbula (2022:417), some of the needs in Africa that require attention include the ability of the 4IR to create digital jobs to reduce unemployment, and the reduction of inequality through equitable access to the internet. This chapter emphasised the significance of digital technology in Africa's developmental agenda as articulated in the AU's Agenda 2063 and the UN's SDGs. It also noted the challenges facing Africa in its endeavours of attaining a digitalised continent.

Africa is less known for innovation in digital technology, yet it plays an instrumental role in providing essential raw minerals that are used in the manufacturing of 4IR and digital technologies. With that being said, the central thesis of this chapter is that African resources are mined to benefit other people other than those who mine them. In the case of 4IR technology which has taken the world by storm, Africa features in the fringes, yet her mineral resources power the 4IR technologies. Using the centre-periphery theory, we made several observations on Africa's contribution to 4IR technology.

First, we noted Africa's contribution to previous revolutions and its key role in 4IR. Second, we observed that Africa's resources are mined legally and illegally to manufacture 4IR technologies. Insofar as the mining of resources illegally is concerned, these have caused several conflicts and exploitation. We referred to the DRC and other countries in Africa whose fragility is partly because of these minerals that have been termed a "curse". Because of issues that they attract and their inability to transform societies wherein they are mined. In the case of the DRC, part of the challenge is that the institutional history of the DRC is profoundly shaped by the extraction of resources—which has historically not benefitted local communities. This and other governance challenges have led to insecurity and poverty, which create the conditions for cycles of violence. More often than not, external players are instigators of these illegal mining by providing the black-market economy and arming of the rivalries in their battle for territorial domination. Third, we argued that Africa fails to reach its full potential in attaining a digitalised continent because of the lack in digital infrastructure and poor education that inspires little innovation.

The chapter acknowledges the potential of 4IR technologies to tackle major global challenges, such as those encapsulated in the SDGs. The authors apply a critical lens to the notion of the 4IR. The chapter highlights the limitations of attaining a digitalised Africa without adequately addressing the preconditions that need to be in place for implementation of the digital technology. For instance, policy strategies to accelerate Artificial Intelligence (AI) adoption in Africa cannot be the same as those in countries with universal, reliable and fast internet access. Instead, a strategic approach to 4I in Africa ought to work in tandem with policies that prioritise support for basic ICT infrastructure. We also noted that deliberate action to create the conditions that will enable the efficient application of 4IR technologies in such a way that they foster inclusion rather than exacerbate inequalities. The chapter acknowledges the potential of 4IR technologies to tackle major global challenges, such as those encapsulated in the SDGs.

The chapter notes that Africa is quite instrumental in 4IR technologies despite it not benefiting enough from the resources that it injects in the making of 4IR infrastructure and digital technologies. Using the centre-periphery theory, the chapter demonstrated that Africa is used as a storehouse of industrialised economies, which occupy the centre while Africa remains in the periphery. Therefore, Africa's role cannot be undermined or be regarded as less important because Africans do not manufacture these sophisticated digital technologies.

REFERENCES

Abrahams, L. (2015). *Framing the digital complexity economy*. Keynote paper delivered at CyberAbuja 2015, the IEEE International Conference on Cyberspace Governance: The Imperative for National and Economic Security, Abuja. 10.1109/CYBER-Abuja.2015.7360506

Abrahams, L. (2016). *ICT regulation for fostering the digital complexity economy in the SADC region 2016-2030*. Unpublished conference paper delivered at the ACER Conference, Livingstone, Zambia. Retrieved from https://static1.squarespace.com/

Akileswaran, K., & Hutchinson, G. (2019). *Adapting to the 4IR: Africa's development in the age of automation*. Tony Blair Institute for Global Change.

Ayentimi, D. T. (2020.) The 4IR and the challenges for developing economies. Developing the Workforce in an Emerging Economy, 18-30.

Ayentimi, D. T., & Burgess, J. (2019). Is the fourth industrial revolution relevant to sub-Sahara Africa? *Technology Analysis and Strategic Management*, *31*(6), 641–652. doi:10.1080/09537325.2018.1542129

Benyera, E. (2022). The Fourth Industrial Revolution and Africa: A Cure Which Kills the Patient. Africa and the Fourth Industrial Revolution, 145-158. doi:10.1007/978-3-030-87524-4_8

Bourreau, M. & Valletti, T. (2015). *Enabling digital financial inclusion through improvements in competition and interoperability: What works and what doesn't?* Academic Press.

Bouzar, K. (2015). *No Peace, No Sustainable Development: A Vicious Cycle that We Can Break*. Academic Press.

Chanias, S., Myers, M. D., & Hess, T. (2019). Digital transformation strategy making in pre-digital organizations: The case of a financial services provider. *The Journal of Strategic Information Systems*, *28*(1), 17–33. doi:10.1016/j.jsis.2018.11.003

Cockburn, A. (2000). Selecting a project's methodology. *IEEE Software*, *17*(4), 64–71. doi:10.1109/52.854070

Ghosh, K., & Dohan, M., & Veldandi, H. (2018). Digital Transformation Strategies for Healthcare Providers : Perspectives from Senior Leadership. *Americas Conference on Information Systems*, 1–5.

Hilson, G., & Maconachie, R. (2008). "Good Governance" and the Extractive Industries in Sub-Saharan Africa. *Mineral Processing and Extractive Metallurgy Review: An International Journal, 30*(1).

Lambert, L. (2017). *The Four Challenges of the Fourth Industrial Revolution.* Market Mogul. Retrieved from https://themarketmogul.com/industry-4-0-challenges/?hvid=2Gt2CE

Melo, D.J., & Solleder, M.J. (2022). *Challenges of digitalisation in MENA and sub-Saharan Africa.* Academic Press.

Mhlanga, D., Ndhlovu, E., & Hofisi, C. (2021). Assessment of the 4IR Challenges of Radical Innovation in Service Delivery in Africa. *Journal of Public Administration, 56*(4.1), 1002-1017.

Murithi, T. (2007). A local response to the global human rights standard: The ubuntu perspective on human dignity. *Globalisation, Societies and Education*, *5*(3), 277–286. doi:10.1080/14767720701661966

Murombedzi, J. C. (2016). *Inequality and natural resources in Africa. World Social Science Report 2016.* UNESCO.

Nalubega, T., & Uwizeyimana, D. E. (2019). *Public sector monitoring and evaluation in the Fourth Industrial Revolution: Implications for Africa* (Vol. 7). Africa's Public Service Delivery and Performance Review.

Ndung'u, N., & Signé, L. (2020). *The Fourth Industrial Revolution and digitization will transform Africa into a global powerhouse.* Foresight Africa Report.

Nyagadza, B., Pashapa, R., Chare, A., Mazuruse, G., & Hove, P. K. (2022). Digital technologies, Fourth Industrial Revolution (4IR) & Global Value Chains (GVCs) nexus with emerging economies' future industrial innovation dynamics. *Cogent Economics & Finance, 10*(1), 2014654. doi:10.1080/23322039.2021.2014654

Schwab, K. (2015). *The Fourth Industrial Revolution: What It Means and How to Respond.* Retrieved from https://www.foreignaffairs.com/articles/2015-12-12/fourth-industrial-revolution

Schwab, K. (2016). *The Fourth Industrial Revolution: What It Means and How to Respond.* World Economic Forum. Available online: https://www.weforum.org/agenda/2016/01/the-fourth-industrial-revolution-what-it-means-and-how-to-respond/

Serumaga-Zake, J. M., & Van der Poll, J. A. (2021). Addressing the Impact of Fourth Industrial Revolution on South African Manufacturing Small and Medium Enterprises (SMEs). *Sustainability (Basel), 13*(21), 11703. doi:10.3390/su132111703

Uleanya, C., & Ke, Y. (2019). Review of preparedness of rural African communities nexus formal education in the fourth industrial revolution. *South African Review of Sociology, 50*(3-4), 91–103. doi:10.1080/21528586.2019.1639074

Uwizeyimana, D. E. (2022). Black Africans' Contributions to Global Industrial Revolutions. *Think (London, England), 90*(1).

van Halm, I. (2023, January 19). *Zimbabwe joins the wave of resource nationalism.* Mining Technology. Retrieved from https://www.mining-technology.com/features/zimbabwe-critical-minerals-resource-nationalism/

Wilk, R. (2004). *The Extractive Economy: An Early Phase of the Globalization of Diet* (Vol. 27). The Environment and World History.

World Bank. (2017). *World development report 2016: Digital dividends.* doi:10.1596/978-1-4648-0671-1

Chapter 8
Digital Technology and Sustainable Development in Africa

Mkululi Gacula
University of Johannesburg, South Africa

ABSTRACT

The role of digital technology in promoting sustainable development and economic growth in Africa has remained a mystery in Africa. Africa is endowed with numerous raw materials and natural resources that should enhance sustainable development in Africa. However, because of coloniality within science and technology, Africa has not benefitted from its indigenous knowledge and technology, and Africa's contributions to the wave of digital technology has not yet been understood nor recognized. The current literature presents that Africa is still behind in technology and science and yet technological innovation started in Africa.

INTRODUCTION

The role of digital technology in promoting sustainable development and economic growth in Africa has remained a mystery in Africa. Africa is endowed with numerous raw materials and natural resources that should enhance sustainable development in Africa. However, because of coloniality within science and technology, Africa has not benefitted from its indigenous knowledge and technology, and the Africa's contributions to the wave of digital technology has not yet been understood nor recognized. This is the current literature presents that Africa is still behind in technology and science and yet technological innovation started in Africa. Due

DOI: 10.4018/978-1-6684-7851-6.ch008

Copyright © 2024, IGI Global. Copying or distributing in print or electronic forms without written permission of IGI Global is prohibited.

to coloniality forces in Africa, the Africa's innovation system has not yet gained speed and this has affected the probability of the continent to achieve sustainable development and economic growth. The role of African indigenous technology to promote technological performance and innovative capacities that would enhance sustainable development in Africa has been suffocated and inhibited by the coloniality forces. of countries. The African innovation system that benefits the continent needs to be informed by the African indigenous technology and this requires a network of institutions that are organized through linkages between African indigenous knowledge and technology. These linkages must relate to each other as elements of a collective system of knowledge creation and use as well as the technologies they use" (OECD,1997:9). These linkages would promote interactive activities between and among institutions to generate and use new products, processes and organizational practices.

In the contemporary Africa, coloniality has permeated within the national system of innovation that include universities, public research institutes, private enterprises, financial institutions such as commercial banks, technology support agencies, policymaking bodies, and the government in general. Indigenous knowledge and technology are not recognized and yet the digital technology the world encounters today started in Africa. This makes the need to decolonize African indigenous knowledge and technology so that one understands the effects of digital technology on sustainable development and economic growth in Africa. Before the arrival of trans-Atlantic slave traders and later colonialists in Africa, the continent had systems of governance that existed within the people's values and traditions. The continent had the then industries, showing the existence of science and technology in Africa. There were systems of administrative justice and local courts that offered justice and administration to the citizens. Science and technology were their means of livelihood, trade form, worship mode, and knowledge level. After colonialists landed in Africa, they colonized Africans and their whole being, they colonized Africans' sources of livelihood, education, and technology. Ibrahim (2014) argued that colonialists came to Africa with a mission to take over technology, natural resources, raw materials, and market outlets. For this reason, they introduced a system of governance that was oppressive and that did not push Science and technology to benefit the continent, rather than colonialists and their countries.

One of the dreadful legacies of the trans-Atlantic slave trade is abolition of Africa's legal institutions and led to traditional courts that were responsible for the administration of justice, rule of law, and governance, colonialists saw the vacuum. They installed an authoritarian system of governance to succeed in their mission. Africans were not given ample time to address the effects of trans-Atlantic slavery, which were to re-install and establish their legal systems and institutions in their respective geographical areas. During the pre-colonial South Africa, there were traditional courts. Traditional courts in many pre-Atlantic slave trade worked well for the benefit of the citizens. Traditional courts in South Africa and the rest of Africa were traditional institutions. Traditional courts were institutions in many pre-Atlantic slave traders that worked well for the citizens' benefit. Africa had a kind of leadership and governance that promoted peace, stability, and technology.

For example, Ojo and Ekhator (2020) argued that in pre-colonial Africa, there were salient novelties in the legal system of the Benin Kingdom and other areas in pre-colonial Africa that promoted justice, peace, and order among people and communities. Special provisions such as collective responsibility in legal personality, the law of primogeniture, the fusion of laws and religion in theory and practice, and the recognition of societal status and political position in legal proceedings amongst other legal concepts were incorporated into the body of laws in many precolonial nations (Ojo and Ekhator, 2020). Past studies have not yet understood that the demolition of these traditional legal systems and traditional courts implied the demolition of governance systems. This chapter aims to assess the role of digital technology in promoting sustainable development in Africa. This assessment requires one to expose coloniality within the legal institutions and systems that were inherited by the post-colonial African leaders from the colonialists. These institutions have promoted coloniality and thus hindering the Africa's contributions to the advancement of digital technology in Africa. This area is highly neglected as it affects African governance systems, showing that Africans cannot govern and lead themselves. It also presents that Africa are not able to contribute to the world of digital technology.

The chapter is thus structured into six main sections. The first section defines digital governance and decolonization concepts. The second section explores the effects of digital technology on sustainable development in Africa, focusing on key areas such as e-governance, e-government, artificial intelligence, and blockchain. The third section explains the characteristics of digital technologies and elements of sustainable development to understand the intersection of digitalization and sustainability Africa. The fourth section examines the effects of digital technology on sustainable development and the need for African governance for digital technology in Africa. The fifth section assesses the factors affecting good governance for digital technology for sustainable development in Africa. third section revisits the effects of colonialism on governance for indigenous knowledge. The section also explains coloniality forces and their effects on digital technology advancement in Africa. The sixth section consists of concluding remarks of the chapter.

DIGITAL GOVERNANCE AND DECOLONIZATION CONCEPTUALIZED

The concept of digital governance is made of two words: digital and governance. Governance is applied in this chapter as an initiative that refers to the process of decision-making and the process by which decisions are made and implemented (Sheng, n.d). the term governance is used in various contexts, such as corporate, international, national, and local (Sheng, n.d). Governance describes the ways in which government links itself to its broader political, social, administrative and environment (Kettl, 2015). As governance is a process of decisions that lead to the establishment of systems that govern people, it depends on who is influencing it and whose values and cultures are adopted in this process. Governance should reflect the values and traditions of the people. It should be embedded in their local traditions and settings. However, in Africa, through proxy leaders installed by the colonialists, strange governance that is not known or recognized by the locals has been imposed on the people.

Governance would mean a system of norms, rules, procedures, and practices, but digital governance means the control and influence over science and technology on rules and practices. The government is one of the actors in governance. Other actors involved in governance vary depending on the government level under discussion. In rural areas, for example, other actors may include influential landlords, associations of peasant farmers, cooperatives, NGOs, research institutes, religious leaders, finance institutions, political parties, the military, etc. The concept of digital or digitized are used in this chapter interchangeably because they all mean the same thing. Digitalization has been identified as the most significant technological trend that is changing both, society and business (Parviainen, Tihinen, Kääriäinen & Teppola, 2017). Digitalization is defined as the way many domains of social life are restructured around digital communication and media infrastructures. In simple terms, digitalization may be defined as the use of digital technologies (Srai and Lorentz, 2019). The integration of the analogue and digital worlds with new technologies that enhance customer interactions, data availability and business process (Eling and Lehmann, 2018). Digitalization refers to the use of digital technology, and probably digitized information, to create and harvest value in new ways (Gobble, 2018). This means that digitalization means the act of transforming analogue ways of doing things into digital ways to improve business relationships between the service providers and receivers.

Digital governance or digitalized governance means public engagement between the services providers and receivers and positions itself on people (Gao & Lee, 2017). It is characterized as the use of information technology in a government operation to increase the production of good and services and as well as enhance the delivery of those goods and services to the citizens (Twizeyimana & Andersson, 2019). Digital governance serves two purposes. It promotes interactions between government, people and government and corporations (Choi & Chandler, 2020). It promotes internal government operations, simplifies, and enhances governance in the political, government and business aspects (Choi & Chandler, 2020; Wang, Medaglia & Zheng, 2018).

The situation in digital governance presents more complexities. There are interconnections between actors involved in digital governance and the nature of governance in terms of defining who contributes to digital technology in today's world. Proxy governance systems, such as the ones we see in today's Africa, would, in many ways, advance the interests of the colonialists. Governance is essential to get digital technology right and enable it to promote sustainable development and economic growth, but in the context of Africa, that is not the case. This digital technology depends solely on Africa's raw materials and natural resources. If you remove Africa from the picture of digital technology and stop all the mining, there would be no digital technology advancement. Africa has 40 percent of the world's gold and up to 90 percent of its chromium and platinum.

Africa has the largest reserves of cobalt, diamonds, platinum and uranium and holds 65 per cent of the world's arable land and ten percent of the planet's internal renewable fresh water source (African Development Bank (AfDB), 2016:3). In Africa, the primary commodities that dominates the world export trade has been petroleum and petroleum products in Libya, Nigeria, Algeria, Egypt, Gabon, the Republic of the Congo, and Angola; iron ore in Mauritania and Liberia; copper in Zambia and the Democratic Republic of the Congo; cotton in Chad (AfDB, 2016:3). With such huge minerals and natural resources that control global world economy, achieving sustainable development in Africa has become a mystery. This shows that Africa is key and crucial in this era of digital technology and yet her contributions are not recognized.

DECOLONIZATION AND AFRICAN INDIGENOUS KNOWLEDGE

In colonial-authoritarianism accounts, colonizers created authoritarian structures by transforming local institutions, which formed the basis of large and decentralized colonial states. As Mamdani (1996) argued, like all colonial powers, the British worked with a single model of customary authority in precolonial Africa. Colonialism presented a colonial model which was highly monarchical, patriarchal, and authoritarian. This model presumed a king at the center of every polity, a chief on every piece of administrative ground, and a patriarch in every homestead or kraal" (wa Thiong'o, 1986). Moreover, scholars have argued and accounted that Britain and other colonizers routinely empowered despotic chiefs by eliminating checks on their authority that were pervasive in the precolonial period (Ranger, 2012; Baldwin, 2015:30–31; Stasavage, 2020: 44). This means that a working, decolonization has to lead Africans recover from colonial impact and restoration of indigenous people's identities, languages and experiences (Datta, 2018). In this manner, indigenous communities can disentangle themselves from the oppressive control of colonizing state government through policy and decolonized strategies (Simpson, 2004). Smith (1999) further described decolonization as a continuous process of anti-colonial struggle that honours indigenous approaches to knowing the world, recognizing indigenous land, indigenous peoples and indigenous sovereignty.

Decolonial thinking has been used to critique the colonial-modern function of assumptions and knowledge forms that are deeply embedded in the discipline and the broader field of Western social science (Seth, 2013). Smith (1999) sees decolonization as a process for conducting research with indigenous communities that places indigenous voices and epistemologies in the center of the research process. For example, interest in research on African indigenous knowledge and technology would be at the centre of that research and be regarded as decolonization process. Smith (1999) suggests that the process of decolonization of indigenous research would help regain control over indigenous ways of knowing and being, and ways in which research can be used for social justice. This is because colonialism had led to the subjugation and suppression of indigenous ways of life (Heleta, 2018), and this was because the colonialists considered indigenous knowledge system as backward and not worthy of any development when compared to other worldviews. However, the subjection and suppression has continued under the coloniality powers and forces.

Coloniality has led to overlooking the remnant of African indigenous and technology on the continent. Mji (2019) argued that there is a general disregard of indigenous knowledge system amongst academics and scientists. As results, Indigenous knowledge and technology in Africa has not been given value and still strategically rejected among academics (Mji, 2019). Even indigenous researchers who were aware of the benefits and superiority of indigenous techniques remarked that they were afraid to admit an interest on this sphere of knowledge, basing this on fear of being ridiculed by Western peers (Mji, 2019). Current western education in Africa presents many challenges to African indigenous knowledge and technology. This therefore makes it difficult for indigenous knowledge promote technologies that would enable indigenous communities and contribute to their development. Western education in Africa present colonised education with western theories which cannot be applied to African context. This kind of education inhibits the usage of acceptable African indigenous research methods and theories. Indigenous decolonizing methods and theories suitable to African contexts must include decolonizing research methodologies and Afrocentrism.

Decolonizing research methodology is an approach that is used to challenge the Eurocentric research methods that undermine the local knowledge and experiences of the marginalized population groups (Smith, 1999). The Ubuntu concept expresses the African philosophy of humanness, that a person is a person through other people (Murithi, 2006). The Ubuntu philosophy is one of the African indigenous knowledge that would promote sustainable development. The Afrocentric is a philosophical and theoretical perspective that suggests cultural and social immersion as the best approach to understand African phenomena as opposed to scientific distance (Mkabela, 2005). Furthermore, Goduka, et al. (2013) argued that for research to be relevant and improve the quality of life of indigenous people, it needs to be rooted in indigenous worldviews, cultural values, and languages.

Scholars have presented that much research on indigenous knowledge has been carried out by researchers without decolonizing the research (Datta, 2018; Keane, Khupe & Seehawer, 2017). However, this research has availed to nothing as coloniality forces have continued to inhibit the role African indigenous technology to sustainable development. Kovach (2010) argued that any research in indigenous without decolonization is referred to as academic oppression towards indigenous communities. Researchers rarely think about sharing their indigenous research with the indigenous knowledge owners and their communities. Lincoln (1994) explains Western research as the rape model of research where the researcher comes in, takes what they want and leave when they feel like it. Indigenous knowledge researchers should thus engage in research not only to produce knowledge but also to make positive change that promote African indigenous knowledge and technology.

As Louis (2007) said, if research does not benefit the communities by extending the quality of life for those in the communities, it should not be done. Knowledge outcomes can be shared in ways that benefit the community through consultations with research participants (Keane, Malcolm, & Rollnick, 2004). However, indigenous researchers in the African context have still not done enough to redress this travesty of conducting research that is not benefitting the indigenous knowledge owners and their communities. Indigenous knowledge thus needs to be decolonized and be shared in ways that benefit indigenous owners and their communities (Sithole, 2007).

Colonialism's impacts include environmental degradation, the spread of disease, economic instability, ethnic rivalries, and human rights violations issues that can long outlast one group's colonial rule. Colonialism not only blocked further political development, but indirect rule made local elites less accountable to their citizens. After independence, even if these states had a coherence others lacked, they had far more predatory rulers (Settles, 1996). The lack of good governance also led to a host of practices that were detrimental to the economy. Nepotism, favouritism, corruption, and tribal politics were commonplace. The result was devastating: a plethora of public servants, a tribal army of thirteen thousand to protect the regime, staggering internal and external debt, lack of basic services, unproductive investments like hotels, bank buildings and a huge budget deficit. The government found it unable to pay the salary of all the bureaucrats and to honour its debts. Instead of adopting sound financial practices, the government decided it was easy to borrow money from international institutions. The International Monetary Fund (IMF) and the World Bank intervened to restructure the economy and balance the budget. These agencies, however, offered assistance on the basis of conditionality and ended with a lot of sacrifices from the population.

Africa's seemingly endemic problems have been attributed to three major factors: inappropriate policies; bad governance; incompetent, corrupt and ineffective leaders. In the last analysis they could all be reduced to leaders and leadership. Given the challenges confronting Africa as it transits towards the 21st Century, it is imperative that Africa creates leaders who are knowledgeable about the ever-changing world, competent, with integrity, vision and commitment. The first generation of African leaders was the product of colonial governance. We can best understand the paucity of African leaders with the requisite capability, integrity, vision, commitment and will power to lead Africa towards its Renaissance by briefly revisiting the manner in which leaders were recruited during the colonial period.

Decolonization as a political, epistemological and economic liberatory project has remained an unfinished business, giving way to coloniality. Coloniality is an invisible power structure that sustains colonial relations of exploitation and domination long after the end of direct colonialism (Maldonado-Torres, 2007: 240-270). Coloniality of power works as a crucial structuring process within global imperial designs, sustaining the superiority of the Global North and ensuring the perpetual subalternity of the Global South using colonial matrices of power (Mignolo, 2007: 155-167). Colonial matrices of power are a set of technologies of subjectivation that consist of four types. The first is control of economy which manifests itself through dispossessions, land appropriations, the exploitation of labour, and control of African natural resources. The second is control of authority which includes the maintenance of military superiority and monopolization of the means of violence. The third is control of gender and sexuality, which involves the re-imagination of 'family' in Western bourgeois terms and the introduction of Western-centric education, which displaces indigenous knowledge. The last is control of subjectivity and knowledge which includes epistemological colonisation and the re-articulation of African subjectivity as inferior and constituted by a series of 'deficits' and a catalogue of 'lacks' (Quijano, 2007:168-187; Grosfoguel, 2007: 214).

COLONIALITY WITHIN DIGITAL TECHNOLOGY IN AFRICA

Colonialism is one experience that has greatly affected Africa in several ways than any other thing. Though many societies of the world had also suffered from colonialism in different senses, but it is African society that was badly hit by colonialism. Colonialism in Africa, like a severe wound, though healed but the scar is still there. Beyond this fact, the scar has also metamorphosed into a dreaded monster called neo-colonialism. Neocolonialism, with all its accompanying side effects, has come to stay in Africa. Though colonialism had since been removed from the continent, but its impact on African mode of reasoning, its effects on African cultures, worldview, identity, languages and history are still felt today. This means, Africans had had an arrangement by which reality was known, transmitted and preserved, and by which traditional societies were organized. But all these were set aside as a strange and difficult situation arose with the experience of colonialism. More than this, African leaders who become the apostles of modern African socio-economic and political philosophies were subjected to colonial rule.

Consequently, this experience also contributed greatly to the wrong orientation and permutation, which they meted out to the Africans during the post-colonial era. But what is coloniality? Coloniality is different from colonialism. Colonialism denotes a political and economic relation in which the sovereignty of a nation or a people rests on the power of another nation, which makes such a nation an empire. Coloniality, instead, refers to long-standing patterns of power that emerged as a result of colonialism, but that define culture, labour, intersubjectivity relations, and knowledge production well beyond the strict limits of colonial administrations. Thus, coloniality survives colonialism. It is maintained alive in books, in the criteria for academic performance, in cultural patterns, in common sense, in the self-image of peoples, in aspirations of self, and so many other aspects of our modern experience. In a way, as modern subjects we breathe coloniality all the time and every day" (Maldonado-Torres, 2007:243). Nkrumah correctly understood imperialism as operating like an octopus in its domination of the world. Global coloniality is also better understood as an octopus constituted by hetararchies of domination, control, and exploitation. Rodney (1972) understood coloniality in terms of 'underdevelopment' and 'dependency' perspectives. This has made Africa to be underdeveloped thus preventing the digital technology in Africa to promote sustainable development. The emphasis in all these articulations of coloniality was on economic domination, which is just one strand of coloniality (Ndlovu-Gatsheni, 2015:487).

The octopus nature of coloniality is well articulated by Grosfoguel who identified its various entangled hetararchies such as complex class formations, core-periphery divisions, male–female hierarchies, heterosexual–homosexual hierarchies, religious/spiritualities hierarchization and divisions, epistemic/knowledge hierarchies, linguistic hierarchies, and aesthetic (Grosfoguel, 2011:16-17). Coloniality is, therefore, an invisible power structure, an epochal condition, and epistemological design, which lies at the center of the present Euro-North American-centric modern world (Ndlovu-Gatsheni, 2015:488). Thus, coloniality is an indirect political, social, and economic subjugation of one political entity by another.

The domination of Africa by the West, which involves direct exercise of political control through the adoption of certain policies aimed at the structural and economic pauperization of the colonized territories. This understanding may be the probable reason that most Marxists perceive colonialism as the monopoly stage of capitalism. Falaiye (1997:32) described vividly the experience of colonialism in Africa as "the European aggressive incursion and subsequent imposition of colonial domination on Africa had both traumatic and destructive consequences on the political, social and cultural situation of the continent". Africa did not only lose her political freedom; but more fundamentally, the structures of political power were weakened, retarded and even dismantled and replaced with those of the metropolis that were considered more "enlightened" and "sophisticated" without any consideration whatever for the people's worldview" (Falaiye, 1997:35). In this way, colonialism destroyed the African indigenous knowledge and technology that was available in Africa.

The above observations may be the probable reason that Jean-Paul Sartre in his "Preface" to Fanon's Wretched of the Earth, informs his fellow whites that they should pick up the book and read to be able to discover the incalculable harm they have done to the black race in colonialism (Fanon, 1968:13). Sartre declared that: "No matter what the whites think about their black victims at colonialism, our victims know us by their scars and by their chains, and it is this that makes their evidence irrefutable" (Fanon, 1968:13). The main reasons for European exploration and subsequent exploitation of Africa were to expand their empires, benefit from Africa's minerals and raw materials and increase knowledge, to spread Christianity and to increase national esteem of established empires. As Khapoya (1971:112, 114) argued that these reasons coincided with the political cum strategic, cultural, and economic reasons for colonization.

CHARACTERISTICS OF DIGITAL TECHNOLOGIES

Digital technology has become an integral part of contemporary society, transforming various aspects of human life, including how we interact, work, and govern. In Africa, the adoption and utilization of digital technologies have the potential to significantly impact sustainable development efforts. Digital technologies encompass a wide range of tools and systems that leverage digital information for various purposes. Understanding the characteristics of these technologies is crucial for assessing their impact on sustainable development in Africa. Digital technology has several characteristics that help achieve sustainable economic development for sustainable development. They include malleability, homogeneity, transferability, sociotechnical maturity, agentic capabilities, and unprecedented economic expansion.

The first characteristic is the fact that digital technology is malleable. It allows continuous evolution and the generation of new forms of technologies (Vărzaru, 2022). The second characteristic of digital technology is homogeneous. This means that it can be applied consistently across different processes to different activities and businesses (Fréour, Fréour, Pohl & Battistelli, 2021). The third characteristic of digital technology is transferable nature of the technologies. This means that digital technology's effects when applied to daily activities and businesses is remarkable and felt within and across processes (Fréour, et al., 2021). The fourth characteristic of digital technology is sociotechnical maturity. At this characteristic, digital technology promotes integration and support other forms of technologies within a sociotechnical environment to promote better results (Vărzaru, 2022). The fifth characteristic of digital technology is the fact that digital technology is also endowed with agentic capabilities, like action selection and protocol development, which if not taken into consideration could disrupt the production of services and goods at workplace (Fréour, et al., 2021). These above characteristics of digital technology contributes to economic growth and exponential social development. This therefore leads economic development and growth that promote the realization of sustainable development.

DIGITAL TECHNOLOGY PROMOTES INNOVATION FOR SUSTAINABLE DEVELOPMENT

For the digital technology to promote economic growth for sustainable development, it converges into digital innovation. Technology innovating digitally means innovating products, processes, or business models using digital technology platforms as a means or end within and across companies or organizations (Ciriello, Richter & Schwabe, 2018). For digital technology to transform into innovation go through three main processes of innovations. The first process is that digitized, information can be stored, transformed, transmitted, and traced by any digital device irrespective of its content (Yoo, Henfridsson & Lyytinen, 2010). The second process is that digital technology is editable through means of re-programming, making digital solutions malleable to changes after deployment by interaction with external systems (Kallinikos, Aaltonen & Marton, 2013). The third process is inherently self-referential. This means that digital technology is needed to create digital technology (de Reuver, Sørensen & Basole, 2018). From the above three processes, one can see that digital technology is both the result of and the basis for developing digital innovations, implying high scalability and low entry barriers and leading to wide participation and democratized innovation (Yoo et al., 2010). This is further explained in the diagram below.

From the above diagram one can see convergence and generativity drive the process of digital technology transforming into innovations to promote sustainable development. Convergence means that the digital technologies combine previously separate components (Ciriello, Richter and Schwabe, 2018). For example, the iPhone brought together features of a music player, a video camera, a GPS unit, a web browser, and a traditional phone, among other things (Ciriello, Richter and Schwabe, 2018). Furthermore, generativity points to the fact that digital technologies are inherently dynamic, extensible, and malleable (Ciriello, Richter and Schwabe, 2018). For example, the iPhone does not only combine different technical features, but it also allows for indefinite expansion by adding features to increase communication or gaming capabilities through its AppStore.

Figure 1.
Adapted from (Ciriello, Richter and Schwabe, 2018).

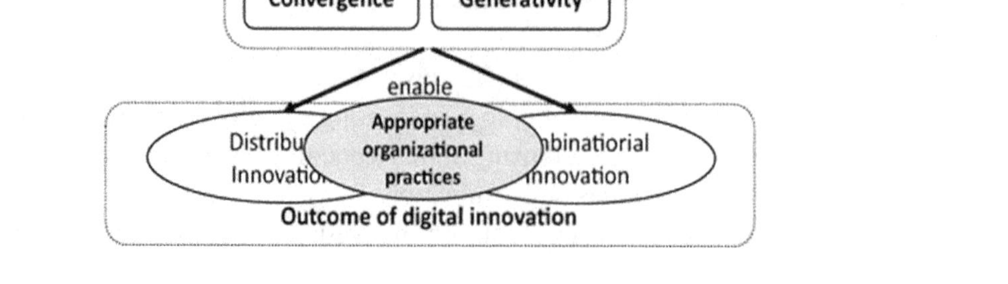

Convergence and generativity make digital technology to be both the basis for digital technology as a means and the result of digital technology as an end digital innovation (Ciriello, Richter and Schwabe, 2018). This means that digital technology enables distributed innovation, combinatorial innovation, and digital technology platforms (Yoo et al. 2012). Distributed innovation means that digital innovations often result from the collection and (re-)combination of digitally encoded information across organizational boundaries (Lusch and Nambisan 2015; Yoo et al. 2012). Open interfaces enable the creation of digital services by using external digital resources while also granting other collaborators access to digital resources generated in the process. For instance, many popular social media applications, such as Instagram, Facebook, and Snapchat, use smartphone cameras to enable users to take, edit, and share photos. In turn, they offer interfaces for providing user data to advertisers (Ciriello, Richter and Schwabe, 2018).

Combinatorial innovation means that new digital solutions are often created by combining existing modules with embedded digital capabilities or blending different modules with the same standard (Ciriello, Richter and Schwabe, 2018). This leads to a wide range of possible combinations of digital technologies, and therefore a wide range of open-ended innovation opportunities enabled by those re-combinable digital technologies (Ciriello, Richter and Schwabe, 2018). Moreover, the flexible character of digital technology enables the modular integration of components into digital technology platforms. Thus, digital technologies enable firms to innovate by creating digital platforms rather than single products (Ciriello, Richter and Schwabe, 2018). A digital technology platform is "a building block, providing an essential function to a technological system which acts as a foundation upon which other firms can develop complementary products, technologies or services" (Gawer, 2014: 2). This means that numerous digital-born companies, such as Facebook, Uber, and Airbnb, provide a digital technology platform with a set of core functionalities that may be extended by complementary contributions from actors outside the organization (Ciriello, Richter and Schwabe, 2018).

Technologies transform entire societies ushering in significant and enduring changes in the way people live and interact with each other, reshaping how governments work and function in digital era and how they deliver public services. Technologies became a catalyst for accelerating human progress, bridging digital divides, developing knowledge societies, and building trust in governance through increased public participation, transparency, and accountability. Digital technologies are at the forefront of breakthroughs giving hope for a sustainable future. However, there is no technological change which is neutral, risk-free, or cost-free. Digital government is now a fundamental pillar of governmental work closely interlinked with shifts in increasingly digital societies and economies, contributing to achievement of the sustainable development goals (SDGs). Below are some of the selected of digital innovation that leads to sustainable development.

E-Governance

E-governance refers to the use of digital technology to enhance the efficiency, transparency, and accountability of government processes. In Africa, e-governance initiatives have the potential to improve citizen engagement, streamline administrative procedures, and combat corruption (Heeks, 2006). For example, electronic voting systems can increase voter turnout and reduce electoral fraud, contributing to more transparent and democratic governance practices (Norris, 2004). E-governance plays promotes sustainable development because entails strategic use of Information Communications Technology (ICT) to support governance processes including ICT-enabled transformation in the relationships between government and citizens, businesses, and other arms of government (Estevez & Janowski, 2013). E-Governance enables the governance of the sustainable development process because it enhances the efficiency of internal government operations. E-governance supports the provision of accessible services needed by the poor and small businesses, delivered at the minimum environmental cost. E-governance increases participation of the poor in government decision- and policy-making processes for sustainable development (Estevez & Janowski, 2013).

E-Government

E-government entails the delivery of government services and information to citizens through digital platforms. By digitizing administrative processes and enabling online access to government resources, e-government systems can enhance public service delivery and promote inclusive development (United Nations Department of Economic and Social Affairs (UNDESA, 2018). In Africa, initiatives such as Kenya's e-Citizen platform have facilitated easier access to government services, including business registration and passport applications, thereby reducing bureaucratic barriers and fostering economic growth (Chigbu, 2019).

Scholars have found that e-government is a tool for improving government and positively impacting good governance, both of which are required for attaining sustainable development (UNDESA, 2016; Alaaraj and Ibrahim, 2014). Von Haldenwang (2004) argued that e-government can increase administrative capability and democratic governance in underdeveloped nations and can better the delivery of public services, reducing costs and saving time. E-government empowers citizens through knowledge sharing with the government, enhancing interactions with business and industry, enhancing effectiveness and efficiency in all areas of government, and boosting revenue growth (Stanimirovich and Vintar, 2013). E-government increases openness, transparency, and decreases corruption (Al-Khatib et al., 2015). Stanimirovich and Vintar (2013) argued that e-government has positive effects of the economic, social and environment thus promotes sustainable development.

Artificial Intelligence

Artificial intelligence (AI) refers to the simulation of human intelligence processes by machines, including learning, reasoning, and problem-solving. In the context of sustainable development, AI technologies offer opportunities to optimize resource management, improve decision-making, and address complex societal challenges (United Nations Educational, Scientific and Cultural Organization (UNESCO), 2020). For instance, AI-powered agricultural systems can enhance crop yield prediction, pest management, and soil conservation practices, contributing to food security and rural livelihoods in Africa (Food and Agriculture Organization (FAO), 2021).

Blockchain

Blockchain is a decentralized digital ledger technology that enables secure and transparent recording of transactions across multiple parties. In the realm of sustainable development, blockchain applications hold promise for enhancing supply chain transparency, facilitating renewable energy trading, and ensuring fair distribution of resources (Swan, 2015). Projects like the BitPesa platform in Kenya demonstrate how blockchain-based remittance services can reduce transaction costs and financial exclusion, thereby promoting economic empowerment and poverty reduction (Hendrickson, Burnett & Haruvy 2019).

The Role of Digital Technologies in Sustainable Development

Current studies present that digital government contribute to greater transparency, openness, and accountability which in turn its government practices increase efficiency and effectiveness needed for sustainable development (Bertot, Jaeger & and Grimes, 2010). Le (2022) argued that governments' digital governance capacity improvements service delivery and thus contribute to achieving sustainable development goals (SDGs). Scholars have noted found that the development of digital government promotes short- and long-term efficiency gains in fossil fuel resources, thereby contributing to high-quality economic growth (Cai et al., 2023; Li et al., 2020; Liu et al., 2022). As digital technologies become more deeply embedded in enterprises' management process, it has been noted that these technologies are gradually improving enterprises to promote economic growth (Xiao & Qi, 2019).

Fitzgerald et al. (2014) proposed that digital technology promote sustainable development because integrating information technology foster services and services. Digital technology is a key foundation for acquiring, utilizing, and managing data, a core means to reduce business costs, reduce duplication of effort, and enhance productivity, and an important tool for resource allocation, organizational boundary reconfiguration, and business model innovation (Liu et al., 2022). Currently, digital technologies have fundamentally changed business strategies, corporate capabilities, and products and services in extended business networks, and largely altered the innovation logic, innovation models, and innovation processes of various innovation agents (Meng & Yu, 2021).

Digital technology applied to business activities improve enterprises' economic performance (ECP) which leads to sustainable growth and development. Digital technology leads to digital economy when applied to industry, but simply acts as a virtual addition to the real traditional economy (Sokolova, 2018). Thus, the growing scale of technological innovations, especially in the field of communications, can transform the system of socio-economic relations and contribute to the spread of digital technologies (Fuchs, 2008). In addition, the digitalization of society provides for the formation of relationships between government, business and society using the latest information technologies, covers socio-economic events on the Internet platform, mobile and sensor networks (Pryshchepa, Kardash, Yakymchuk, Shvek, Parlov & Kramarenko, 2020). Semjachkov (2017) argued that to produce digital products and services, digital technologies are the key and core of the digital economy. However, for the digital technology to lead to digital economy, people must be given ability to access the Internet (Pryshchepa et al., 2020). This therefore makes it possible for digital economy to increase labor productivity, business competitiveness and the well-being of citizens (Maksymiv, 2016).

Digital technologies applied to business enable enterprises to optimize their organizational structure, promote rapid business service-oriented changes, and create a flexible and efficient governance environment for their production and operations (Li, Wu, Zhang & Yang, 2022). The accurate use of digital technologies by enterprises can help optimize their operational processes and improve the efficiency of organizational operations (Mubarak & Petraite, 2020), and further shorten the product development and production cycle. The use of digital technology and the platforming of digital innovation achieve more efficient production management (Huang, Yu & Zhang, 2019). Digital technology improves the resource efficiency and information transfer in the process of digital management (Qi & Xiao, 2020), break the geographical limitation of digital innovation agents (Yan, Ji & Xiong., 2021), and reduce the enterprises' market transaction costs, communication and coordination costs, and operation costs.

More and more enterprises are applying digital technologies in their business management practices to effectively solve various problems encountered in innovation (Wang et al., 2021). The converging applications of digital technologies can break through enterprise boundaries and industry barriers, provide digital value cyberspace for enterprises through information structure adjustment, organizational structure change, and advanced production, and realize intelligent transformation management innovation (Liu, Dong & Wei, 2020). Therefore, the emergence and wide application of digital technologies has largely changed the intrinsic nature of innovation (Yan et al., 2021). Most enterprises in developed world are applying digital technologies to promote their economic performance in order to improve the sustainable development performance (SDP). Digital technology have been applied to emerging industries and transforming people's income and leading to sustainable economic growth which serves as the basis for sustainable development (Wang & Du, 2021).

Even though digital technology improves enterprise management efficiency, there must be appropriate systems such as strategies and capabilities to control and regulate their functionalities as in to reduce their effects on people's lives. The development of digital technologies has reconfigured the process and coordination mechanism of enterprise innovation (Wang, Li & Tan, 2022). It is becoming a critical endeavor for many states applying digital technology in their enterprises to rapidly enhance innovations that promote digital economy (Wang et al., 2022). However, for digital technology to promote digital economy, it must promote digital technology capability (DTC) which crucial for achieving digital innovation a prerequisite for enterprises to gain competitive advantage and sustainable economic performance (SEP) (Annarelli et al., 2021). Sabai (2019) argued that digital technology capabilities are important factors that transform economies and businesses. Innovation resources continue to be digitized, enterprises' innovation have combined their own characteristics of digital technologies (Nambisan et al., 2019) and are having the innovation ability to promote their economies (Nasiri et al., 2020).

The usage of digital technologies affects digital innovation, and the mechanism by which digital technology capabilities affect business enterprises to promote sustainable economy which then increase the probability of achieving sustainable development (Tortora et al., 2021). In addition, DTA brings about digital strategy (DS) which increases the usability of digital technologies to affect daily business (Nambisan et al., 2019). Morgan and Anokhin (2020) argued that DS improves the use of digital technologies to lead to economic performance and sustainable income that promote people's livelihoods. It have been equally argued that DS enables enterprises to cope with changing economic conditions and improves utilization of resources, thus achieving sustained competitive advantage through enhanced innovation potential.

For instance, applying cloud computing and big data technology to business enterprises enables the collection of massive volume of available product (Blichfeldt & Faullant, 2021). It also enables the collection of market information and as well as predicting market trends and movements to respond to dynamic market changes (Blichfeldt & Faullant, 2021). Through the use of digital technology, the focus on consumers' potential needs, competitors' dynamics, and industry frontier information are strengthened (Jiao, 2020). Through the digital technology, the enterprises can achieve accurate capture of users' diverse needs and promote product innovation toward a more personalized and differentiated trend (Li et al., 2022). Moreover, digital technology enhances the ability of enterprises to transfer and reorganize knowledge and absorb external heterogeneous knowledge in an open posture (Wang & Du, 2021). It enables internal knowledge to be reorganized by analyzing and organizing massive data to realize the transformation from knowledge discovery to value creation, which enables scientific management decisions on development programs (Liu et al., 2020). Eventually, this leads to the transformation of enterprises from traditional innovation networks to innovation ecosystems with complementarities, thus affecting their competitive advantage (Ren & Deng, 2022). The diagram below explains the interactions between DS, DTA and DTC to promote economic performance that in turn promote sustainable development.

By applying digital technology, enterprises are guided to continuously improve their resource and energy efficiency and productivity, and to reduce the impact of various production activities on environmental quality. At the same time, by effectively regulating and assessing the environmental impact of the production process, the emission of additional environmental pollutants is reduced or even curbed, thus bringing great potential for achieving sustainable development (Ou, 2021). Nambisan et al. (2019) argue that pursuing a digital technology direction has an impact on the development and acquisition of new skills, competencies and knowledge, which are important resources that contribute to the introduction of new products or processes. Therefore, deploying a DS can help enterprises effectively manage scarce resources such as new skills and knowledge, which can in turn

Figure 2.
Adapted from: (Li, Wu, Zhang & Yang, 2022).

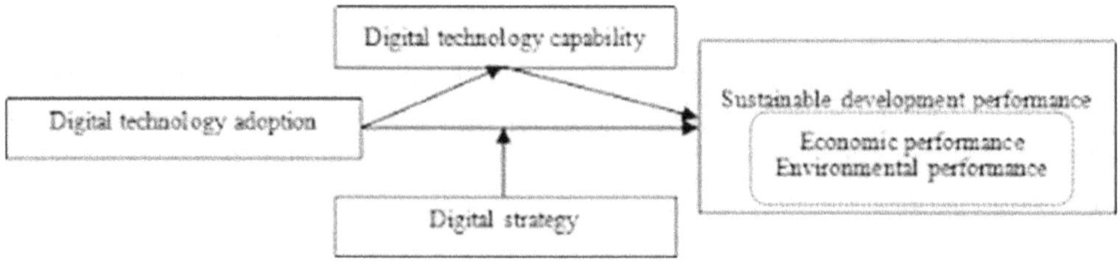

improve internal workflows and advance product or service innovation, thereby positively impacting corporate innovation (Wang et al., 2022). A DS will promote the application of digital technology for value creation and innovation activities because of its strategic innovation, which can in turn enhance the innovation capability of enterprises and significantly contribute to innovation performance (Xiao & Li, 2019).

The key examples of digital technologies that are profoundly promoting sustainable and economic development in the world include e-governance, e-government systems, and service delivery through artificial intelligence. For example, e-governance enable greater transparency in government processes through open data initiatives, online portals for public consultation, and real-time monitoring of government activities (United Nations E-Government Survey, 2020). It enhances citizen participation through digital platforms such as social media, mobile apps, and online forums empower citizens to participate in decision-making processes, hold governments accountable, and advocate for their rights (Bertot et al., 2010). It also improves service delivery, through digitizing administrative processes and automating routine tasks, e-governance initiatives can improve the efficiency and effectiveness of public service delivery, reducing bureaucratic inefficiencies and enhancing citizen satisfaction (UNDP, 2019).

The use of e-government systems promote access to services as they provide citizens with convenient access to government services, such as education, healthcare, and financial assistance, regardless of their geographical location or socioeconomic status (Alemayehu & Sun, 2017). They serve as cost savings thus enable the achievement of sustainable income. Digitalization of government services can lead to cost savings through reduced paperwork, streamlined processes, and increased operational efficiency, enabling governments to reallocate resources to priority areas such as infrastructure development and social welfare programs (Yu, 2012). E-government system strengthens governance and facilitate data-driven decision-making, performance monitoring, and policy evaluation, enabling governments to improve governance outcomes, enhance public trust, and foster sustainable development (UNDESA, 2016).

The other digital technology that plays a key role in sustainable development and economic development is Artificial Intelligence. The effects of Artificial Intelligence in service delivery. Artificial Intelligence technologies such as remote sensing, predictive analytics, and precision farming tools can improve agricultural productivity, optimize resource use, and mitigate climate risks, thereby contributing to food security and rural development (FAO, 2020). Artificial Intelligence powered educational platforms offer personalized learning experiences, adaptive assessment tools, and interactive content, enhancing educational outcomes, promoting lifelong learning, and bridging the digital skills gap (UNESCO, 2021). Artificial Intelligence applications to healthcare, such as telemedicine, medical imaging analysis, and predictive diagnostics, expand access to quality healthcare services, particularly in underserved rural areas, improving health outcomes and reducing disparities (World Health Organization (WHO), 2021).

The other digital technology that is promoting economic growth for sustainable development is blockchain technology. Niyitunga (2023) argued that blockchain technology promotes and enhances economic security today, blockchain technology enhances trust and security across various sectors. From business and economic activities, banking systems and financial transactions, property rights, and intellectual property to all public essential services needed by the citizens to better their lives. Blockchain restores and reforms all governmental services increase transparency and accountability and prevents corruption and corrupt arrangements while providing those services. Niyitunga (2022) argued that blockchain technology can help government officials or service providers establish a more secure, transparent, and trustworthy environment for digital asset ownership and public services. For example, financial transactions and payment transfers have led to economic insecurity in South Africa, and through blockchain technology, money transfers and payments, property registries, contractual agreements, and identity confirmation are made possible and protected (Ljutic & McPhee, 2017).

Digital technology used in business activities promote sustainable development because they curb corruption, that has become rampant in the public sphere. For instance, e-government initiatives in the form of digitized record-keeping, web-based procurement systems, and electronic service delivery can enhance transparency and accountability in public service provision (Baniaman 2015; Davies and Fumega 2014). Moreover, digital technology through e-government can strengthen reform-oriented actors, limit government officials' opportunities for discretionary behavior through improved monitoring of behavior. It can equally enhance citizens' access to information about government processes, and their ability to hold government accountable (Shim and Eom 2008; Andersen 2009). Digital technology can also promote governance and management of natural resources and raw materials in transparent and accountable ways. Kolstad and Søreide (2009) argued that natural resource sectors are susceptible to corruption, and corruption is one of the primary ways through coloniality forces get natural resources from Africa and failed at low cost.

Corruption within the extraction and production of natural resources are serious. It has led to natural resource and raw material depletion at higher rates, heightened costs of extraction and production and lead to revenue loss, thus entrenching poverty and underdevelopment in Africa. Technological initiatives have the potential to reduce corruption and thereby increase the likelihood that resource wealth will lead to positive development outcomes in developing countries. For example, technology-facilitated information flows and participatory mechanisms can help to ensure that oil and mining revenues are spent for public rather than private interests. Digital technologies can also remove the information asymmetries that enable illegal fishing and wildlife trafficking. It can enable individuals involved in decision-making about the use of land, forest, and water resources to become accountable. However, unsustainable exploitation and management of these resources pose significant challenges to sustainable development. Extractive industries often lead to environmental degradation, land degradation, and displacement of indigenous communities (Ovadia, 2019). For instance, the extraction of oil in Nigeria's Niger Delta region has caused widespread pollution, deforestation, and social unrest, undermining the region's sustainable development prospects (Akpor, Ohiembo & Olaolu, 2018).

Coloniality: A Hindering Factor to Sustainable Development in Africa

The promulgation and endorsement of Sustainable Development Goals (SDGs) in 2015 came with the mantra of "Leave No-One Behind" (Klasen & Fleurbaey, 2018). The proponents of the slogan of not leaving anyone behind did not take into consideration the coloniality forces and their effects on promoting inequality between the North and South thus hinder the aim of achieving sustainable development. The legacies of trans-Atlantic slave trade, colonialism and now the ongoing decoloniality have had impact on the continent of Africa and have made the confident unbale to achieve sustainable development. For instance, the coming of SDGs resulted from the failures of the Millennium Development Goals (MDGs) as many states in the global system were marginalized and thus left behind. Marginalization increased inequality which in turn served as hindrance to sustainable development for all. As the MDGs scholars and political leaders failed to understand the coloniality forces in Africa, the same ignorance still exists under the endorsement of the SDGs thus making the probability of not leaving anyone behind in the process of achieving sustainable development minimum.

One of the consequences of coloniality is the presence of inequality in the global system which has made African states be put at the periphery. Inequality whether at local and continental levels is the barrier to development and economic growth. The Millennium Development Goals (MDGs), which failed to incorporate the issue of inequalities (Stuart & Samman, 2017). Numerous development efforts and initiatives reproduced and worsened existing inequalities by failing the most-marginalized and benefitting so-called low-hanging fruits, meaning those who were comparatively somewhat better off (Bhatkal et al., 2015). For not leaving anyone behind, SDGs proponents and scholars in Africa must therefore pay particular attention to addressing the effects of coloniality in Africa. This is because coloniality has increased the level of inequality between Africa and other world states and has also led to inequality within the African states themselves.

The effects of coloniality in Africa prove particularly relevant to reduce the probability of digital development in Africa and are affecting the use of digital technologies in pursuit of the SDGs. They are promoting digital technology inequality (DTI) because these effects have permeated within African indigenous knowledge and technology to promote traditional inequalities. Scholars have argued that digital inequalities is a hindrance to sustainable development because it shapes marginalization which is actually an enemy economic growth (Helsper, 2021; Ragnedda, 2018; van Dijk, 2005). This means that traditional inequalities would lead certain groups disadvantaged and promote exclusion that affect sustainable development. This therefore means that the promise to leave no-one behind requires the recovery from the trans-Atlantic slave trade, colonialism and coloniality. This recovery would enable the view of the Africa's contribution to the wave of digital technology and the growth of industrialization on the continent. Moreover, the recovery must start with Western education also known by colonial education in Africa so that African indigenous knowledge and technology is given an upper hand as in to bring on board an inclusive development programs in Africa. Oba and Onyije (2011) quoted in Niyitunga (2021:171) highlight that:

It is an educational system designed to promote the most rational use of material and social resources. It consists of studies about Africans but not for Africans. It is never designed to give young people confidence and pride as members of African society. It instils a sense of deference towards all that is European and capitalist. African history is considered either non-existent or unimportant. This has resulted in African history's not going beyond the transatlantic slave trade, and many Western scholars refer to Africa as the dark continent. The colonial system also stimulated values and practices that amounted to informal education. The great events and historical developments that are taught are European and of colonialism.

Niyitunga (2021) further argued that western education in Africa has promoted subordination, exploitation, created mental confusion and development of underdevelopment. It is an education which far from giving people confidence in their ability and capacities to overcome obstacles (Niyitunga, 2021). It tends to make them feel their inadequacies and their inability to do anything about the conditions of their lives. It promotes corruption of the thinking and sensibilities of the African and filling them with abnormal complexes (Niyitunga, 2021). Ngũgĩ wa Thiong'o (1981) added that colonial education destroyed African culture, philosophy and epistemology. Ndlovu-Gatsheni argues that, in contemporary Africa, children and youth begin a journey of alienation from their African culture and context the very moment they step into the school, church and university (Ndlovu-Gatsheni, 2017). He further argues that Africans begin the painful path of learning to hate their progenitors as demons; they begin to be taught that all the knowledge they possessed before coming to school was nothing but folk knowledge, barbarism and superstitions that must quickly be forgotten (Ndlovu-Gatsheni, 2017). This means that the existence of the Western universities in Africa constitutes a major barrier to the development of digital technology and growth of industrialization that would lead to sustainable development. Therefore, positive recovery and reforms need establishment of African universities, universities that reflect Africa's realities.

Colonial education in Africa has led to digital literacy and skills gap, which undermines the effective utilization of digital technologies for sustainable development. Many African countries face shortages of skilled professionals, such as software developers, data scientists, and digital entrepreneurs, who can harness the potential of digital technologies for innovation and economic growth (UNCTAD, 2019). Additionally, low levels of digital literacy among policymakers, civil servants, and the general population hinder the adoption and implementation of digital solutions in key sectors such as education, healthcare, and agriculture (Chigbu, 2020). Without adequate skills and knowledge, individuals and organizations struggle to leverage digital technologies effectively, limiting their impact on sustainable development outcomes (World Bank, 2016).

Colonial education has also eradicated African indigenous knowledge and technology on which the development of digital technologies and industrialization in Africa is based upon. It has retarded the development of digital technology in Africa and yet digital technologies have the potential to significantly advance sustainable development. They enhance good governance, improve service delivery, and foster innovation thus promote economic growth. The lack of adequate infrastructure and connectivity, lack access to reliable electricity, internet connectivity, and digital infrastructure, particularly in rural and remote areas prevents the achievement of sustainable development (International Telecommunication Union (ITU), 2020). As a result, millions of people are excluded from the benefits of digital technologies, hindering efforts to promote inclusive and equitable development (Kende & Gillwald, 2017). For example, in countries like Nigeria and Kenya, inadequate broadband infrastructure and high internet costs limit access to online services and digital opportunities, perpetuating digital divides and widening existing inequalities (Global System for Mobile Communications Association (GSMA), 2019).

Rodney and Ndlovu-Gatsheni found that education did exist in pre-colonial Africa and was uprooted and destroyed by the trans-Atlantic Slave trade, colonialism and coloniality which exist in modern era. This Africa education was rooted in the African indigenous knowledge and was for the better of the continent. For example,

Rodney (1972) quoted in Niyitunga (2021:170) identified different features of formal education systems in precolonial Africa that:

It had close links with social life in both a material and a spiritual sense. It was collective in nature, it was development oriented, it was also available at later stages of life, such as on the occasion of passing from one grade to another or joining a new brotherhood. There was a specific programme and a conscious division between teachers and pupils, educational practices were based on communal value, it was directly connected with the purposes of the society, there was no separation of education and productive activity, or any division between manual and intellectual education.

Undeniably, precolonial African educational systems matched the realities of the people in the society and produced well-rounded personalities to fit into that society (Niyitunga, 2021). In precolonial Africa, the literacy rate was high, to the extent that there were universities in Egypt, Morocco and Mali that testified to the standard of education achieved in Africa before the trans-Atlantic slave trade and colonial invasions (Rodney, 1972). The precolonial education reflected the division of labour that made it necessary to create guilds for passing down the technologies of ironmaking, leather-making, cloth manufacture, pottery and professional trading (Rodney, 1972).

The effects of coloniality on preventing the role of digital technologies to promote economic for sustainable development are enormous. The first effect is on the political governance for digital technologies and industrialization in Africa. Pre-trans-Atlantic slave trade and pre-colonialism political governance and structure were intact and were able to lead to economic development by then. The post-colonial federal authority, like the colonial authority before it, lacked the administrative apparatus and political clout to integrate these local authorities, even when they wanted to (Whatley, 2012). From this perspective, state-failure in West Africa may be rooted in a political and economic history that is unique to Africa in many respects, a history that dates at least as far back as the era of the transatlantic slave trade (Whatley, 2012). In some of the African states the remnant of the African indigenous knowledge and technology continue to influence local outcomes in Africa (Englebert, 2000; Gennaioli and Rainer, 2007; Michalopoulos and Papailoannou, 2010), but coloniality forces in the unaddressed colonial legacies have hindered its influence on digital technologies in Africa. Coloniality has made African states incapable from benefiting from the digital technologies in order to promote economic growth and development.

Farah, Ibrahim, Kiamba, Sylvia, Kesegofetse and Mazongo (2011:2) argued that Africa had democratic culture in the pre-colonial era which later dismantled through the influence of slavery, colonialism, and neo-colonialism. Wyk (2007:3) noted "contemporary state in Africa is a remnant of a colonially imposed governance system." African states adopt the colonialists' centralized state system which produced ethnic, and authoritarian based political culture (Wyk (2007:7). Coloniality is not interested to flourish African governance for digital technology which through regime change instals proxy leaders who are not prepared to administer themselves effectively, instead, they are used as stooges, and dictated what to do and what to say as far digital technology is concerned. It becomes difficult thus for digital technologies to lead to economic development for sustainable in a situation of coloniality because the colonialists who controls the global economic system becomes the beneficiaries of technologies. It is clearly noted that "the colonial state in Africa is still an authoritarian bureaucratic apparatus of control (Copson, 2001:12) and has killed any interests in African indigenous knowledge in Africa.

The lack of interest in African indigenous knowledge has given an upper hand for coloniality forces to exploit the continent and leave it undeveloped. This means that coloniality forces have used the state as instrument of exploitation throughout the modern era, and the Africans have remained ignorant. The coloniality forces has produced coloniality administrators and political elites that are not interested in the African affairs, African indigenous knowledge and technology. These coloniality leaders are autocratic and dictatorial in the eyes of coloniality forces and make decisions on behalf of the indigenous people without consulting them. Dictated by coloniality forces, the current African political parties have become a top-down organizational structure and have largely produced autocratic leaders who have failed to nurture a good governance for digital technologies to achieve economic growth. Salih et al (2007:41) argued that "African political parties originated in the illegitimate and non-democratic colonial domination and rule have become a coloniality tool in modern era. Although, multiparty system that leads to democratic governance is introduced in Africa it is not genuinely exercised as it is brought coloniality agents for their own benefits. Political parties are allowed to emerge and be formulated but the environment is full of colonial forces that makes it unconducive to use digital technologies for economic growth in Africa.

Due to coloniality forces, African political leaders' class are highly corrupt and corrupted, because they act as colonialists in the black helmets and are responsible for the coloniality agents that put them on political power. This argument is supported by Njoku (2005:99) who argued that colonial powers' evil socio-political culture produced the habit of corruption in public service of the contemporary Africa. From this, therefore, one can deduce that the existing corrupt behavior of the contemporary leaders of Africa is the continuation of the policies and administration of imperialists. It is obvious that corruption is one of the most attributing factors for poverty and poverty in turn leads for corruption creating vicious cycle. This situation is true in the contemporary Africa, and it is attributed to effects of coloniality. Currently, many

African leaders on the continent enjoy political power as a means of accumulating wealth (Wyk, 2007:5). Thus, state has got its instrumental value for the wealth of the few, leaders. The characteristics of greedy, selfish, and rent-seeking nature of contemporary African political leaders can therefore be traced back to the deed of colonizers and the continuation of those deeds by the coloniality forces.

The second factor is the economic retardation that results from the legacies of trans-Atlantic slave trade and colonialism in Africa. Underdevelopment and poverty status of the African states have hindered the development of digital technologies. It has also affected growth of industrialisation that would use digital technologies to promote sustainable development in Africa. in the economic retardation results from trans-Atlantic slave trade and colonialism and has persisted in this modern era of coloniality thus making it difficult for Africa to benefit from its raw materials and digitalization process. As argued earlier, digital technology promotes economic growth which in turn leads to the achievement of sustainable development. However, this has been made impossible because Africa has not yet recovered from trans-Atlantic slave trade and colonialism legacies which destroyed and dismantled African indigenous knowledge that would promote digital technology and making it to promote sustainable development.

For instance, Africa prior to colonialism was not economically isolated from the rest of the world and had a huge voice and contributions to the world economy. Settles (1996) argued that African states had engaged in international trade from the time of the pharaohs of ancient Egypt, and west Africa specifically had developed extensive international trading systems during the eras of Ghana, Mali, and Songhai (Settles, 1996). These huge empires relied heavily on the taxing of foreign trade to finance goveffilllent expenditures. The wealth of these nations was dependent largely on the trade in gold, but also on the levying of customs, taxes, booty from foreign expeditions, and fees associated with administrative offices (Diop, 1988). Unfortunately, the trans-Atlantic trade and colonialism have caused harm to African economic growth. Colonialism has influence in the contemporary African economic performance. The economic underdevelopment of Africa is rooted from the European colonization (Nunn, 2003:2).

The Atlantic slave trade introduced to the continent sophisticated systems of credit and exchange (Settles, 1996). Credit, debt in currency, and interest all developed apace with the trade (Vansina, 1990) and this resulted in the increase in international trade was that indigenous African economy was subordinated to the interests of Europe (Vansina, 1990). Settles (1996) argued that slave trading system did this by substituting European manufactured products for those products which normally would have been made locally. Thus, Africa was made a source of raw materials (needed for digital technology), in this case labor, and as only a consumer of goods (Settles, 1996). This change of international trade, therefore hindered the

Africa's contributions to the wave of digital technologies. Even innocuous imports such as textiles, competed damagingly with local products, with the result that the technological gap between Africa and Europe, confined in the 15th century to shipping and firearms, widened to such an extent that African technology failed to progress (Herlehy & Wickins, 1983). This pattern of unbalanced trade continued into the period of colonialism and has remained in modern era of coloniality.

As it is obvious that the European colonizers heavily exploited the resources of Africa, its negative effect on the contemporary Africa's underdevelopment is a well-established fact. As Rodney (1973:146) clearly expressed Europeans taken away the young, healthy and the most productive forces of the continent leaving the children, aged, and unhealthy section of the population. Thus, it is easy to deduce from this that the European powers contributed a lot to the economic underdevelopment of contemporary African states. The colonizers left African poor and oppressed states. Hence, Africans started their economic performance at zero level. This, in turn, caused unequal trade transactions (Farah et al., 2011:3). As a result of the exploitation of necessary resources, the continent has been forced to rely on the importation of foreign products (Thomson, 2010:192). This compelled African states to import very expensive manufactured goods from outside by what they gain from export of their cheaper primary products. This leads to unequal exchange of market products and trade deficit. Hence, Africa's integration into the world capitalist economy with its less valuable products is plainly unequal and usually makes it risk taker.

Financial constraints and limited investment in digital infrastructure and innovation are also major challenges. African countries face funding gaps for digital infrastructure projects, research and development initiatives, and capacity-building programs (World Bank, 2020). While there has been a growing interest from international donors and private investors in supporting digital initiatives in Africa, financing remains insufficient to meet the scale of demand and address systemic challenges (African Development Bank (AfDB), 2017). Moreover, the lack of access to affordable financing options and venture capital impedes the growth of digital startups and innovative enterprises, limiting their ability to scale and contribute to sustainable development (UNCTAD, 2020).

The third factor is the failure to govern raw materials and natural resource in Africa. one of the consequences of coloniality under the neocolonialism era is the creation and installation of proxy leaders in Africa. The lack of political governance has influenced the failure to govern raw materials and natural resources for digital technologies in Africa. Africa's abundant natural resources, including minerals, forests, and arable land, play a crucial role in driving economic growth and human development. Robinson (2002) argued that we should "...hypothesize that slavery-induced predatory institutions had significant adverse influence on development paths not just in the Americas where the slaves were used, but also in Africa where the slaves originated. Inikori (1982, 2003) argues that the slave trade encouraged the formation and spread of banditry and small militarized states. Nunn and Wantchekon (2011) presented evidence showing that centuries of slaving in Africa fomented mistrust among people and towards their institutions, and that this slavery-induced mistrust persists till today. This means that slave trade dismantled Africans ways of governance and instead instilled mistrust and fear which has brought about inferiority complex within the Africans. Whatley and Gillezeau (2011b) presented that the slave trade increased ethnic fragmentation in Africa which was reinforced by colonialism and has persisted today under coloniality forces that control modern era. Ethnic fragmentation has impeded the growth of digital technology and has on the other prevented the achievement of sustainable development in Africa.

Nunn (2008) finds that trans-Atlantic slave trade hampered long-term governance for economic progress in Africa. Among the many possible channels is its impact on the governance of economic and political institutions. Colonial institutions were authoritarian, unaccountable, and repressive at the national level. the central strain in traditional African political thought by which Africans sought to maintain the "internal autonomy of the local community" (Vansina 1990, 119; see also Hanson 2022), which prevented any single individual from accumulating too much power. Existing research often conceives of British indirect rule as a coherent ideology that imposed homogeneous and authoritarian local institutions. Nonetheless, we concur with the general contention that the colonial period contributed to authoritarian rule in Africa after independence. Colonial rule set bad precedents of mass coercion (Young 1994), favoring resource extraction over public good provision (Cooper 2002), and reifying ethnic identities (Mamdani 1996; Pengl, Roessler and Rueda 2022).

Colonialism was not just about economic subjugation, but about the ability to wrest control of the local economy from African rulers (Bolt, Leigh, Kohler, Paine, Jack and Robinson, 2022). Improving the production methods or strengthening the economy was not important. Colonial powers instituted trade controls that limited colonial imports to those from the colonizing power and restricted exports to that same market. This reduced the freedom of choice in marketing goods that was previously available to commodities producers. Taking this power away under the guise of colonial development made traditional rulers weaker because their power base was destroyed. Since the colonialism era to this era, Africans are still not allowed to improve their technological methods of production or to market their goods freely. They are still forced to abide with colonial system dictations which hamper the development of digital technology and industrialization in Africa. Digital technologies are driven by outside forces or coloniality forces and the locals Africans have no ability to decide for themselves what to trade and how to use digital technology to increase economic performance. The forces behind trans-Atlantic slave trade, colonialism and coloniality made Africans to be unequal partners in the arena of international trade and economics, thus promoting underdevelopment in Africa.

CONCLUSION

Governments worldwide are increasingly embracing digital technologies to tackle diverse challenges, marking a global consensus on the urgent need for digital government advancement. Digitalization serves as a catalyst for innovation, transforming governance and bolstering national and international competitiveness. In Africa, digital technologies hold promise for fostering sustainable development by bolstering governance, service delivery, and innovation. However, reaping these benefits necessitates addressing hurdles like the digital divide, data privacy, and environmental concerns. Sustainable development in Africa is intricately linked to factors such as population growth, natural resource management, and governance quality. To effectively tackle these challenges, holistic approaches prioritizing environmental sustainability, social equity, and economic growth are essential. Furthermore, investing in renewable energy, resilient infrastructure, and sustainable agriculture can mitigate development's adverse impacts. Indigenous African knowledge systems also play a crucial role in sustainable development across various domains. Moreover, overcoming challenges such as infrastructure gaps, skills shortages, and regulatory barriers requires collaborative efforts from governments, private sectors, and international partners. By addressing these hurdles, African nations can harness digital technologies to achieve sustainable development goals and foster inclusive, resilient, and prosperous societies.

REFERENCES

African Development Bank. (2016). *Catalyzing growth and development through effective natural resources management.* African Development Bank (AfDB).

African Development Bank. (2017). African *Economic Outlook*. Entrepreneurship and Industrialization. African Development Bank.

Akpor, O. B., Ohiembor, M. O., & Olaolu, T. D. (2018). Environmental degradation in the Niger Delta: A human security challenge. *Journal of Human Security*, *14*(2), 11–22.

Alemayehu, D. A., & Sun, Y. (2017). E-government service delivery in Ethiopia: Achievements and challenges. *International Journal of Information Management, 37*(5), 378–386.

Andersen, T. B. (2009). E-government as an anti-corruption strategy. *Information Economics and Policy, 21*(3), 201–210. doi:10.1016/j.infoecopol.2008.11.003

Annarelli, A., Battistella, C., Nonino, F., Parida, V., & Pessot, E. (2021). Literature review on digitalization capabilities: Co-citation analysis of antecedents, conceptualization and consequences. *Technological Forecasting and Social Change, 166*(3), 120635. doi:10.1016/j.techfore.2021.120635

Bai, D.B., Hu, J., Irfan, M., & Hu, M.J. (2023). Unleashing the Impact of Ecological Civilization Pilot Policies on Green Technology Innovation: Evidence from a Novel SC-DID Model. *Energy Economics, 106813*. https://doi.org/.eneco.2023.106813. doi:10.1016/j

Baldwin, K. (2016). *The paradox of traditional chiefs in democratic Africa*. Cambridge University Press.

Baniaman, H. M. (2015). *Controlling corruption through e-governance: Case evidence from Bangladesh U4 Brief.* Bergen, Norway: U4 Anti-Corruption Resource Centre.

Bayeh, E. (2015). The Political And Economic Legacy Of Colonialism In The Post-Independence African States. International Journal in Commerce, IT & Social Sciences, 2(2).

Bertot, J.C., Jaeger, P.T., & Grimes, J.M. (2010). Using ICTs to create a culture of transparency: E-government and social media as openness and anti-corruption tools for societies. *Govern. Inf. Q., 27*(3), 264–271. https://doi.org/.giq.2010.03.001. doi:10.1016/j

Bi, S., Shao, L., Tu, C., Lai, W., Cao, Y., & Hu, J. (2023). Achieving carbon neutrality: The effect of China pilot Free Trade Zone policy on green technology innovation. *Environmental Science and Pollution Research International, 30*(17), 50234–50247. doi:10.1007/s11356-023-25803-1 PMID:36790713

Blichfeldt, H., & Faullant, R. (2021). Performance effects of digital technology adoption and product & service innovation: A process-industry perspective. *Technovation, 105*, 102275. doi:10.1016/j.technovation.2021.102275

Cai, D., Hu, J., Jiang, H., Ai, F., & Bai, T. (2023). Research on the relationship between defense technology innovation and high-quality economic development: Gray correlation analysis based on panel data. *MDE. Managerial and Decision Economics*, *44*(7), 3867–3877. Advance online publication. doi:10.1002/mde.3925

Castro, C., & Lopes, C. (2022). Digital government and sustainable development. *Journal of the Knowledge Economy*, *13*(2), 880–903. doi:10.1007/s13132-021-00749-2

Chang, A. (2018). Analysis on corporate governance compliance standards in New Zealand–a qualitative study on disclosures using content analysis and interviews. *Journal of Financial Regulation and Compliance*.

Chigbu, U. E. (2019). Digital technologies, youth entrepreneurship, and sustainable development in sub-Saharan Africa: Prospects, challenges, and policies. *Journal on Sustainability*, *11*(4), 1032–1062.

Chigbu, U. E. (2020). Digital technologies, youth entrepreneurship, and sustainable development in sub-Saharan Africa: Prospects, challenges, and policies. *Journal on Sustainability*, *12*(1), 354–375.

Choi, T., & Chandler, S. M. (2020). Knowledge vacuum: An organisational learning dynamic of how e-government innovations fail. *Government Information Quarterly*, *37*(1), 101416. doi:10.1016/j.giq.2019.101416

Ciriello, R. F., Richter, A., & Schwabe, G. (2018). Digital Innovation. *Business and Information Systems Engineering*. https://link.springer.com/article/10.1007%2Fs12599-018-0559-8

Davies, T., & Fumega. S. (2014). *Mixed incentives: Adopting innovations for transparency, accountability, and anti-corruption. U4 Issue*. Bergen, Norway: U4 Anti-Corruption Resource Centre.

de Reuver, M., Sørensen, C., & Basole, R. C. (2018). The digital platform: A research agenda. *Journal of Information Technology*, *33*(2), 124–135. doi:10.1057/s41265-016-0033-3

Diop, A. (1988). *Precolonial Black Africa*. Lawrence Hill.

Eling, M., & Lehmann, M. (2018). The impact of digitalization on the insurance value chain and the insurability of risks. *Geneva Pap. R. I.*, *43*(3), 359–396.

Englebert, P. (2000). Pre-Colonial Institutions, Post-Colonial States, and Economic Development in Tropical Africa. *Political Research Quarterly*, *53*(1), 7–36. doi:10.1177/106591290005300101

Farah, I., Kiamba, S., & Mazongo, K. (2011). Major challenges facing Africa in the 21st century: A few provocative remarks At the International Symposium on Cultural Diplomacy in Africa - Strategies to confront the Challenges of the 21st Century: Does Africa have what is required? *International Symposium on Cultural Diplomacy in Africa*.

Fitzgerald, M., Kruschwitz, N., Bonnet, D., & Welch, M. (2014). Embracing digital technology-a new strategic imperative. *MIT Sloan Management Review*, *55*(2), 1–12.

Food and Agriculture Organization. (2020). Overcoming water challenges in agriculture. Food and Agriculture Organization. *The State of Food and Agriculture*.

Food and Agriculture Organization. (2021). *Digital technologies in agriculture and rural areas: Status report.* Food and Agriculture Organization (FAO) of the United Nations.

Fuchs, C. (2008). The implications of new information and communication technologies for sustainability. *Environment, Development and Sustainability*, *10*(3), 291–309. doi:10.1007/s10668-006-9065-0

Gawer, A. (2014). Bridging differing perspectives on technological platforms: Toward an integrative framework. *Research Policy*, *43*(7), 1239–1249. doi:10.1016/j.respol.2014.03.006

Gennaioli, N., & Rainer, L. (2007). The Modern Impact of Precolonial Centralization in Africa. *Journal of Economic Growth*, *12*(3), 185–234. doi:10.1007/s10887-007-9017-z

Gil-Garcia, J.R., & Flores-Zúniga, M.A. (2020). Towards a comprehensive understanding of digital government success: integrating implementation and adoption factors. *Govern. Inf. Q.*, *37*(4), 101518. . doi:10.1016/j.giq.2020.101518

Global System for Mobile Communications Association. (2019). *The Mobile Economy Sub-Saharan Africa 2019*. GSMA Headquarters.

Gobble, M. (2018). Digitalization, digitization, and innovation. *Research Technology Management*, *61*(4), 56–59. doi:10.1080/08956308.2018.1471280

Grosfoguel, R. (2011). Decolonizing Post-Colonial Studies and Paradigms of Political-Economy: Transmodernity, Decolonial Thinking, and Global Coloniality. *TRANSMODERNITY: Journal of Peripheral Cultural Production of the Luso-Hispanic World, 1*(1).

Heeks, R. (2006). E-governance for development: A focus on rural areas. *Journal of Urban Technology*, *13*(2), 123–137.

Hendrickson, J., Burnett, N., & Haruvy, E. (2019). Bitcoin and the charity of digital giving. *Journal of Political Economy*, *127*(6), 234–261.

Herlehy, T. J., & Wickins, P. (1983). An Economic History of Africa from the Earliest Times to the Partition. New York: Oxford University Press, 1981. 323 pp. Maps, tables, biblio., index. $29.95, hardcover. *African Studies Review*, *26*(1), 157–158. doi:10.2307/524615

Hu, J., Hu, M., & Zhang, H. (2023). Has the construction of ecological civilization promoted green technology innovation? *Environmental Technology & Innovation*, *29*, 102960. doi:10.1016/j.eti.2022.102960

Huang, Q. H., Yu, Y. Z., & Zhang, S. L. (2019). Internet development and manufacturing productivity improvement: Internal mechanism and China's experience. *China's Industrial Economy*, (8), 5–23.

International Telecommunication Union (ITU). (2020). *The State of Broadband Report 2020*. International Telecommunication Union HQ.

Kallinikos, J., Aaltonen, A., & Marton, A. (2013). The Ambivalent Ontology of Digital Artifacts. *Management Information Systems Quarterly*, *37*(2), 357–370. doi:10.25300/MISQ/2013/37.2.02

Keane, M., Malcolm, C., & Rollnick, M. (2004). What is science education doing for rural South Africans, and what would they like it to do. In *Proceedings of the 12th Annual Southern African Association for Research in Mathematics, Science and Technology Education Conference* (pp. 453-458). Cape Town: University of Cape Town.

Kende, M., & Gillwald, A. (2017). The impact of mobile broadband in Sub-Saharan Africa: Policy considerations for digital diversity. *Journal of Global Information Technology Management*, *20*(2), 63–87.

Kettl, D. F. (2015). The job of government: Interweaving public functions and private hands. *Public Administration Review*, *75*(2), 219–229. doi:10.1111/puar.12336

Le, T. H. (2022). Are digital business and digital public services a driver for better energy security? Evidence from a European sample. *Environmental Science and Pollution Research International*, *29*(18), 27232–27256. doi:10.1007/s11356-021-17843-2 PMID:34978034

Li, M., Hu, J., Liu, P., & Chen, J. (2023). How can digital finance boost enterprises' highquality development?: Evidence from China. Environ. Sci. Pollut. *Control Ser. 1–, 15*. Advance online publication. doi:10.1007/s11356-023-28519-4 PMID:37440136

Li, X., Wu, T., Zhang, H., & Yang, D. (2022). Digital Technology Adoption and Sustainable Development Performance of Strategic Emerging Industries: The Mediating Role of Digital Technology Capability and the Moderating Role of Digital Strategy. *Journal of Organizational and End User Computing*, *34*(8), 1–18. doi:10.4018/JOEUC.315645

Li, X. Y. (2022). *Research on the impact and mechanism of digital technology driving enterprise innovation*. East China Normal University.

Liu. (2020). Digital innovation management: Theoretical framework and future research. *Management World*, *36*(7), 198–217.

Liu, N., Teng, L., Tian, W., & Li, Y. (2023). Does digitalization enhance fossil fuels resources efficiency? *Resources Policy*, *85*, 103878. doi:10.1016/j.resourpol.2023.103878

Liu, X. X., Yang, Y. Q., & Sun, Z. J. (2022). *The construction and evolutionary development of enterprise digital capabilities: A multi-case exploratory study based on leading digital*. Enterprises Reform.

Liu, Y., Dong, J. Y., & Wei, J. (2020). Digital innovation management: Theoretical framework and future research. *Management World*, *36*(7), 198–217.

Ljutic, A., & McPhee, C. (2017). Editorial: Blockchain. *Technology Innovation Management Review*, *7*(10).

Lusch, R. F., & Nambisan, S. (2015). Service innovation: A service-dominant logic perspective. *Management Information Systems Quarterly*, *39*(1), 155–175. doi:10.25300/MISQ/2015/39.1.07

Ma, D., & Zhu, Q. (2022). Innovation in emerging economies: Research on the digital economy driving high-quality green development. *Journal of Business Research*, *145*, 801–813. doi:10.1016/j.jbusres.2022.03.041

Maksymiv, Y. V. (2016). Reporting as an important tool in ensuring interaction between stakeholders. *Actual Problems of Economics*, *178*(4), 304–310.

Maldonado-Torres, N. (2007). On the coloniality of being. *Cultural Studies*, *21*(2-3), 2–3, 240–270. doi:10.1080/09502380601162548

Meng, Q. S., & Yu, J. (2021). Impact mechanism of digital technology innovation on new generation of information technology industry upgrading. *R & D Management*, *33*(1), 90–100.

Michalopoulos, S., & Papailoannou, E. (2010). *Divide and Rule or the Rule of the Divided? Evidence from Africa.* Discussion Paper Series No. 8088. Center for Economic Policy Research. doi:10.1016/j.jbusres.2019.06.019

Mubarak, M. F., & Petraite, M. (2020). Industry 4.0 technologies, digital trust, and technological orientation: What matters in open innovation? *Technological Forecasting and Social Change, 161*, 120332. doi:.techfore.2020.120332 doi:10.1016/j

Nambisan, S., Wright, M., & Feldman, M. (2019). The digital transformation of innovation and entrepreneurship: Progress challenges and key themes. *Research Policy*, *48*(8), 103773. doi:10.1016/j.respol.2019.03.018

Nasiri, M., Saunila, M., & Ukko, J. (2020). Shaping digital innovation via digital-related capabilities. *Information Systems Frontiers*, 1–18.

Niyitunga, E. B. (2022). The Current and Potential Role of Blockchain-based Technology in Managing Medical Records in Africa. *Digital Policy Studies Journal*, *1*(1), 52–68. Retrieved December 01, 2023, from. doi:10.36615/dps.v1i1.1253

Niyitunga, E. B. (2023). Economic Insecurity Threatens Public Safety and Social Order in South Africa. *Can Blockchain Technology Provide a Sustainable Solution.*

Njoku, U. J. (2005). Colonial Political Re-Engineering and the Genesis of Modern Corruption in African Public Service: The Issue of the Warrant Chiefs of South Eastern Nigeria as a Case in Point. *Nordic Journal of African Studies*, *14*(1), 99–116.

Norris, P. (2004). *Democratic phoenix: Reinventing political activism.* Cambridge University Press.

Oba, N. F., & Onyije, E. L. (2011). Colonialism and education. *Mediterranean Journal of Social Sciences*, *2*(4), 41. Retrieved September 28, 2019, from https://www.richtmann.org/journal/index.php/mjss/article/view/10879

OECD. (1997). *National Innovation Systems.* Organisation for Economic Co-operation and Development.

Ou, Y. Y. (2021). *Research on the impact of digital maturity of construction enterprises on environmental performance.* Chang'an University.

Ovadia, J. S. (2019). Resource extraction and security in Africa: What role for the private sector? *African Security Review*, *28*(2), 147–163.

Parviainen, P., Tihinen, M., Kääriäinen, J., & Teppola, S. (2017). Tackling the digitalization challenge: How to benefit from digitalization in practice. *IJISPM*, *5*(1), 63–77. doi:10.12821/ijispm050104

Pryshchepa, O., Kardash, O., Yakymchuk, A., Shvec, M., Pavlov, K., Pavlova, O., & Kramarenko, I. (2020). Optimization of multi-channel queuing systems with a single retail attempt: Economic approach. *Decision Science Letters*, *9*(4), 559–564. doi:10.5267/j.dsl.2020.8.002

Qi, Y. D., & Xiao, X. (2020). Enterprise management reform in the digital economy era. *Management World*, (6), 135–152.

Ren, Z. Z., & Deng, F. (2022). Digital technology, factor structure transformation, and high-quality economic development. *Soft Science*, (9), 1–10.

Rodney, W. (1972). *How Europe Underdeveloped Africa*. Bogle-L'Ouverture Publications.

Sabai, K. (2019). Digital technology, digital capability, and organizational performance: A mediating role of digital innovation. *International Journal of Innovation Science*, *11*(2), 177–195. doi:10.1108/IJIS-08-2018-0083

Semjachkov, K. A. (2017). The digital economy and its role in the management of modern socioeconomic relations. *Modern Management Technology, 8*(80). https://sovman.ru/article/8001/

Settles, J. D. (1996). *The Impact of Colonialism on African Economic Development*. Chancellor's Honors Program Projects. https://trace.tennessee.edu/utk_chanhonoproj/182

Shim, D. C., & Eom, T. H. (2008). E-Government and anti-corruption: Empirical analysis of international data. *International Journal of Public Administration*, *31*(3), 298–316. doi:10.1080/01900690701590553

Sokolova, H. B. (2018). Some aspects of the development of the digital economy in Ukraine. *Economic Bulletin of Donbass*, *1*(51), 92–96.

Srai, J., & Lorentz, H. (2019). Developing design principles for the digitalisation of purchasing and supply management. *Journal of Purchasing and Supply Management*, *25*(1), 78–98. doi:10.1016/j.pursup.2018.07.001

Stasavage, D. (2020). *The decline and rise of democracy: A global history from antiquity to today*. Princeton University Press.

Swan, M. (2015). *Blockchain: Blueprint for a new economy*. O'Reilly Media.

Tortora, D., Chierici, R., Farina Briamonte, M., & Tiscini, R. (2021). "I digitize so i exist:" Searching for critical capabilities affecting firms' digital innovation. *Journal of Business Research*, *129*, 193–204. doi:10.1016/j.jbusres.2021.02.048

Twizeyimana, J. D., & Andersson, A. (2019). The public value of E-Government–A literature review. *Government Information Quarterly*, *36*(2), 167–178. doi:10.1016/j.giq.2019.01.001

United Nations Conference on Trade and Development (UNCTAD). (2019). *Digital Economy Report 2019: Value Creation and Capture: Implications for Developing Countries*. Geneva: United Nations Conference on Trade and Development.

United Nations Conference on Trade and Development (UNCTAD). (2020). *World Investment Report 2020: International Production Beyond the Pandemic*. Geneva: United Nations Conference on Trade and Development.

United Nations Department of Economic and Social Affairs. (2016). E-government survey 2016: E-government in support of sustainable development. New York: United Nations Department of Economic and Social Affairs (UNDESA).

United Nations Department of Economic and Social Affairs. (2018). E-government survey 2018: Gearing e-government to support transformation towards sustainable and resilient societies. New York: United Nations Department of Economic and Social Affairs (UNDESA).

United Nations Development Programme. (2019). Digital governance in Zimbabwe: Analysis of strategies, actors and institutions. New York: United Nations Development Programme (UNDP).

United Nations E-Government Survey. (2020). *Digital government in the decade of action for the Sustainable Development Goals: E-government for building back better*. New York: United Nations Department of Economic and Social Affairs (UNDESA).

United Nations Educational, Scientific and Cultural Organization. (2020). Artificial intelligence: A human rights perspective. Paris: United Nations Educational, Scientific and Cultural Organization (UNESCO).

United Nations Educational, Scientific and Cultural Organization. (2021). AI in education: Smart learning environments. Paris: United Nations Educational, Scientific and Cultural Organization (UNESCO).

wa Thiongó N. (1981). *Decolonizing the mind: The politics of language in African literature*. Portsmouth, NH: Heinemann.

Wang, C., Medaglia, R., & Zheng, L. (2018). Towards a typology of adaptive governance in the digital government context: The role of decision-making and accountability. *Government Information Quarterly*, *35*(2), 306–322. doi:10.1016/j.giq.2017.08.003

Wang, H. H., & Du, M. (2021). Digital technology, employee participation, and enterprise innovation performance. *Research and Development Management*, (1), 138–148.

Wang, H. H., Li, Y., & Tan, Y. Y. (2022). The impact of digital transformation on enterprise performance based on meta-analysis. *Journal of System Management*, (1), 112–123.

Whatley, W. (2012). *The transatlantic slave trade and the evolution of political authority in West Africa.* Academic Press.

World Bank. (2016). *World Development Report 2016: Digital Dividends*. World Bank Group.

World Bank. (2020). *Digital Infrastructure Investment: Financing the Transformation of Emerging Economies*. World Bank Group.

World Health Organization. (2021). Artificial intelligence for health. Geneva: World Health Organization (WHO).

Xiao, X., & Qi, Y. D. (2019). Value dimension and theoretical logic of industrial digital transformation. *Reform*, (8), 61–70.

Xie, W., Zheng, D., Luo, J., Wang, Z., & Wang, Y. (2022). Research of cooperation strategy of government-enterprise digital transformation based on differential game. *Open Mathematics*, *20*(1), 791–808. doi:10.1515/math-2022-0480

Xie, X., Huo, J., & Zou, H. (2019). Green process innovation, green product innovation, and corporate financial performance: A content analysis method. *Journal of Business Research*, *101*, 697–706. doi:10.1016/j.jbusres.2019.01.010

Yan, J. Z., Ji, W. Y., & Xiong, Z. (2021). Overview and prospect of digital innovation research. *Scientific Research Management*, (4), 11–20.

Yoo, Y., Boland, R. J. Jr, Lyytinen, K., & Majchrzak, A. (2012). Organizing for innovation in the digitized world. *Organization Science*, *23*(5), 1398–1408. doi:10.1287/orsc.1120.0771

Yoo, Y., Henfridsson, O., & Lyytinen, K. (2010). Research commentary-The new organizing logic of digital innovation: An agenda for information systems research. *Information Systems Research*, *21*(4), 724–735. doi:10.1287/isre.1100.0322

Yu, V. F. (2012). E-government for good governance in developing countries: Empirical evidence from the eFez project in Morocco. *Government Information Quarterly*, 285-315.

Yuan, Y.-P. (2023). Government digital transformation: understanding the role of government social media. *Govern. Inf. Q., 40*(1), 101775. https://doi.org/. giq.2022.101775. doi:10.1016/j

Zhang, W., Li, G., & Guo, F. (2022). Does carbon emissions trading promote green technology innovation in China? *Applied Energy, 315*, 119012. doi:10.1016/j. apenergy.2022.119012

Zhao, S., Peng, D., Wen, H., & Wu, Y. (2022). Nonlinear and spatial spillover effects of the digital economy on green total factor energy efficiency: Evidence from 281 cities in China. *Environmental Science and Pollution Research International, 30*(34), 81896–81916. Advance online publication. doi:10.1007/s11356-022-22694-6 PMID:36029445

Chapter 9
Harnessing Digital Technology for Economic Development in Africa

Victor H. Mlambo
https://orcid.org/0000-0002-7705-8863
University of Johannesburg, South Africa

Xolani Thusi
University of Limpopo, South Africa

Muzi Shoba
https://orcid.org/0000-0002-0370-6856
Nelson Mandela University, South Africa

Hv Mlambo
University of Zululand, South Africa

Hlanganani Mnguni
Hlanganani Mnguni University of Zululand, South Africa

Lethiwe Mbongwa
Daleview Secondary School, South Africa

ABSTRACT

According to Ambassador Albert Muchanga, the African Union's Commissioner for Economic Development, Trade, Tourism, Industry, and Minerals, digital technologies could present Africa with a significant opportunity to open new avenues for swift, equitable economic growth and the creation of jobs. For Africa, digital transformation has the potential to be revolutionary. It is a chance to advance industrialization and economic growth while reducing poverty and enhancing people's quality of life. Utilizing digital services and technology will support Agenda 2063 of the African Union. Digital technology can spur innovation, economic growth, and employment creation in many critical economic sectors. It also makes African markets more interconnected with the rest of the world and with each other.

DOI: 10.4018/978-1-6684-7851-6.ch009

Copyright © 2024, IGI Global. Copying or distributing in print or electronic forms without written permission of IGI Global is prohibited.

INTRODUCTION

According to Ambassador Albert Muchanga, the African Union's Commissioner for Economic Development, Trade, Tourism, Industry, and Minerals, digital technologies could present Africa with a significant opportunity to open new avenues for swift, equitable economic growth and the creation of jobs (African Union, 2023). For Africa, digital transformation has the potential to be revolutionary. It is a chance to advance industrialization and economic growth while reducing poverty and enhancing people's quality of life. Utilizing digital services and technology will support Agenda 2063 of the African Union. Digital technology can spur innovation, economic growth, and employment creation in critical economic sectors. It also makes African markets more interconnected with the rest of the world and with each other (African Development Bank Group, 2023). It can improve everyone's access to the market and to finance, especially in underserved areas that traditional financial institutions tend to overlook. In line with more considerable good governance and development policies and programs, promoting digitalization in Africa will optimize the impact on health, energy, transportation, agriculture, and education, making essential social services more accessible. Reducing disparities and closing the digital divide depends on an easily accessible, safe, and dependable internet. It has the potential to significantly improve human capital and give young people in Africa new opportunities. However, the digital divide will most likely impact women and girls significantly, particularly those poor and residing in rural areas (African Development Bank Group, 2023).

Between 2017 and 2025, the number of mobile connections will double. In many of the region's nations, the percentage of people with internet access is also rising quickly. For instance, since 2010, internet penetration has increased by 123 percent in Cameroon and 114 percent in Kenya. Due mainly to the COVID-19 pandemic, millions of Africans used the internet and e-commerce for the first-time during lockdowns to obtain consumer goods, vital health supplies (Runde & Bryja, 2023), and critical information. Nonetheless, one unrecognized obstacle preventing internet access in Africa may be the absence of a dependable power source. For instance, it is estimated that 645 million Africans lacked access to electricity in 2016.2. In sub-Saharan Africa, just 40% of adults had access to mobile internet services in 2022. A lack of affordability and other access barriers kept 44% of the population unconnected in those areas with mobile connectivity.

Furthermore, it is projected that in 2022, merely 20 million households in sub-Saharan Africa will have a wired internet connection, with the most prevalent type being Asymmetric Digital Subscriber Lines, or ADSL. Considering this, this chapter examines the difficulties in using digital technology to advance Africa's economy. The chapter also examines how digital technology can be harnessed for African development.

METHODOLOGY

As with any other research paper, for a literature review to be considered a proper research methodology, the necessary steps and actions must be taken to ensure the review is accurate, precise, and reliable. In light of the research question for this review/paper, a semi-systematic review was adopted to understand the challenges of digital development in Africa. A semi-systematic review, also known as the narrative review approach, is a type of review that follows the partial as opposed to full guidelines of a systematic review in terms of a literature survey and selection but also includes steps for framing a written discourse about literature that can be used as part of the research process (Zunder, 2021). This type of review is designed for topics conceptualized differently and studied by various groups of researchers within diverse disciplines (Snyder, 2019).

Information Sources and Data Collection Process

A search strategy for identifying relevant literature was developed and conducted between November 2022 and April 2023 under three electronic databases: Google Scholar, Sabinet and the Social Science Citation Index (Web of Science). Table 1 below presents the total results for each database during the preliminary literature search. The selected search terms included were "technology," "Africa," "colonialism," post-colonialism," OR "neo-colonialism." These terms are directly related to the research question of this review. To examine and track trends and challenges of inclusive development in Africa, we searched for literature within the past 30 years, as we needed the old literature to find trends of inclusive development that explain the latest developments of this phenomenon.

Table 1. Results of the preliminary literature search using databases.

Databases	Total Number of Results	Peer-Reviewed Papers	Included for Review
Google Scholar	587	203	30
Sabinet	860	434	45
Social Science Citation Index (Web of Science)	146	38	3

Source: Authors own compilation

Article Selection

As shown in Table 1, our search yielded 1,872 articles, then duplicates and studies irrelevant to inclusive development in Africa and the post-colonial era were removed, leaving a total of 811 articles. After a further rigorous review process and consideration of all the inclusion criteria (discussed in detail below), only 89 articles were selected for analysis. Data extraction was in the form of reading abstracts first and making selections, followed by reading full-text articles later before making the final selection. Once this step was completed and the initial articles (or other relevant literature) had been collected, the texts were screened in full to ensure they met the inclusion criteria. As an additional strategy, references in the selected articles were scanned to identify other potentially relevant articles. Findings were synthesized through thematic analysis, a method commonly used for identifying, analyzing and reporting patterns in the form of themes within a text (Snyder, 2019). The data abstracted was based on findings and conceptualization of inclusive development. The information of interest was themes in literature, perspectives and a historical analysis of inclusive development within a post-colonial era.

Inclusion and Exclusion Criteria

Regarding research quality, deciding on inclusion and exclusion criteria is one of the most critical steps when conducting a review (Snyder, 2019). The inclusion criteria were the year of publication (between 1998 and 2023), the language of the article (strictly English), the type of articles were conceptual and empirical papers, and media sources as well as reports from institutional organizations. Studies considered were qualitative, quantitative and mixed methods. The sources that did not talk about digital development in Africa were excluded.

LITERATURE REVIEW

Conceptualization of Digital Technology in Africa

Digital technology refers to any systems and equipment that encode and utilise the binary number system to represent data. These products span digital watches and TVs to advanced robotics and artificial intelligence. Digital technology has altered how we study, communicate, work, and much more (Berman, 2021). Digital technologies include various technologies, tools, services, and applications that use different hardware and software (Rice, 2003). Digital technologies enable electronic services or activities such as information creation, storage, processing, transmission, and presentation. Personal computers, digital television, radio, mobile phones, and robotics are all examples of digital technology (Tulinayo et al., 2018). Digital technology comprises electronic apparatus, devices, and systems that are utilised to manipulate, distribute, and retain data in binary format. Digital technology, as opposed to analogue technology, which transmits information via electromagnetic signals, encodes data as either true or false (Berman, 2021).

Digital technologies have progressed from standalone initiatives to networks of tools and programmes that connect people and things worldwide and help address personal and global concerns. Digital technology has shown to be essential for the growth of modern economies. They have altered people's communication styles, decreased market information frictions, increased inclusiveness (particularly in finance), and resulted in more efficient service delivery (Mensah & Mensah, 2023). Digital technology is highly significant in today's contemporary culture for communication, entertainment, information, and education (Habibi et al., 2023). Digitalisation is not widely practised in Africa. Only 28% of Africans utilise the internet, resulting in a digital divide that prohibits the continent from entirely using the potential of digital technology. Africa lacks more than simply internet connectivity.

Other basics include electricity, literacy, financial inclusion, and legislation. While digital technology has the potential to accelerate development, many African countries continue to have inadequate access to technology and are falling behind. As a result, consumers cannot employ existing digital solutions (IMF, 2021). Digitalisation is widely recognised as one of the most effective methods for implementing the 2030 Agenda for Sustainable Development and Africa's Agenda 2063 (UNCTAD, 2020). With Africa's proportion of the global workforce expected to be the greatest by 2100, African countries must increase their use of digital technology to create job development for the continent's more than 22 million people (World Bank, 2023). Digital technologies could provide Africa an excellent opportunity to open new avenues for rapid, inclusive economic growth and employment creation (African Development Bank, 2023). Digital technology can potentially be a transformative force in Africa, promoting economic growth, deepening democratic government, and increasing civil society engagement (UNDP, 2022). Across Africa, the inexorable spread of networks, sensors, artificial intelligence, and automation is propelling a revolution to an uncertain future. Emerging technologies like CCTV cameras with facial recognition systems, drones, robotics, and "smart cities" are rising. Digitisation improves government income collection and reduces corruption (Allen,2021). More African countries, however, risk being destabilised or restricted in their ability to reap the benefits of the digital revolution. Nigeria boasts more tech hubs than any other African country but has also become a global hotspot for cybercriminals (Allen, 2021).

Digital Technology in a Pre-Colonial Era

The imposition of colonialism permanently changed the history of Africa. African ways of life, cultural development patterns, and modes of thought were permanently altered by the shift in political power brought about by colonialism. Through the process of imperialism and the economic policies that accompanied colonisation, the Atlantic slave trade had a significant impact on the African economy (Dwayne, 1996). The official division of Africa by the major European powers, known as the "Scramble for Africa," occurred before the African countries progressed in all spheres, especially trade. Exploiting a region's natural, human, and financial resources for the benefit of the colonising country was the primary goal of colonialism. To achieve this goal, European powers promoted the growth of cash crop agriculture, a trading network connecting a region's entire economic output to the requirements of the colonising state, and commodity-based trading systems (Dwayne, 1996). The African economic system's natural development was halted by the rise of colonialism and the division of Africa by European colonial powers.

Africa was not economically cut off from the rest of the world before colonisation. African nations have been involved in international trade since the Egyptian pharaohs. In particular, west Africa established sophisticated international trading networks under the governments of Ghana, Mali, and Songhai. To fund government spending, these massive empires mainly relied on taxing foreign trade (Dwayne, 1996). These countries' wealth was mainly derived from the gold trade but was also augmented by taxes, customs, loot from overseas expeditions, and fees for using government offices.

In addition to being shaped by their diverse post-colonial experiences and pre-colonial heritage, modern African societies still bear the scars of colonialism. The continent's current understanding reflects African societies' political, economic, and historical intrusions and constructions due to colonial contact with Europeans in the 18th and 19th centuries. The development of Africa is evaluated from the viewpoints of its former colonisers, especially in the fields of science and technology. One has a propensity to view Africa as having no science or technology before the arrival of European settlers. Indigenous African communities and societies were heavily influenced by colonial ideas, which were used to portray indigenous peoples as uniform and unchanging (Brownlie, 2005).

In pre-colonial Africa, scientific and technological techniques were developed to improve the quality of life for indigenous societies. Traditional skills and techniques were applied in art creation (Shizha,2016). Pre-colonial Africans are widely known for their traditional iron smiting and smelting, wood and ivory crafting, fabric weaving, ceramics, and indigenous traditional medication manufacturing and administration skills. When exploited, these can assist in developing poor indigenous societies (Akuamoah, 2017). During the pre-colonial period in West Africa, wheeled transport technology was well understood but used sparingly. Beginning in the 15th century, West Africa gained access to wheeled transport technologies via Atlantic commerce with Western Europe. Several Europeans brought wheeled vehicles to the West (Law,1980). Pre-colonial African technology discussions generally focus on why standard Eurasian technologies like carts, ploughs, and potter's wheels did not spread to Africa despite centuries of contact with outsiders. To better understand the barriers to adopting various technologies, it is more effective to approach them as technological systems rather than individually. A technical system consists of shared resources, energy sources, labour and social organisations, tools and processes, and outputs of products and services (Austen & Headrick, 1983). While ancient Greece is frequently recognised as the birthplace of science and technology, classical agricultural and metal-using technologies emerged and expanded rapidly in pre-colonial Africa during the second millennium B.C. (The Confucian Weekly Bulletin, 2020). Austen and Headrick (1983) agree that.

In contrast, agriculture and metallurgy emerged in Africa roughly a millennium later than in the Middle East and Europe; these technologies were independent and revolutionary for the continent's peoples. One of the reasons why technologies such as domesticating food production evolved later was because African populations could sustain themselves by foraging vegetation, hunting wildlife, and fishing. History demonstrates that Africans displayed technological brilliance prior to their encounter with Europeans. However, colonial policies undermined indigenous businesses, with blacksmiths bearing the brunt of the consequences. For example, colonial reforms prohibited possessing and manufacturing weapons and transferring land and resources to the British, denying smelters access to land and minerals (Mises Institute, 2022). During the pre-colonial period, Africa's technical progress was natural and methodical. However, Europeans disturbed this process. The disruption began in the fifth century, first with the Arabs and then with the Atlantic slave trade (Oloidi, 2014).

Digital Technology in the Post-Colonial Era in Africa

Literature shows that digital technology is spreading all over the African continent at an inexorable speed, as Ndemo and Weis (2017) described. There has been an unprecedented focus on the potential of digital technology to drive and accelerate efforts toward the realisation of the United Nations Agenda 2030 – of sustainable development goals and Africa's Vision 2063. This focus is mirrored by the increasing interest in information and communication technologies (ICT) within the development context of African countries (African Union, 2023). ICT, or digital technologies, are viewed here as a crucial enabling infrastructure for driving economic growth and development in the continent. There is a strong sentiment that ICT infrastructure attracts much-needed foreign direct investments and generates significant fiscal revenues and work opportunities in Africa. The World Bank (2013) has already described technological progress as a considerable driving force behind economic growth and has implied developing countries should take advantage of technology.

Therefore, it is not surprising that many African countries, such as South Africa and Nigeria, have developed their own ICT policies. Notably, the ICT policies of both South Africa and Nigeria considerably highlight the developmental aspirations of the countries and how they view digital technologies as an opportunity to catalyse and sustain socio-economic development critical to their respective countries' needs. In the context of South Africa, the government's national policy known as the National Development Plan (NDP) asserts the country's commitment to building an ethical and capable developmental state characterised by inclusiveness, tolerance and harmony. The view is that a capable developmental state would eliminate inequality and reduce inequality in the country by 2030, which aligns with the United Nations Sustainable Development Goals.

Accordingly, the government considers digital technology an essential role player that can facilitate all the objectives of the country set out in the NDP. The South African White Paper on ICT expresses the government's commitment to ensuring that everyone in South Africa, regardless of who they are, where they live or their socio-economic status, can improve the quality of their lives by accessing the benefits of participating in the digital society" (South African ICT White Paper, 2016, 1a). Furthermore, the White Paper postulates that the government views information and communication technologies as a means to facilitate the inclusive socio-economic transformation of South Africa (South African ICT White Paper, 2016, 1b). Digital technology has become an integral part of how people live their daily lives, as people now require digital addresses as they would equally require physical addresses. Indeed, the surge of technology in the post-colonial era is occurring at an inexorable pace (Ndemo & Weis, 2017). Some argue that digital technology has already significantly shaped various aspects of African communication and connectivity. The widely cited data shows that the proliferation of digital technologies, such as the internet and mobile phones, has facilitated greater access to information and communication networks, bridging geographical and social divides within the continent. The following section focuses on the role of digital technology in African development.

The Role of the Impact of 4IR in Other Developing Regions

The Fourth Industrial Revolution (4IR) is causing significant disruptions to traditional manufacturing methods and presenting new obstacles and growth prospects. There appears to be a concentration of innovations related to the Fourth Industrial Revolution (4IR) in a few prominent countries. Despite the aforementioned fact, implementing

Fourth Industrial Revolution (4IR) technologies presents a potential avenue for emerging economies to expedite their progress by bypassing certain intermediary phases of industrialization. According to Carlota and Luc (1988), individuals can transition directly to utilizing more sophisticated technologies. This phenomenon is frequently exemplified by the swift uptake of cellular phone technology in nations without extensive landline coverage (Keun & Chaisung, 2001). When considering the industrial sector, the Fourth Industrial Revolution (4IR) presents potential prospects for developing nations that have not made substantial investments in previous manufacturing technologies. Instead of gradually retooling existing factories, it is theoretically possible to construct new manufacturing facilities from scratch, following the principles of smart factories. The realization of this potential is contingent upon various factors, encompassing technological accessibility, the proficiency of workers, and prevailing market conditions. These factors are examined to ascertain how countries can effectively utilize the technological disruptions linked to the Fourth Industrial Revolution (4IR) to enter the manufacturing sector (Keun, 2019). Like other developing regions, Africa must promptly establish and strengthen policy frameworks pertinent to ensure the continent can effectively leverage development opportunities. The Fourth Industrial Revolution (4IR) encompasses several significant features that have profound implications for education, employment, and the future of work. These features include the rapid advancement of digitalization, artificial intelligence (AI), cloud computing, robotics, and 3D printing (Morsy, 2020). This is particularly accurate for African nations. In the last ten years, the proportion of the continent's population under the age of 20 has increased by over 25% and is expected to become the largest age group on the continent by 2070. With the advent of the Fourth Industrial Revolution (4IR), Africa's youth will emerge as a crucial resource. To harness the demographic dividend, African nations must undertake comprehensive reforms of their education systems in anticipation of the forthcoming technological revolution (Morsy, 2020). Automation has the potential to raise skills premiums and worsen income inequality, but it can also enhance productivity and generate new job opportunities. Therefore, the Fourth Industrial Revolution (4IR) presents a distinctive prospect for African nations to overcome developmental obstacles through the utilization of technology. The Fourth Industrial Revolution (4IR) will significantly impact the skills required in the job market (Morsy, 2020). Globally, there is a shift in demand towards flexible social, behavioural, and non-repetitive cognitive abilities rather than focusing on repetitive tasks and specialized skills associated with occupations. The African continent is experiencing an increasing demand for professionals in various fields, such as software engineering, marketing, writing, and financial advice. Conversely, there is a declining demand for mechanical technicians, administrative assistants, and accountants. Nevertheless, Africa and other countries considered 'followers' encounter numerous overarching obstacles when enhancing connectivity to exploit emerging technologies fully. According to Markowitz (2019), the primary limitations on the supply side pertain to the insufficiency of facilitating ICT infrastructure and the absence of policy frameworks and policy agility required to implement emerging GPT investments. However, it may also lead to social and economic upheaval, including exacerbating inequality, if appropriate measures are not taken (Fox & Signe, 2022). Given that the significant proportion of the workforce is engaged in informal employment, which is projected to persist for many years, Africa's education and industrial policies must find a middle ground. This involves promoting private investment necessary for establishing new formal jobs through advanced technology while guaranteeing that all new workers possess the fundamental skills and infrastructure to earn a satisfactory livelihood (Fox & Signe, 2022).

From the onset of the First Industrial Revolution in England in the latter half of the 18th century, swift technological and scientific breakthroughs have significantly transformed the lives of millions of individuals globally. In the northern hemisphere, there has been a notable rise in life expectancy and a significant increase in the availability of food and goods. Simultaneously, it is noteworthy that a substantial portion of the prevailing environmental challenges faced by the planet, including climate change, biodiversity loss, air pollution, and oceanic waste accumulation, can be attributed to the unintended repercussions stemming from prior industrialization (Siekmann et al., 2023). Furthermore, the advantages of technological advancements are progressively unevenly allocated, both on a global and national scale, resulting in a heightened wealth accumulation among a smaller group of individuals. Amidst the prevailing environmental, social, and economic challenges, the global community is poised to confront a significant transformation in the forthcoming years, commonly called the Fourth Industrial Revolution (4IR) (Siekmann et al., 2023). Previous industrial revolutions predominantly depended on fossil resources as the primary energy source, whereas the forthcoming revolution is expected to rely on renewable energy sources (RES) primarily. Several interconnected trends, including decarbonization, digitalization, and decentralization, characterize the Fourth Industrial Revolution (4IR). These trends are expected to significantly impact the fundamental power and data infrastructures, business models and industries within and outside the energy market (Siekmann et al., 2023). Considering the favourable effects on the quality of life and the associated social and environmental repercussions is crucial. Managing these intricate advancements entails significant ambiguity and peril, necessitating enduring strategic choices amidst rapidly evolving conditions. Prior studies have demonstrated that in such situations, the progress of technology is enhanced by having a clear objective and a guiding principle to ensure stability (Siekmann et al., 2023).

The Digital Divides in Africa: Questions Around Consolidating the 4IR

According to Afrobarometer, most Africans possess a mobile phone and can connect to a mobile telecom network. Nevertheless, less than 50% of them possess a mobile phone with Internet connectivity, and an even smaller proportion can obtain a computer (Malephane, 2022). This phenomenon results in diminished levels of Internet utilisation across numerous nations. Women and individuals who are rural, older, less affluent, and less educated are significantly more prone to experiencing a lack of Internet access and usage. A significant disparity exists between less developed countries (LDCs) and more developed nations (Malephane, 2022).

Africa necessitates a digital revolution to expedite economic expansion and foster employment generation. According to the World Bank, achieving the African Union's objective of providing universal and affordable internet access will result in a yearly increase of 2 percentage points in Africa's GDP growth. Moreover, the likelihood of securing employment, irrespective of one's level of education, rises by 6.9 to 13.2 percent when fast internet becomes accessible. This is due to its ability to establish new businesses and enhance productivity and exports. According to Ghanem (2020), digital technologies present a distinct opportunity for African nations to reshape multiple sectors of their economies profoundly.

Nevertheless, realising this potential is contingent upon Africa's ability to tackle its substantial deficiencies in digital infrastructure effectively. Africa is significantly falling behind in terms of digital advancements. For instance, the level of internet penetration, quality, and affordability in this region is significantly lower compared to other regions globally. In 2019, Africa's internet penetration rate was 39.6 percent, while the global average was 62.7 percent. However, there is significant variation among countries, with Kenya having 89.8 percent and Burundi having 5.3 percent (Ghanem, 2020). In addition, Africa accounted for a mere 1 percent of the global international internet bandwidth in 2017. The average mobile broadband download speed in Africa is 2.7 megabits per second (Mbits/s), approximately half of the global median of 5.2 Mbits/s.

Additionally, the monthly expenditure for a fixed broadband connection account for 36.6 percent of the gross national income, in contrast to the global average of 14.5 percent. African nations allocate approximately 1.1 percent of their Gross Domestic Product (GDP) towards digital investment, whereas advanced economies allocate an average of 3.2 percent. Therefore, maintaining the current situation is not viable, as it will perpetuate the digital divide's expansion and exacerbate Africa's marginalisation (Ghanem, 2020). According to Thompson (2023), Africa has the lowest internet penetration rate among the continent's population, at 39%, in the context of the Fourth Industrial Revolution (4IR). This is significantly lower than the global average of almost 60%.

Moreover, substantial disparities exist in internet accessibility between rural and urban regions, with urban areas surpassing rural areas by nearly 200 percent in certain countries regarding smartphone usage. Cost is the predictable cause of Africa's internet deficiency. Thompson (2023) states Africa has the highest internet connectivity cost globally. Based on data from the Alliance for Affordable Internet, it is observed that Africans allocate an average of 8.8% of their monthly income towards the acquisition of 1GB of data, in contrast to Latin America's average of 3.6% and Asia's average of 1.5%. In certain instances, such as in Chad, the DRC, and CAR, it was observed that 1GB incurred a cost equivalent to one-fifth of the earnings. De Feydeau, Menski, and Perry (2022) provide additional evidence in favour of this proposition and contend that while enhanced internet accessibility and associated infrastructure could yield significant consequences throughout the continent, Africa lags significantly behind other regions in terms of a fibre network and broadband connectivity, spectrum availability, and data centre processing capabilities.

Nevertheless, the digital infrastructure in sub-Saharan Africa is experiencing rapid growth, primarily driven by robust investment incentives (de Feydeau, Menski, & Perry, 2022). The potential for digital expansion in Africa is immense, but it will hinge on guaranteeing affordable and widespread digital accessibility. Consequently, equity investors and financiers will need a substantial amount of capital. The impact of the Fourth Industrial Revolution (4IR) on African societies and economies will ultimately depend on the specific policies implemented by African governments in response (Du Preez, 2019). To attain genuine success in the Fourth Industrial Revolution (4IR), African nations must develop comprehensive policy frameworks that address two significant challenges: infrastructure backlogs and insufficient investment in human capital. Another issue in Africa is the persistent scarcity of essential skill sets. Skills shortages have hindered the developmental potential of numerous African nations since gaining independence. If these issues are not effectively addressed, they will be further exacerbated by the Fourth Industrial Revolution (4IR) (Du Preez, 2019).

Undoubtedly, the successful integration of Fourth Industrial Revolution (4IR) technologies necessitates the collaborative engagement of various stakeholders. By actively involving these stakeholders in developing and deploying these technologies, we can leverage their potential to generate significant social impact. According to the World Economic Forum (202), the full potential of the Fourth Industrial Revolution (4IR) can only be realised if public and private stakeholders act. Without the involvement of both parties, technology will not be expanded to achieve the Global Goals. Maintaining the status quo is not a viable choice: opting for a "wait and see" approach will impose unsustainable environmental and social burdens on individuals and our planet. South Africa's conversation surrounding the Fourth Industrial Revolution (4IR) is typically centred around government and private sector involvement. However, it is essential to acknowledge that other stakeholders in civil society, including academia, women, and youth, also have a crucial role in ensuring widespread support from all members of society. The Fourth Industrial Revolution (4IR) is evident in technological advancements and has a wide-ranging impact on all aspects of society. Therefore, it is necessary to adopt a more comprehensive perspective and approach (Department of Telecommunications and Postal Services, 2018). Establishing partnerships and fostering a collaborative mindset among various stakeholders will be crucial in shaping novel governance frameworks for the Fourth Industrial Revolution (4IR), ensuring its long-term viability. To advance the Fourth Industrial Revolution (4IR) agenda in Africa, it is imperative for the private sector to actively collaborate with various stakeholders, such as government entities, organisations, and policymakers. While technological advancements primarily characterise 4IR, it is also marked by the prevalence of digital collaboration. Organisations have the potential to create opportunities and reshape the narrative of the Fourth Industrial Revolution (4IR) in Africa through the establishment of partnerships (Madiba, 2023).

The Role of Digital Technology in African Development

Thus far, the chapter has already established that African governments view new and emerging technologies as an essential enabling infrastructure for Africa's economic growth and development. ICT is seen, for example, as a vehicle for Africa's attainment of San's (1999) development indicators that include, among other things, economic opportunities, political freedoms, social facilities, and transparency guarantees, as well as the construction of knowledge societies as imperative drivers of development. A strong consensus is that ICT or digital technology will transform Africa into a knowledge-based and globally competitive society. Digital technology features significantly in the Africa Agenda 2063 as one of the essential drivers and critical enablers for Africa's advancement.

For instance, the African Continental Free Trade Area (AfCFTA) is expected to boost African economies, promote trade liberalisation, and transform Africa's industrial landscape internally and globally. The AfCFTA is a flagship project of the AU Agenda 2063, which is expected to improve continental trade that is presently approximated at 16%, with Africa's share of world trade valued at merely 3% (The Department of Trade, Industry and Competition, 2023). Furthermore, it is believed that AfCFTA will build an African integrated market of over 1.4 billion people with a combined gross domestic product of about US$3.4 trillion (Congressional Research Service, 2023). In this context, digital technologies will be expected to enable

AfCFTA institutions to detect trading abnormalities faster and more accurately and improve outcomes rapidly. The AfCFTA will boost industrialization and economic growth in Africa by instituting digital technology policies, principles and practices. Recognizing the significance of digital technology in economic advancement, the AU Assembly decided on 10 February 2020 to mandate negotiations for an AfCFTA E-commerce protocol.

The negotiations on the AfCFTA E-commerce protocol are still ongoing. However, the protocol is expected to regulate and facilitate e-business and trading on digital platforms. It should boost and transform Africa's digital economy and enhance continental economic integration. All this combined remarkably demonstrates how Africa has the role that digital technological progress could play in the continent's advancement. However, Africa still has a long way to go in achieving meaningful development. The current political and economic landscape that is characterized, inert alia, by challenges of corruption, maladministration, political intolerance, and instability remains a huge stumbling block for the continent. Such challenges have been accelerated by weak state institutions that have been infiltrated and politicized by the ruling elites so that these institutions serve their interests rather than those of the populace. In the preceding part of the chapter, we documented five Sen's (1999) development indicators. It is general knowledge these development indicators are still a dream to many on the continent.

Fukuyama (2019), cited in Shoba (2022), suggests strong evidence that peace and security are essential elements of economic growth and development. Similarly, Sen (1999) underlines the import of freedom for development. Sen's (1999: 1a) point of view is that "Development can be seen as a process of expanding the real freedoms that people enjoy." Simply put, "development requires the removal of major sources of unfreedom." Such primary sources entail, among other things, "poverty as well as tyranny, poor economic opportunities as well as systematic social deprivation, neglect of public facilities as well as intolerance or overactivity of repressive states." (Sen, 1999: 1b). In the African context, it is extensively documented that poverty, tyranny, limited economic opportunities, and restricted political participation persist as detrimental realities due to the pervasive influence of repressive leadership across the continent. Accordingly, this chapter contends that whilst technological progress "can substantially contribute to expanding human freedom" (Sen, 1999:1c), which is central to the development process, Africa must work toward attaining peace and security for the continent's development to take place.

The Question of Regulation of 41R in Africa

Implementing innovative policy and legislative reforms is crucial in facilitating digital transformation. Governments are granted the power to implement measures and allocate resources to address the challenges and capitalize on the opportunities presented by the digital age. The advent of the fourth industrial revolution has introduced novel challenges, including trade restrictions, enterprise data security, liability concerns, and threats to personal data privacy. These challenges necessitate the implementation of stringent regulations through establishing standards, legislation, and policies (Zhou & Zhou, 2015). The role of policy is significant in the governance of the intricate smart environment (Scholl & Scholl, 2014). The European Commission has formulated an industrial policy addressing the fourth industrial revolution to guide Europe in its transformation process. According to the European Commission (2015), the policy is anticipated to address various aspects such as skills, infrastructure, funding, and regulation. The policy establishes a conducive environment for the flourishing of "smart industries." Governments must formulate economic, industrial, and labour market policies that are adaptable and capable of adequately equipping industry, citizens, and government for the prospects of the fourth industrial revolution.

According to Njuguna and Signe (2020), the centrality of innovation in the Fourth Industrial Revolution (4IR) necessitates strengthening state and institutional capabilities to foster and facilitate innovation and establish a conducive business environment to achieve success. One significant regulatory obstacle pertains to the augmentation of cybersecurity measures. Many African nations face a deficiency in terms of a comprehensive legal framework and institutional capacity to effectively combat cybercrime. In recent years, there has been a noticeable rise in initiatives aimed at mitigating cybercrime, primarily at the local level or through the active involvement of private sector entities (Njuguna & Signe, 2020). Between 2015 and 2016, there was a notable surge of 73 percent in the number of companies that obtained certification for Information Security Management Systems. Specifically, the number of such companies rose from 129 in 2015 to 224 in 2016, with South Africa, Nigeria, and Morocco being the primary contributors (International Organization for Standardization, 2018). Implementing universally recognized and suitable standards and regulations, such as these, is an initial measure towards enhancing cybersecurity. Simultaneously, companies should allocate resources to improve their employees' cybersecurity competencies and incorporate cyber risk mitigation into their decision-making procedures. The African Continental Free Trade Agreement presents a distinctive prospect for bolstering governance during the Fourth Industrial Revolution (4IR). By implementing coordinated policies and procedures, the continent can effectively adjust to the swift transformations brought about by the Fourth Industrial Revolution (4IR) and utilize it to expedite engagement in global value chains.

According to Oloruntade and Omoniyi (2023), regulating artificial intelligence (AI) and its associated technologies can be approached from three perspectives.

The initial aspect pertains to legislation passed by a legislative body, followed by the formulation of approaches to embrace artificial intelligence and, ultimately, the formulation of policies. The issue of regulation, particularly within the African context, pertains to the inadequacy of comprehensive regulation on artificial intelligence (AI) due to the comparatively lower level of technological advancement in Africa compared to developed nations. The utilization of AI in Africa is a subject of disagreement. The representation of African content is insufficient, resulting in the perpetuation of specific stereotypes and biases.

Additionally, Africans and their data are vulnerable to exploitation (Musoni, 2023). One notable instance of the utilization of African data pertains to the initiation of Worldcoin's artificial intelligence responsibility, wherein individuals from Kenya were extended an invitation to scan their eyeballs on the crystal-ball-shaped Worldcoin Orb in return for Worldcoin tokens, which were valued at approximately $49 (Musoni, 2023). The emergence of AI has sparked concerns regarding privacy, security, regulation, and ethics. As a result, Kenyan authorities have temporarily halted Worldcoin operations while investigating. Uniform or antiquated regulations can hinder innovation and investment in sectors related to the Fourth Industrial Revolution (4IR). The establishment of a conducive regulatory framework is imperative. Although there is a diffusion of e-government policy elements from Europe and the United States to Africa, these policies are often not adequately tailored to specific local contexts. The proposed measures do not evaluate risks or potential consequences, nor do they consider the feasibility of local implementation (Signé, 2023).

Moreover, the political leadership in numerous African nations is deficient in the motivation required to instigate transformation in the public sector and seldom regards e-government as a top priority. Most African nations lack the legal structure to facilitate and promote digital governance or disseminate emerging technologies (Signé, 2023). It is still necessary to give priority, revamp, and reframe e-government initiatives in Africa to meet the requirements of the majority of citizens. If governments exhibit a sluggishness in embracing novel technologies, they will be unable to achieve the necessary efficiency improvements required to provide adequate public services.

The Challenges of Digital Technology in Africa

Scholars such as Enaifoghe (2021) and Baku (2022) have indicated the importance of digital technology in various African sectors, including the economy, education, and health. However, these scholars have been concerned about the challenges that deny the smooth implementation of digital technology in Africa. Such challenges continue to raise questions about what must be done to digitalise Africa better. Therefore, this section highlights some significant challenges that affect the effective use of digital technology in Africa, particularly for economies. According to Finsense Africa (2023), African digital technology is currently affected by challenges such as lack of necessary resources and limited budgets, legacy systems, and resistance to change, cyber security threats, lack of digital skills, lack of leadership support and regulatory challenges, and lack of collaboration.

Lack of Resources and Limited Budgets

Vogelsang et al. (2019) point out financial limitations as a significant challenge affecting digitalisation in Africa. These scholars say limited financial support blocks Africa's digital infrastructure and transformation. Most countries cannot provide enough budgets to support the implementation. As a result, most communities, companies, and other organisations struggle with digital technology. This assertion is also supported by Finsense Africa (2022), who states that almost 60% of the chief information officers (CIOs) lack the financial means to support the proper implementation of digital technologies for their organisations. Manny et al. (2021) argue that African governments must try to budget for implementing digital technology resources. This will assist in enhancing the complex instrumentation and highly accurate measurements for the growing use of sensors and data transmission technologies. Again, such budgets will also assist in producing massive volumes with incoming real-time data streams and calls for sophisticated analytics or algorithms.

Legacy Systems and Resistant to Change

Change is vital in human lives, but people find it challenging to adapt in most instances. Introducing a specific new digital technology piece automatically requires people to change how they live or do things. Then, this process is not loved by everyone. In other words, it is challenging to integrate new digital technologies with legacy systems, which many African institutions and businesses still use. Many organisations are comfortable using their old systems, which are not competent and fast for today's circumstances (Sebastian et al., 2020). According to these scholars, new digital technology comes with game-changing opportunities and existential threats that most old businesses fear. This fear primarily arises from the fact that the rules and culture of business have to change (Zissis, 2020).

Cyber Security Threats

Deloitte (2021) pointed out the issue of cyber security threats in the African continent. This institution has noted that in 2022, about 89% of African organisations will face some cyber security attack incidents. Typically, the goals of these cyberattacks are to disrupt regular corporate operations, obtain, alter, or delete sensitive data, or use ransomware to demand money from customers. They are unsuitable as they affect the organisations' regular business and smooth operations. In the education sector, cyber security threatens losing valuable information for their students and affecting daily practices. Combating cyber security threats is still a real challenge in Africa. Hence, only 10,000 cybersecurity professionals are estimated to be available for the entire continent (Deloitte, 2021). This number is deficient compared to the risk that is available and cases that have been reported. Therefore, strong cybersecurity measures are highly recommended to be implemented to reduce cybersecurity threats, such as firewalls, encryption, and multi-factor authentication. Deloitte Africa advises African businesses to address cybersecurity from a risk-based perspective, concentrating on their most important assets.

Lack of Digital Skills

Advanced skills play a crucial role in digital technology. People need various skills to operate such pieces of technology. It has been observed that many African organizations lack the digital skills to implement digital technology properly. Mckinsey Africa (2022) observed that around 64% of African CIOs lack digital skills for their organizations. For example, finding a local specialist for obsolete versions of new fundamental economic systems is impossible. This further affected the productivity of the organizations. Perifanou et al. (2021) indicate the importance of digital skills in the economy, indicating that individuals participating in sectors such as businesses, customers, and suppliers must have some digital skills for their respective roles. López Peláez (2020) indicates that digital skills are essential to avoid the digital divides between organizations and societies. Therefore, organizations need to provide their staff with training and upskilling. Additionally, they can collaborate with educational institutions and training centres to create talent pipelines for digital jobs.

Lack of Leadership Support and Regulatory Challenges

Effective leadership always plays a crucial role in any organization. This further implies the rules and regulations that are in place to regulate that organization. According to Matyana and Thusi (2023), good leadership is crucial in driving the organization in a practical direction while promoting productivity. Due to regulatory obstacles, African organizations may struggle to implement digital transformation activities.

Mckinsey Africa (2022) observed that most African organizations often need more leadership support for digital transformation initiatives; around 42% of the Chief Information Officers need more effective leadership to implement successful digital transformation. Therefore, senior executives must engage in digital transformation and ascertain the strategic significance of digital technology. To support digital activities, there should also be a dedicated leader for digital transformation.

Other challenges involve a lack of collaboration between the stakeholders, such as the community, departments and other stakeholders that might be affected. Secondly, in Africa, there is an issue of Inadequate infrastructure that cannot support the proper implementation of digital technology. This further implies that modern technologies are incompatible with outdated data centres and servers. Thirdly, data privacy concerns also affect the proper implementation of digital technologies in Africa.

Digital Technology in Africa: Prospects

Based on the above challenges facing the proper implementation of digital technology, the following strategies are recommended by Finsense Africa (2022). They are useful to every African organization but in this regard, they are linked particularly with economic operations.

1. **Evaluation of digital technology-** There is no one-size-fits-all set of technologies, so CIOs must carefully consider the ones most appropriate for their unique problems and organizational objectives. When putting these technologies into practice, they need to consider the regulatory environment and talent availability.
2. **Managed services-** Digital technology allows companies to monitor and maintain their services if they have planned for the services.
3. **Outsourcing Talent-** Utilizing outside knowledge and resources boosts productivity, lowers expenses, and quickens innovation. Outsourcing provides specialized skills and knowledge while filling talent shortfalls. Small enterprises and those with minimal IT resources can benefit from it. They can outsource software developers, digital project managers, cybersecurity experts, quality assurance engineers, and many more.
4. **Involvement in Policy Making-** Organizations may guarantee that their interests are represented and help shape the technology landscape by participating in the policymaking and regulation-development process. This fosters creativity and reduces risk as well. Companies could shape industry policies and speak up for their interests. Interacting with legislators also gives companies information about new rules and compliance specifications.
5. **Private and Hybrid Cloud-** Organizations can benefit from increased scalability, flexibility, and cost savings by migrating resources to the cloud. Remote collaboration is another benefit of cloud computing. It is only feasible if cloud migration is planned for. The workload associated with routinely updating and maintaining antiquated, unpatched on-premises technology can

be lessened with cloud infrastructure. Thanks to its flexibility and scalability, businesses can quickly and simply add or remove resources as needed without altering the underlying technology. Moreover, monitoring your cloud can obtain faster disaster recovery choices and increased data security.

6. **Collaboration with FinTech Consultants-** Before making any technology investments, consulting professionals may help save money and guarantee that companies get the most out of their resources. Best practices, anticipated difficulties, and opportunity identification can all be aided by consulting. Alignment with business goals is ensured by consulting with technology specialists.

Although the Fourth Industrial Revolution (4IR) has been regarded as crucial for progress, there have been concerns regarding its ethical implications. For instance, artificial intelligence (AI) has progressively assumed a significant role in our daily existence, revolutionizing our work methods, communication, and cognition. The emergence of the Fourth Industrial Revolution (4IR) has propelled artificial intelligence (AI) to a prominent position as a highly disruptive technology in modern society. AI facilitates automation, machine learning, and advanced data analytics. Nevertheless, the ethical implications of AI have emerged as a crucial matter that necessitates attention if we are to guarantee a society that is equitable and impartial in the times to come (Mbiza, 2023). An urgent ethical issue related to AI is the possibility of bias and prejudice. The impartiality of machine learning algorithms is contingent upon the quality of the training data. According to Mbiza (2023), if the data exhibits biases or mirrors societal inequalities, the resultant algorithms will likewise demonstrate bias and prejudice. This issue is particularly problematic in domains such as recruitment and lending determinations, where artificial intelligence (AI) has the potential to sustain preexisting biases and prejudices in manners that are frequently challenging to identify or contest. An additional crucial ethical concern pertains to the capacity of AI to mechanize occupations and displace labourers. AI can generate novel employment opportunities and sectors, yet it also can disrupt established ones and result in the displacement of numerous workers. The phenomenon above can intensify preexisting economic disparities and foster social turmoil and instability (Mbiza, 2023).

Additionally, the matter of accountability and responsibility must be considered. With AI's increasing prevalence and potency, attributing accountability for its actions and consequences is becoming progressively challenging. This phenomenon prompts significant inquiries regarding the allocation of responsibility in instances where an AI system commits errors or inflicts harm, as well as how we can guarantee equitable distribution of the advantages and drawbacks associated with AI within society (Mbiza, 2023).

CONCLUSION

Providing Africa's youth with meaningful work opportunities is already a top development policy concern. Given the population bulge in Africa and the wave of young Africans who are anticipated to enter the workforce during the next 20 years, it will continue to be an issue. Emerging technologies in the Fourth Industrial Revolution (4IR) have been hailed by recent research as a game-changer that can hasten the economic transformation of developing nations. African governments are advised to equip their labour force with skills to organise and invest in this revolution.

Adopting 4IR technologies might result in significant economic growth and welfare benefits for sub-Saharan Africa. Yet, it could also cause social and economic disruption, asymmetry in opportunities, earnings, and incomes between workers with lower and higher levels of education, and an acceleration of trends in inequality. What countervailing measures should African policymakers take to ensure that all new workers have the fundamental skills and resources to earn a living wage while fostering the private investment necessary to create jobs utilising cutting-edge technology?

Furthermore, how probable is it that African producers will use the new technology? Due to high costs and the fact that many technologies fall short of addressing the specific challenges that African producers face in raising productivity and profitability, the continent of Africa has yet to embrace new, productive technologies quickly. Africa's digital innovation faces several obstacles. Realising the full benefits of digitalisation is hampered by several factors, including poor infrastructure, low human capital, low stakeholder engagement, and a lack of user education and training. Institutional, infrastructure, and sociocultural constraints restrict digital innovators' ability to participate in and impact the ecosystem of the circular plastic economy. Frontier technologies and innovations (FTI) hurt Africa because of the continent's poor adoption of technological innovations and favourable ICT policies, culture, ethics, and values. Digital businesses in Africa face additional difficulties due to the unfriendly business environment.

Nonetheless, Africa has welcomed the Fourth Industrial Revolution (4IR) and has many tech hubs that attract outside investment. Africa must prioritise education, capacity building, transformative leadership, increased telecommunications investments, the development of necessary infrastructure, and related industries to reap the full benefits of disruptive technologies. At the national and local levels, there is a strong correlation between the availability and utilisation of digital technologies and economic growth, innovation, job creation, and inclusion. However, Sub-Saharan Africa still has a long way to go in terms of digital development.

It has been suggested that the government should prioritise developing and enhancing the resilience of both public and private institutions. This can be achieved through investments in education related to the Fourth Industrial Revolution (4IR), improvements in infrastructure, and the establishment and enforcement of legal frameworks that safeguard the security and privacy of data.

Furthermore, governments must create a conducive environment that encourages the adoption of 4IR technology and fosters innovation and entrepreneurship. Examples of this include the reduction of red tape, the expediting of business registration, and the simplification of bureaucratic procedures. Formulating policies prioritising the growth and improvement of digital infrastructure, such as broadband connectivity and 5G networks, is crucial to achieving the Fourth Industrial Revolution (4IR). Authorities can establish explicit objectives and schedules for the advancement of infrastructure. Furthermore, governments should encourage public-private partnerships (PPPs) to stimulate private-sector investment in 4IR projects. By leveraging the resources and expertise of the corporate sector, these collaborations have the potential to propel public agendas forward. It is crucial to prioritise investing in science, technology, engineering, and mathematics (STEM) education from early childhood to tertiary levels to develop a workforce with the necessary skills for the Fourth Industrial Revolution (4IR). Governments must revise educational curricula to integrate digital literacy, coding, data analytics, and other skills associated with the Fourth Industrial Revolution (4IR) into formal educational systems. Implementing technical and vocational training programs can offer alternative avenues for individuals who may not have the opportunity to pursue conventional higher education to acquire skills relevant to the Fourth Industrial Revolution (4IR).

REFERENCES

African Bank. (2023, May). *Digital technologies key to inclusive growth in Africa - African Union Commissioner*. https://www.afdb.org/en/news-and-events/digital-technologies-key-inclusive-growth-africa-african-union-commissioner-60818

African Development Bank Group. (2023, May 3). *Digital technologies key to inclusive growth in Africa*. https://www.afdb.org/en/news-and-events/digital-technologies-key-inclusive-growth-africa-african-union-commissioner-60818

African Union. (2023, May 3). *Digital technologies key to inclusive growth in Africa - African Union Commissioner*. https://www.afdb.org/en/news-and-events/digital-technologies-key-inclusive-growth-africa-african-union-commissioner-60818

Ahmed, S. (2022, January 25). *Leveraging the potential of digitalisation, disruptive technologies, and cross-border data flows to facilitate aspirations of a Pan-African Digital Single Market.* https://researchictafrica.net/2022/01/25/leveraging-the-potential-of-digitalisation-disruptive-technologies-and-cross-border-data-flows-to-facilitate-aspirations-of-a-pan-african-digital-single-market/v

Akuamoah, K.D. (2017). *History of Technological innovations and the changing trends in African societies: Pottery as a case study*. Academic Press.

Allen, N. (2021, March 12). *The promises and perils of Africa's digital revolution.* https://www.brookings.edu/articles/the-promises-and-perils-of-africas-digital-revolution/

Amaury de Feydeau, A., Menski, M., & Perry, S. (2022, May 26). *Africa's digital infrastructure transformation.* https://www.whitecase.com/insight-our-thinking/africas-digital-infrastructure-transformation

Austen, R. A., & Headrick, D. (1983). The role of technology in the African past. *African Studies Review*, *26*(3-4), 163–184. doi:10.2307/524168

Baku, A. A. (2022). Digitalisation and new public management in Africa. *New Public Management in Africa: Contemporary Issues*, 299-316.

Berman, M. (2021, October 28). *What is Digital Technology? Programming Insider.* Programming Insider. https://programminginsider.com/what-is-digital-technology/

Congressional Research Service (CRS). (2023). *Report on, African Continental Free Trade Area (AfCFTA): overview and issues for congress.* CRS.

Department Of Telecommunications and Postal Services. (2018). *Invitation To Nominate Candidates for The Presidential Commission On Fourth Industrial Revolution*. https://www.gov.za/sites/default/files/gcis_document/201812/42078gen764.pdf

Development Bank of Southern Africa. (2024). *How digital transformation will secure a sustainable future for Africa.* https://www.dbsa.org/article/how-digital-transformation-will-secure-sustainable-future-africa#:~:text=This%20is%20because%20Africa%20is,a%20state%20of%20cataclysmic%20pause

Doloitte. (2021). *Future of Cyber Survey: Complexity is the new normal.* Accessed: 17 February 2024. https://www.deloitte.com/za/en/services/risk-advisory/services/2021-future-of-cyber-survey.html

Du Preez, J. (2019, June 24). *The 4th industrial revolution in Africa: The next great frontier.* https://www.inonafrica.com/2019/06/24/the-4th-industrial-revolution-in-africa-the-next-great-frontier/

Dwayne, S. J. (1996). The Impact of Colonialism on African Economic Development. *University of Tennessee Honors Thesis Projects*.

Enaifoghe, A. (2021). Digitalisation of African Economies in the Fourth Industrial Revolution: Opportunities for Growth and Industrialisation. *African Journal of Development Studies*, *11*(2), 31.

Fox, L., & Signe, L. (2022, September 23). *Inclusion, inequality, and the Fourth Industrial Revolution (4IR) in Africa*. https://www.brookings.edu/articles/inclusion-inequality-and-the-fourth-industrial-revolution-4ir-in-africa/#:~:text=Adoption%20of%20Fourth%2DIndustrial%2DRevolution,discussed%20in%20our%20recent%20report

Fukuyama, F. (2019). *Dimension of development. Global Empowerment Conference*. Harvard University. Available from https://www.youtube.com/watch?v=5R-dty5V9DM&t=458s

Ghanem, H. (2020). *Shooting for the moon: An agenda to bridge Africa's digital divide*. https://www.brookings.edu/articles/shooting-for-the-moon-an-agenda-to-bridge-africas-digital-divide/

Habibi, A., Sofyan, S., & Mukminin, A. (2023). Factors affecting digital technology access in vocational education. *Scientific Reports*, *13*(1), 5682. doi:10.1038/s41598-023-32755-6 PMID:37029180

IMF. (2021, March 1). *Africa Goes Digital – IMF F&D*. https://www.imf.org/en/Publications/fandd/issues/2021/03/africas-digital-future-after-COVID19-duarte

International Organization for Standardization, ISO Survey of Management System Standard Certifications. (2018). *International Telecommunications Union, Measuring the Information Society Report 2018* (Vol. 1). International Telecommunications Union.

Lashitew, A. (2022, November). *Digital technologies open vast business opportunities in Africa*. Academic Press.

Law, R. (1980). Wheeled transport in pre-colonial West Africa. *Africa*, *50*(3), 249–262. doi:10.2307/1159117

Lee, K. (2019). *The Art of Economic Catch-Up. Barriers, Detours and Leapfrogging in Innovation Systems*. Cambridge University Press. doi:10.1017/9781108588232

Lee, K., & Lim, C. (2001). Technological Regimes, Catching-up and Leapfrogging: Findings from the Korean Industries. *Research Policy*, *30*(3), 459–483. doi:10.1016/S0048-7333(00)00088-3

López Peláez, A., Erro-Garcés, A., & Gómez-Ciriano, E. J. (2020). Young people, social workers and social work education: The role of digital skills. *Social Work Education*, *39*(6), 825–842. doi:10.1080/02615479.2020.1795110

Madiba, R. (2023, July 19). *Connective collaboration key to the 4IR agenda in Africa*. https://www.bcx.co.za/insights/connective-collaboration-key-to-the-4ir-agenda-in-africa/

Manny, L., Duygan, M., Fischer, M., & Rieckermann, J. (2021). Barriers to the digital transformation of infrastructure sectors. *Policy Sciences*, *54*(4), 943–983. doi:10.1007/s11077-021-09438-y PMID:34751195

Markowitz, C. (2019). Harnessing the 4IR in SADC: Roles for Policymakers. *Occasional paper, 303*. https://saiia.org.za/research/harnessing-the-4ir-in-sadc-roles-for-policymakers/

Matyana, M., & Thusi, X. (2023). Unemployment and poverty in South Africa: Assessing the National Development Plan 2030 predictions. *International Journal of Development and Sustainability*, *12*(6), 212–226.

Mbiza, M. (2023, June 21). *The Ethics of AI in the Fourth Industrial Revolution: Ensuring Responsible and Fair Use*. https://www.linkedin.com/pulse/ethics-ai-fourth-industrial-revolution-ensuring-fair-use-mbiza/

Mensah, A. D. J. T., & Mensah, A. D. J. T. (2023, October 2). *Ten Facts About Digital Technology Adoption in Ghana*. World Bank Blogs. https://blogs.worldbank.org/africacan/ten-facts-about-digital-technology-adoption-ghana

Mises Institute. (2022, August 12.). *Yes, Precolonial Africa Had Technology and Economic Life before*. https://mises.org/wire/yes-precolonial-africa-had-technology-and-economic-life-colonialism

Morsy, H. (2020, August 18). *How Africa Can Harness the Fourth Industrial Revolution*. https://www.project-syndicate.org/commentary/how-africa-can-close-education-skills-gap-by-hanan-morsy-2020-08

Musoni, M. (2023, September 18). *Looking into the crystal ball: Artificial intelligence policy and regulation in Africa*. https://ecdpm.org/work/looking-crystal-ball-artificial-intelligence-policy-regulation-africa

Ndemo, B., & Weiss, T. (2017). Making sense of Africa's emerging digital transformation and its many futures. *Africa Journal of Management*, *3*(3-4), 328–347. doi:10.1080/23322373.2017.1400260

Njuguna, N., & Signe, L. (2020, January 8). *The Fourth Industrial Revolution and digitization will transform Africa into a global powerhouse.* https://www.brookings.edu/articles/the-fourth-industrial-revolution-and-digitization-will-transform-africa-into-a-global-powerhouse/

Oloidi, J. (2014). African traditional technology: The Ekiti of southwestern Nigeria experience. *International Journal of Social Science and Humanities Research, 2*(3), 310–315.

Oloruntade, G., & Omoniyi, F. (2023, May 26). *Where is Africa in the global conversation on regulating AI?* https://techcabal.com/2023/05/26/where-is-africa-in-the-global-conversation-on-regulating-ai/

Perez, C., & Soete, L. (1988). Catching-up in technology: entry barriers and windows of opportunity. In G. Dosi, C. Freeman, R. Nelson, G. Silverberg, & L. Soete (Eds.), *Technical Change and Economic Theory*. Francis Printer.

Perifanou, M., Economides, A. A., & Tzafilkou, K. (2021). *Teachers' digital skills readiness during COVID-19 pandemic*. Academic Press.

Rice, M. F. (2003). Information and communication technologies and the global digital divide: Technology transfer, development, and least developing countries. *Comparative Technology Transfer and Society, 1*(1), 72–88. doi:10.1353/ctt.2003.0009

Runde, D., & Bryja, B. (2023, November 7). *The Role of AGOA in Accelerating Africa's Digital Transformation.* https://www.csis.org/analysis/role-agoa-accelerating-africas-digital-transformation

Scholl, H. J., & Scholl, M. C. (2014). Smart governance: A roadmap for research and practice. *iConference 2014 Proceedings.* http://www.effra.eu/attachments/article/129/Factories%20of%20the%20Fut ure%202020%20Roadmap.pdf

Sebastian, I. M., Ross, J. W., Beath, C., Mocker, M., Moloney, K. G., & Fonstad, N. O. (2020). How big old companies navigate digital transformation. In *Strategic information management* (pp. 133–150). Routledge. doi:10.4324/9780429286797-6

Sen, A. (1999). *Development as freedom*. Oxford University Press.

Shizha, E. (2016). African indigenous perspectives on technology. In *African indigenous knowledge and the sciences* (pp. 47–62). Brill. doi:10.1007/978-94-6300-515-9_4

Shoba, M. (2022). The Phenomenon of China Shops in South Africa: Development, Entrepreneurship and Social Cohesion. *African Renaissance, 19*(1).

Siekmann, F., Schlör, H., & Venghaus, S. (2023). Linking sustainability and the Fourth Industrial Revolution: A monitoring framework accounting for technological development. *Energy, Sustainability and Society*, *13*(1), 26. doi:10.1186/s13705-023-00405-4

Signé, L. (2023, June 8). *What Does the Fourth Industrial Revolution Mean for Africa?* doi:10.1017/9781009200004.005

South Africa's White Paper. (2016). *National Integrated ICT Policy White Paper*. Available at https://www.gov.za/sites/default/files/gcis_document/201610/40325gon1212.pdf

The Confucian Weekly Bulletin. (2020, February 17). *Technological Developments in Pre-Colonial Africa*. https://confucianweeklybulletin.wordpress.com/2020/02/17/technological-developments-in-pre-colonial

The Department of Trade, Industry and Competition (DTIC). (n.d.). *A Guide to the African Continental Free Trade Area*. Available at http://www.thedtic.gov.za/wp-content/uploads/Guide-to-the-ACFTA_2023.pdf

The World Bank. (2013). *Information and Communication Technologies: Results Profile*. Available at https://www.worldbank.org/en/results/2013/04/13/ict-results-profile#

Thompson, M. (2023, May 31). *Impact of the Internet in Africa*. https://unconnected.org/blog/impact-of-the-internet-in-africa-2021#:~:text=Africa%20is%20the%20continent%20with,global%20average%20of%20nearly%2060%25

Tulinayo, F. P., Ssentume, P., & Najjuma, R. (2018). Digital technologies in resource constrained higher institutions of learning: A study on students' acceptance and usability. *International Journal of Educational Technology in Higher Education*, *15*(1), 1–19. doi:10.1186/s41239-018-0117-y

UNCTAD. (2020, September 7). *The Role of Digitalization in the Decade of Action for Africa*. https://unctad.org/news/role-digitalization-decade-action-africa

UNDP. (2022, June 13). *Digital Technologies Key to Unlocking Transformation in Africa*. https://www.undp.org/africa/news/digital-technologies-key-unlocking-transformation-africa

UNESCO. (2024, February 9). *Digital learning and transformation of education*. https://www.unesco.org/en/digital-education

Vogelsang, K., Liere-Netheler, K., Packmohr, S., & Hoppe, U. (2019). *Barriers to digital transformation in manufacturing: development of a research agenda.* Academic Press.

World Bank. (2023, April 25). *Accelerating the Use of Digital Technologies is Key to Creating Productive Jobs and Boosting Economic Growth in Africa.* https://www.worldbank.org/en/news/press-release/2023/03/13/accelerating-the-use-of-digital-technologies-is-key-to-creating-productive-jobs-and-boosting-economic-growth-in-africa

World Economic Forum. (2020, January). *Unlocking Technology for the Global Goals.* https://www.pwc.com/gx/en/services/sustainability/publications/accelerating-sustainable-development.html

Zhou, K., Liu, T., & Zhou, L. (2015, August). Industry 4.0: Towards future industrial opportunities and challenges. In *Fuzzy Systems and Knowledge Discovery (FSKD), 2015 12th International Conference on* (pp. 2147-2152). IEEE

Zissis, G. (2020). Power Distribution Systems: An Old Business With Modern Challenges [President's Message]. IEEE Industry Applications Magazine, 26(4), 5-6.

Chapter 10
Africa's Contributions to Digital Technology:
A Case of M-Pesa Technology in Kenya

Patrick Makau Musembi
https://orcid.org/0009-0005-4003-7216
Daystar University, Kenya

ABSTRACT

Historical literature shows that Africa is the cradle of mankind and has the world's oldest record of science and human technological achievements. The oldest stone tools in the world have been recorded in Eastern Africa, and later evidence for tool of technological production by humans' hominin ancestors was recorded in West, Central, Eastern, and Southern Africa. For example, some of the earliest inventions and African technology include hand axes, fishing hooks, bows and arrows, and even boats. Many of the African inventions in history pre-date the use of the written word by tens of thousands of years. These earliest technological advancements became the basis and foundations of the digital technologies the world experiences today.

INTRODUCTION

History literature shows that Africa is the cradle of mankind and has the world's oldest record of science and human technological achievements. The oldest stone tools in the world have recorded in eastern Africa, and later evidence for tool of technological production by humans' hominin ancestors recorded in West, Central, Eastern and Southern Africa. for example, some of the earliest inventions and African technology include hand axes, fishing hooks, bows, and arrows, and even

DOI: 10.4018/978-1-6684-7851-6.ch010

Copyright © 2024, IGI Global. Copying or distributing in print or electronic forms without written permission of IGI Global is prohibited.

boats. Many of the African inventions in history pre-date the use of the written word by tens of thousands of years. These earliest technologies advancements became the basis and foundations of the digital technologies the world experiences today. Digital technology refers to tools, systems and devices that can generate, create, store or process data (Edwards, 2014). The data processing and logic capabilities of digital technologies are enabled through microprocesses that are programmed to perform various functions (Edwards, 2015). Digital technologies refer to devices such as personal computers and tablets, tools such as cameras, calculators and digital toys, systems such as software and apps, augmented and virtual reality, and less tangible forms of technology such as the Internet (Danby, Fleer, Davidson & Hatzigianni, 2018).

Despite the effects of trans-Atlantic slave trade, colonialism and coloniality forces in modern era to have demolished and prevented the growth of indigenous knowledge and technology and hindered that Africa's contributions to the wave of digital technology, the mobile money (pesa) (M-PESA) is successful digital technology created by Africans in the post-colonial and post-cold War states in Africa. Digital money circulation in the form of Mobile Money (MM) is a mobile phone-based electronic funds storage and transfer service. The whole development was enabled by the process of changing cash into electronic units of money, e-money, and store in the mobile phone. A user can deposit, store, transfer and withdraw funds from their mobile money account, much like a bank deposit. The firm providing the service is normally a mobile network operator or another type of non-bank, called a MM firm in this chapter. Users can deposit and withdraw funds through cash merchants or agents. These are corner stores, petrol stations, and other outlets operating on behalf of MM firms.

The aim of this chapter is to explain the current Africa's contributions to the digital technology by assessing the impacts of M-PESA digital technology on economic development and financial inclusion. The chapter is divided into five sections. While the first section explains the origins of M-PESA, the second section discusses the structure of M-PESA and assesses various ways Kenyans are using M-PESA to realise their basic and essential needs. The third section explains the socioeconomic effects of M-PESA on the quality of life in Kenyan. The fourth part examine M-PESA 's role as a money transfer service and assesses unpleasant effects of using M-PESA on the Kenyan society. The fifth section explains the role of M-PESA in financial inclusion and sustainable development. The fifth section provides concluding remarks of the chapter.

FACTORS INFLUENCING THE ORIGINS OF M-PESA IN KENYA

Kenya was the first developing nation to launch a mobile-banking service based on text messaging (Shaikh, & Karjaluoto, 2014). M-PESA started in 2007 as an Electronic Money (EM) transfer product that enables users to store value on their mobile phones. It is based on a platform of electronic units of money that can be used for multiple purposes including transfers to other users, payments for goods and services, and conversion to and from cash. The lack of financial inclusion in Kenya is fuelled by an array of adverse policies and conditions. Many Kenyans live in isolated, rural areas, locations where banks see traditional establishments as extremely unprofitable (Mbiti and Weil, 2011). Moreover, Kenyan culture embraces strong familial connections and a larger sense of community. This leads to the belief in informal lending practices, fostered by the trust individuals have with their relations. Transferring money between friends, family, and the larger community can be expensive, time consuming, and often dangerous. Furthermore, even those with access to formal financial institutions often do not open accounts due to their lack of trust that decades of questionable banking practices have created (Mbiti and Weil, 2011).

These factors created a fissure between the ease and comfort associated with informal financing and the security and reliability that came with the use of formal financial institutions. The executives at Vodafone and Safaricom recognized that a mobile financial service could address many of the problems present in Kenya's financial sector. Unsurprisingly, this breakthrough came from the mobile phone industry, an industry built on connecting people through new ideas and technology (Mbiti and Weil, 2011). M-PESA, sought to address the gap that separated the unbanked from the advantages of financial inclusion (Hughes and Lonie, 2007). Although this was a novel idea, in no way did it seek to overhaul the existing payment structure. Jack and Suri (2011) have instead posited that "M-PESA is not designed to replace all payment mechanisms but has found and filled a niche in the market in which it provides significantly enhanced financial services".

M-PESA was originally branded as an alternative remittance system. Kenyan cities are full of individuals moving in from small towns, looking to make money and send it back to assist their families in their hometowns. Although these individuals may have left their ancestral home, they are expected to maintain a strong relationship and still contribute to their extended families, leading to approximately 21% of Kenyan adults relying on a form of money transfers for survival (Johnson et al., 2012). At its core MPESA's goal, and corporate slogan, is to allow individuals to "send money home" (Mas and Ng'weno, 2010). In March 2007, when M-PESA was officially launched, there were only "poor alternatives for making domestic money transfers, particularly in the absence of technology-enabled or retail-based alternatives with a broad network" (Mas and Morawczynski, 2009). At that time, there were two options to sending remittance payments: using a secure, but very expensive formal service or delivering the money in person, a low-cost, but risky method. M-PESA 's service was marketed as the best of both worlds. MPESA promotions claimed that with the punching of a few keys, one could now easily transfer monthly remittances from the comfort of their couch. This service had bridged the gap between the two alternative forms, providing the benefits of each.

Bank fees were also unreasonably high by any standard. In 2012, the "withdrawal fee of Kshs 30 (around $.40) from an ATM represented the price of a kilo of unmilled maize which would feed three people for one meal" (Johnson et al., 2012). Yet, the alternatives weren't much better. If one saves their excess money in their house, they run a substantial risk of loss or theft. There is an estimated $3.4 billion being stored by the unbanked, "stuffed in jars or mattresses" (Economist, March 30, 2013). Customers were looking for a way to have liquidity available, but not pay the high fees associated with the traditional financial sector or risk keeping their savings in cash and making transfers through often unscrupulous agents. Like in the remittance industry, M-PESA sought to provide a middle ground, essentially to secure, short term liquidity at a reasonable price. This solution was not a process intended to eliminate cash altogether. In fact, "Safaricom has not set out to replace cash in day-to-day life, they simply offered a new solution" (Mas and Ng'weno, 2010).

M-PESA is a mobile money transaction service that allows users to deposit, withdraw, transfer money, pay bills, and purchase goods and services and access credit and savings all using a mobile devise (Wikipedia, 2023). Developed by Safaricom, a Kenyan telecommunications company, M-PESA was launched in 2007 and has since become one of the most successful mobile money transfer technologies in the world (Mbiti & Weil, 2014). M-PESA in Kenya stands out as a key so-called success story of mobile money. While limited examples of mobile money existed in other countries, M-PESA was the world's first major mobile money service. Launched in 2007, there are now 30,530,500 active registered mobile money subscriptions in Kenya of which M-PESA subscribers are 30.2 million while Airtel money and T-Kash are 0.344 million (Communications Authority of Kenya - CAK, Q4 Statistical Report 2020) as of June 2020. The evolution that is critical to its success is associated with the development of a retail electronic payments platform that was real time and that has become a game changer in digital financial services ecosystem.

M-PESA TECHNOLOGY IN THE WORLD OF DIGITAL TECHNOLOGY

M-PESA is Africa's leading international digital international money transfer and transactions. M-PESA global is a service that enables M-PESA registered customers to send and receive money globally. M-PESA customers will be able to send funds; To East Africa: Rwanda, Tanzania & Uganda. M-Pea is a global digital technology recognized and used by different people in transacting and transferring money electronically. The technology has millions of Bank accounts and is overtalking Western Union in Africa which has over 500,000 Western Union locations globally (Donovan, 2012). M-PESA had started as a pilot project of the multinational United Kingdom (UK) mobile operator, the Vodafone Group (di Castri, 2013:6-7). The technology received huge support from a the United Kingdom (UK) Department of International Development (DFID) grant, the Vodafone M-PESA team had chosen Kenya as the country in which to pilot a microfinance institution (MFI)-based loan disbursal and repayment system (di Castri, 2013:6-7).

At a global market, M-PESA has become an international electronic market as it expanded its money transfer and transactions to the global world. The company's major stakeholder, Vodafone, has recognized this as M-PESA is no longer restricted to the Kenyans only (Ntara, 2015). It has been extended beyond Kenya's borders and onto the international scene. M-PESA activities have been launched and made operational in several countries outside Africa, such as Afghanistan, Fiji, and India. Additionally, Safaricom's partnership with Western Union permits M-PESA customers to receive money from 45 nations outside of African continent, including the UK, United States (USA), Italy and Canada (CISCO, 2013; Ntara, 2015).

M-PESA technology has permeated the entire continent. The technology is used in North Africa, West Africa, Central Africa, Horn of Africa, and Southern Africa. Millions of poor people are using basic mobile phones to transfer money; pay for goods; and access sophisticated financial services, such as credit, insurance, and savings accounts (Donovan, 2012). It is a money transfer system operated by Safaricom, Kenya's largest cellular phone provider, and allows users to exchange cash for electronic float e-float on their phones, to send e-float to other cellular phone users, and to exchange e-float back into cash. In a financialized economy, capital markets aiming for growth require innovation (Ismail Ertürk, Julie Froud, Sukhdev Johal, Adam Leaver, and Karel Williams 2013; Ertürk 2020). These innovations depend on the digital technology. Hence the narrative of the hero entrepreneur as the key agent for technological change and economic development (Mariana Mazzucato, 2014; Deirdre McCloskey and Alberto Mingardi 2020).

M-PESA was first developed as bank product in partnership between Safaricom, a telecommunication (telco) company and the Commercial Bank of Africa (CBA).

In subsequent years, it has further evolved into a platform for a wide range of financial services such as virtual savings accounts in commercial banks. Harnessing the power of big data, M-PESA now also operates as a channel of credit supply by commercial banks, microfinance institutions and cooperatives (SACCOs). In addition to domestic financial services, M-PESA allows users to send and receive cross border remittances using their mobile phones. M-PESA and similar digital financial services represent a significant improvement in the national payments technology, reducing transaction costs and lowering the barriers to entry into the formal financial system (Ntara, 2015). Financial institutions in Kenya have embraced M-PESA as a platform to manage micro accounts, build customer deposits, and broaden their customer network. Consequently, Kenya has emerged as a leader in financial inclusion in Sub-Saharan Africa. In 2016 over 75% of the adult population in Kenya. This is particularly true in Kenya, where M-PESA has had a significant impact on the country's economy and society as will be discussed below.

The significance of M-PESA is underscored by its geographical spread and the value of monetary transactions moved through M-PESA. According to Safaricom (2003) the service has spread to cover a total of 12 countries including Kenya, Tanzania, Uganda, Rwanda, South Africa, Afghanistan, Lesotho, DRC, Ghana, Mozambique, Egypt, and Ethiopia with a total of 52.4 million active users in 2022 from an initial 20, 000 customers in 2007. The Central Bank of Kenya (CBK) estimated the value of mobile money transactions at 56.8 percent of Kenya's Gross Domestic Product (GDP) in 2021 an increase from 48.7 percent in 2020 and the value is projected to reach 68 percent of Kenya's GDP by the end of 2023 (Business Daily, 2023; Statista, 2023).

One of the success factors of M-PESA is the fact that the product concept is very simple. An M-PESA customer can use his or her mobile phone to move money quickly, securely, and across great distances, directly to another mobile phone user (Wikipedia, 2023). The customer does not need to have a bank account. The customer is simply required to register an M-PESA account with Safaricom using their national identity and nothing more. Customers turn cash into e-money at Safaricom dealers who are spread across the country, and then follow simple instructions on their phones to make payments through their M-PESA accounts. The system provides money transfers as banks do in the developed world (Safaricom, 2023). The account is very secure, PIN protected, and supported with a 24/7 service provided by Safaricom and Vodafone Group (Hughes & Lonie, 2007).

Another reason for the success of M-PESA in Kenya is its ability to reach people who were previously unbanked. Prior to the introduction of M-PESA, many Kenyans did not have access to formal banking services due to a lack of infrastructure and the stringent requirements for opening and maintaining bank accounts (Mbiti & Weil, 2014). A survey conducted in early 2007 revealed that 38 per cent of Kenyans had no access to formal financial services, according to a national survey, and only 19% had access to formally regulated financial institutions such as banks with only 400 branches and slightly over 600 ATMs in Kenya at the time (Munga, 2007). From the early 2000s banks required customers to maintain minimum balances ranging from Kenya shillings 5000 to Kenya shillings 50, 000 in their bank accounts in addition to providing proof of permanent residential address. This was in response to the threat of non-performing loans (Waweru & Kalani, 2009). Such stringent conditions cut off many Kenyans from accessing banking services leading to the emergence and strengthening of existing local banks. These banks, however, continued the same trend of requiring stringent requirements creating a gap in the banking sector. The success of M-PESA was a clear sign that M-PESA filled a gap in the market (Suri, T & Jack, W. 2016). M-PESA allows users to transfer money using their mobile phone, which is a device that many Kenyans own. This has made it possible for many Kenyans to participate in the formal financial system for the first time.

THE WORKINGS OF M-PESA

There are three basic transactions that customers conduct with M-PESA. First of all, a customer may deposit money at an M-PESA outlet in return for e-float (called a "cashin" transaction) (Mbiti & Weil, 2011). The customer is required to show a valid identification document, and his identity and the amount of the deposit are logged in a book kept at the outlet (Mbiti & Weil, 2011). Upon receipt of the money, the M-PESA agent enters the customer's telephone number and deposit information into his/her cell phone, and the customer waits at the outlet window until he/she receives a confirmation text message that e-float has been deposited (Mbiti & Weil, 2011). Unless the system is running slowly (which happens occasionally), the whole transaction takes about a minute or less.

A customer may exchange e-float for cash at an M-PESA outlet (called a "cash out" transaction) (Mbiti & Weil, 2011). Again, the customer must show a valid identification document, and the transaction is logged. The customer tells the shop clerk how much cash he/she wants, then chooses "withdraw cash" on the M-PESA menu on his phone, enters the amount to be withdrawn (plus the relevant fee), and enters the agent number (Mbiti & Weil, 2011). The agent then receives a text indicating that the transaction is complete, and the agent then gives the appropriate amount of cash to the customer. This whole transaction takes about one minute.

A user may transfer e-float from their phone to another phone. This chapter refers to such a transfer as a digital technological transfer of a "person-to-person transfer." Moreover, during the transactions, one or both of the parties may be either a local or an international institution or firm (Mbiti & Weil, 2011). The user enters the phone number of the recipient and the amount to be transferred on their Cellphone. The sender and recipient each receive a text message stating that money has been transferred (Mbiti & Weil, 2011). These three basic transactions can be combined in several ways. For instance, a user may deposit cash and send the full amount deposited to another user, who can then withdraw the full amount transferred. The chapter thus refers to this use as "digital technology deposit-transfer-withdraw." Instead, a user who receives a transfer from one person may transfer the e-float to some other user instead of withdrawing cash (Suri & Jack, 2011). E-float could circulate in this manner indefinitely, like conventional cash (Mbiti & Weil, 2011). A third usage possibility is where a user deposits cash and then later withdraws it themselves without having transferred it (Suri & Jack, 2011). It has been argued that people use this digital technology of transferring and transacting money for safety for example when they are traveling (Vaughan, 2007; Morawczynski, 2009).

Moreover, the application M-PESA technology patterns described above can be mixed in varying ways. A user may receive a transfer and withdraw some of the value while transferring some of the remaining amount elsewhere and leaving some e-float in his account for future transactions (Mbiti & Weil, 2011). This is a pattern in which a user might receive a transfer and not withdraw it right away for several reasons. The first reason is to economize on transaction fees, the second reason to economize on the effort of going to an M-PESA outlet, and third to benefit from the safety of storing value on a mobile phone rather than in cash or in the banks. M-PESA is safer than cash because a PIN is required to perform any transaction (Mbiti & Weil, 2011). If a phone is stolen or lost, the M-PESA funds are safe unless the PIN has been compromised. If the PIN is compromised and funds are transfered to another account, the legitimate account holder can recover his/her funds if they have not been withdrawn by the fraudulent recipient by initiating a transfer reversal through the customer service department (Mbiti & Weil, 2011).

In its workings, M-PESA technology is simple and easy to use by everyone regardless their education or technological awareness. The simplicity of the M-PESA user interface complements the service's value proposition and operates on an application on the user's mobile phone (Ntara, 2015). People using the application often find the M-PESA services by launching it right from the main menu of their phones where it resides. The technology is so simple such that people using it do not need to download it from the network every time it is required. Ntara (2015) argued that these people also are guided by several simple prompts regarding the number of the destination phone, amount of money to be transferred and sender's personal identification number. Upon the reception of a phone text message that confirms electronic transactions, the information is sent to the M-PESA server using a single text message (Safaricom, 2015). This therefore reduces messaging costs and the risk of interruption of the transaction request (Safaricom, 2015). M-PESA is primarily a simple money transfer service and other uses of M-PESA suggest other functions (Mbiti & Weil, 2011). To the extent that electronic money (e-money) circulates among several users between an initial cash-in transaction and a final cash-out transaction, it can be seen as an evolving alternative to currency. Similarly, to the extent that people hold e-float balances on their phones for significant periods of time, M- Pesa can be seen as having aspects of banking (Ntara, 2015).

All M-PESA e-float is backed 100% by deposits held at three commercial banks in Kenya (Mbiti & Weil, 2011). Interest earned on these deposits is donated to a charity, which allows Safaricom to avoid being regulated as a bank. An extensive description of the arrangements between Safaricom and the network of agents who service M-PESA users can be found in Eijkman, Kendall, and Mas (2010) and Suri and Jack (2011). In Kenya, M-PESA has several uses. These uses include storing, transferring and withdrawing money electronically, topping up airtime, buying goods and services, offering savings and loan products (Safaricom, 2015). They also include paying school fees, utility bills and international remittances (Safaricom, 2015). Companies and government institutions have also used M-PESA to disburse employee salaries and collect bill payments (Safaricom, 2015).

HOW KENYANS ARE USING M-PESA TO IMPROVE THEIR LIVES

M-PESA has transformed the lives of millions of Kenyans by providing them with access to financial services, even in areas with limited banking infrastructure. The model which started as a means of sending money between friends and relatives has since evolved to include various forms of payment and transactions. M-PESA provides safe, secure, domestic, and international transfers of money, top-up airtime, bill payments, salary deliveries, short-term loans and much more (Vodafone, 2023). M-PESA customers can send money in real time to any other M-PESA customer with an account registered in the same country as well as M-PESA Global where customers can receive and send money across borders in real time (Vodafone, 2023). M-PESA has continued to revolutionize the retail business sector and other sectors of the Kenyan economy.

M-PESA is pioneering a cashless economy in Kenya. For example, the service is used in church and in public transport. It is used in the payment of college and school fees, healthcare services, micro-saving, and access to credit (Business Daily, 2023). M-PESA services are performed in church for worship when congregants pay tithes and offerings and any other form of transactions such fundraisings, purchasing goods and services from different service providers. For example, in the church matters, during offering time, the church media team displays the church Paybill number on a big screen with account details for members to give their donations (Business Daily, 2023). The use of M-PESA has lessened the security challenges faced by mega city churches that used to collect huge sums of money in cash on Sundays when banks are usually closed (Business Daily, 2023). Churches are no longer worrying about hiring private security firms to guard Sunday collections. The use of M-PESA facilitates donations to the church at any time of the week from the comfort of their homes or offices (Business Daily, 2023).

Correspondingly, the use of the M-PESA cashless model is preferred in public service vehicles where customers do not have to worry about being robbed of their cash money at night. In the transportations industry, the passengers have reverted to the M-PESA services as they are secure and safe to use. M-PESA services are electronic and cashless model and have created sanity in the transport industry. Moreover, even those willing to invest in the transport have resulted to the use of M-PESA services because they do not require crew to handle cash money (Business Daily, 2023). The use of M-PESA services has to some extend reduced the endemic corruption in the traffic department of the police which has consistently been highlighted by various reports. According to the 2018, 2020, and 2021 graft survey reports released by the Ethics and Anti-Corruption Commission (EACC), the Kenya police and particularly the traffic department was perceived as the most corrupt public institution (EACC corruption report, 2018, 2020, 2021). The gravity of corruption in the traffic police department was highlighted by the case of a traffic police officer in 2021 who agreed to forfeit Sh26 million linked to bribes received from transport industry operators on Garissa-Thika highway to the State (Business Daily, 2021).

The use of M-PESA has been linked to commercial banks has reshaped the way Kenyans send, receive, and store money. Cook and McKay (2017) while discussing the strategies banks have adopted to respond to mobile banking argue that bank accounts in Kenya today outnumber mobile money accounts by more than 30 percent as most banks have collaborated with M-PESA . As a result, Kenyans can send and receive money directly from their bank accounts. M-PESA electronic transactions and transfers have made it easier for people to send money to their loved ones, pay bills and buy goods and services. Some of other M-PESA services performed by locals via electronic transactions are electricity, water, and internet bills. M-PESA services have made people or users not to travel to physical payment points. These services also often have included payment of school fees whereby parents pay the school fees for their children from the comfort of their homes without traveling or make long queues to the banks. Thus, M-PESA electronic services save time and money that would otherwise be spent on travel and queuing.

The governmental institutions have not been left behind in making use of MPESA. In 2013 the Kenya Revenue Authority (KRA) in a bid to improve service delivery and expand the scope of payment channels available to taxpayers launched the

Lipa Ushuru na M-PESA (Pay your taxes via M-PESA) to allow taxpayers to pay taxes swiftly and conveniently using their mobile phones (KRA, 2023). In the past, taxpayers had to physically visit banks or KRA offices to make tax payments. Since the launch of this service an average tax amounting to USD 1.1 million is collected via mobile money monthly and there is an upward growth in the amount of tax paid via mobile money (OECD, 2022). According to available data from KRA the financial year 2019/2020 saw a 95.2 percent increase in mobile money payments from USD 20.9 million collected in 2018/2019 financial year to USD 41.9 million collected via MPESA (OCED, 2022). To appeal to many Kenyans to use MPESA, the taxman has made the service simple and easy to follow and provided easy steps to pay taxes on the KRA website (KRA, 2023).

Furthermore, M-PESA has enhanced the digitization of government services through the Electronic Citizen (eCitizen) digital payments platform that enables Kenyan citizens, residents, and visitors to access and pay for government services electronically. The eCitizen platform covers services such as applications and payments for electronic passport (e-passport), temporary permit and electronic vis (e-visa), registration of marriages, business names and companies, driving licenses, search for a Title deeds, land rates, and the repayment of student loans advanced by the higher education loans board among others (eCitizen, 2023). Payment of such services via M-PESA has eliminated cartels, middlemen, and unnecessary delays. An applicant for a service pays for the required service from the comfort of their home or office and acquires a reference number through which one can track progress of applied service (eCitezen, 2023). As a result, M-PESA has eliminated the handles eased the struggle of setting up companies in Kenya. For example, is establishing a trading company, one is required to apply for a name search, and pay for the service via M-PESA . This process takes about three days after which one can lodge to register the proposed company and in less than a week a company can be registered and ready to do business without the applicant having to step foot in the physical office of the registrar of businesses and companies.

Another use of M-PESA is the ability for Kenyans to access credit. M-PESA has enabled Kenyans to access credit to start businesses or to expand existing businesses. The service has made it possible for lenders to assess creditworthiness based on M-PESA transaction history, which has increased access to credit for Kenyans who would otherwise have been excluded from traditional banking services. For example, in 2012 Safaricom in partnership with the Commercial Bank of Africa (CBA) of Kenya launched M-Shwari Kenya. M-Shwari is a savings and loan service which enables M-PESA customers to instantly access loans and contribute to savings directly from their phone (Safaricom, 2023). M-Shwari allows subscribers to save as little as Ksh. 1, and access credit from Ksh. 1,000. In addition, one can earn interest of up to 6.3% per annum on savings balance, save for a fixed period and access loans instantly, credited to M-PESA account from a minimum of Kshs. 1,000 and up to Kshs. 1million account charged at 9% (Safaricom, 2023). In addition, a customer can access *Fuliza* service, a continuous overdraft service that allows Safaricom's M-PESA customers to complete their M-PESA transactions even when they do not have enough funds in their M-PESA account (Safaricom, 2023). Safaricom sets the maximum credit limit on one's Fuliza account based on the customer's saving habits.

Another use of M-PESA is the *Changa na* M-PESA Service which is a fundraising platform for crowdfunding patterned after the popular Kenyan harambee spirit in which friends come together to pool resources for a given cause. This is like, "GoFundMe" and can be used for such causes as raising school fees, medical bills, funeral support, emergency funds, and other forms of charity among others. Customers can make E-Money payments to the client for a specific period and for specific purposes. All you need to do is consistently share the fundraiser with your friends, family, and online community. In no time, support will start pouring in (M-Changa Africa, 2023). Closely related to M-Changa is the next of kin claim payment of beneficiary dues. Next of kin is a term used to describe a person's closest living blood relative whom Safaricom may contact in case of any eventuality. This is the legal representative who is authorized to deal with the estate of a deceased person whether through appointment under a will or by right and thereafter application to a court as per the law. Safaricom requires that the claimant should visit any of the retail centers and provide certain documents for inspection and approval before any funds can be released (Safaricom, 2023).

SOCIO-ECONOMIC AND CULTURAL EFFECTS OF USING M-PESA SERVICES

There is a growing body of literature examining the effects of mobile money on welfare outcomes in the global south (Van Hove & Dubus, 2019; Suri & Jack, 2016; Mbiti & Weil, 2015). It has been indicated that mobile electronic banking holds the prospect of offering a low cost, accessible transaction banking platform for currently unbanked and the poor; it also has the potential to expand financial services to those who have been previously systematically excluded (Hughes & Lonie, 2007). The benefits of mobile banking in terms of economic development include poverty reduction, power to access loans and insurance, the ability to save for a future productive investment and enabling the poor to take control of their own livelihoods (Chib, Leon & Rahim, 2015). In Kenya research suggests that M-PESA has had significant social, economic, and cultural effects on the lives of Kenyans. Kenyans have used M-PESA in a variety of ways to enhance the quality of their lives including financial inclusion, improving access to healthcare and education, supporting the growth of small businesses, boosting women's economic opportunities, reducing poverty, and giving rural women the power to take charge of their lives (Chib, Leon & Rahim, 2015). Since its inception M-PESA has been a strong driver of socioeconomic and political growth in Kenya; it has empowered more Kenyans to be in control over their lives including health, education, economic and political situations, and self-respect, dignity, and community engagement.

M-PESA has increased access to healthcare in rural communities in Kenya. A study conducted by the center for Individuals with Disabilities Education Act (IDEA) and Research for Inclusive Settings (IRSIS) of Vanderbilt University found that the per capita expenditure on health in Kenya was $68 of which 46 percent was borne by households (Haas & Nagarajan, 2011). The study also established that 44 percent and 18 percent of the sick did not seek healthcare due to high costs and long distance to the nearest health facility respectively (Haas & Nagarajan, 2011). Lack of adequate medical care begins from birth for many Kenyans. For example, in 2009, only 44 percent of births were attended by a skilled health professional compared to 70 percent in 2016 (World Bank, 2023; WHO 2009). With little access to funds to afford preventative care and high incidence of diseases, many Kenyans find themselves in a situation where emergency medical care is necessary but not unaffordable (Haas & Nagarajan, 2011).

However, M-PESA has enabled users to share risk by facilitating the prompt mobilisation of funds from friends and relatives who live miles away to pay for medical emergencies. A study conducted among low-income people in Kenya which analysed 18, 000 M-PESA transactions of 92 low-income people in 2011 found out that 35 per cent of hospital bills were paid using M-PESA remittance pooled from family and friends (Mbiti & Weil, 2011). This means that M-PESA is providing users with prompt access to cash that they can use to pay for needed healthcare. Moreover, Kenya's social health insurance, the National Hospital Insurance Fund (NHIF), experienced a 500 per cent increase in voluntary payment subscribers between 2009 and 2017 after the organisation started receiving payments via M-PESA (Colbourn, Seal & Obadha, 2018). M-PESA made payment of insurance premiums easier, quicker, and safer through Paybill and *Lipa na* M-PESA services. It has also been argued that such payments likely reduced workload on the NHIF staff who could in turn focus on processing new registrations (Nyaronga, 2015).

The M-PESA technology has improved savings practices among Kenyans as it reduces transaction costs, provides a safe and trusted place to store money, and facilitates transfers to and from other savings vehicles such as mobile savings and credit accounts, bank accounts, and informal financial groups. Colbourn, Seal and Obadha (2018) found out that three per cent of mobile money users saved money in *M-shwari* for medical emergencies while 33 per cent saved to increase the amount of credit they could qualify for in future. This future credit could be used for anything including paying for medical expenses when needed, home improvement or purchase of livestock such as goats or a cow. The survey also showed that eight per cent of *M-shwari* users who took loans to pay for medical expenses with the average loan amount being 1,280 Kenyan shillings (almost US$13) (Obadha, Seal, & Colbourn, 2018). Therefore, products such as KCB M-PESA and *M-shwari* are shifting borrowing from social networks because of the ease of how one can access credit faster without having to negotiate and convince social networks who might not come through in times of medical emergency.

It has been argued that economic growth is the most powerful instrument for reducing poverty and improving the quality of life in developing countries particularly through financial inclusion of the poor and marginalized groups (Moin, 2009). M-PESA has enabled the poor and other vulnerable groups including women to participate fully in the opportunities unleashed by mobile money. Suri and Jack (2016) estimated that access to M-PESA increased per capita consumption levels and lifted 194,000 households, translating to 2 percent of Kenyan households out of poverty. They concluded that impact of M-PESA on poverty was more pronounced in female-headed households because of changes in financial behavior including increased financial resilience and saving, occupational choice, especially for women, who moved out of agriculture and into business (Suri and Jack, 2016). The reality is that women simply transitioned their agricultural business to M-PESA (Suri and Jack, 2016).

However, financial inclusion goes beyond access to and usage of appropriate, and affordable financial services (Van Hove, & Dubus, 2019). Triki & Faye (2013) for example, argued that financial inclusion involves access and availability of formal, regulated financial services, the physical proximity, actual usage of financial services and products including frequency and duration of time used, and whether the products are well tailored to suit the client needs. This understanding of financial inclusion implies that what may have been considered by scholars such as Suri & Jack may not have necessarily been financial inclusion. First, the explosion of digital lending occasioned by M-PESA has opened the flood gates for expensive and unregulated digital lending applications. As a result, Kenyans unable to access bank loans are being exploited, harassed and 'debt-shamed' by unlicensed online lenders if they default (Kimeu, 2022). In October 2021 Kenyan authorities introduced new laws to regulate digital lenders. However, by September 2022, only 10 out of 288 operating digital lenders qualified for a license from the Central Bank of Kenya (CBK) (Kimeu, 2022).

M-PESA has empowered more Kenyans to be in control over their lives including health, economic and political situations, and self-respect, dignity, and community engagement. New innovations continue to be developed out of M-PESA (Kimeu, 2022). They include M-Tiba, a mobile health wallet, enables people to save money for medical treatment and benefit from health insurance which now has 2.8 million users. M-Tiba enables people to control their medical benefits directly from their phone via a PIN (*M-tiba*, 2023). They can initiate medical treatments for themselves and their loved ones from their comfort homes through their mobile phones (*M-tiba*, 2023). This therefore means that M-Tiba's mobile technology connects everyone in the healthcare space and makes it more affordable for millions of Kenyans. This is because M-Tiba gives loans to health practitioners to finance needs of healthcare facilities such as purchasing ambulance, ultrasound, power backup in collaboration with Safaricom, major commercial banks, and insurance companies (*M-tiba*, 2023).

M-PESA 'S ROLE AS A DIGITAL FINANCIAL MARKETS

M-PESA has become the benchmark for successful mobile money launches and operations. New mobile financial service operators seek to emulate the success and market penetration Safaricom has been able to achieve. M-PESA has been inextricably woven into the daily life of Kenyans, rich and poor, rural and urban. It would be hard to find a person in Kenya who was unaware of M-PESA . According to di Castri (2013), over 18 million Kenyans use the mobile service, with an 86% penetration rate amongst families. Between the years 2008 and 2011, M-PESA grew at 88% annually (Deb and Kubzansky, 2012.). Now, upwards of two-thirds of the adult population transfers a proposed $1.6 billion a month on the mobile platform (di Castri, 2013).

The transactions can even amount to "as much as 60% of the country's GDP" (The Economist Intelligence Unit, May 13, 2014). Never before has there been such a take up of mobile financial technology. Jack and Suri's survey (2011) revealed that only 2% of the population in Kenya claims that the closing of M-PESA would have no effect on them at all. M-PESA is also maintaining growth at an astonishing rate, with competitors having a negligible effect on their vast market penetration. M-PESA has leveraged its status as a mobile phone operating network to create a profitable mobile financial system by fundamentally altering the credit system in place. The spreading of "informal credit" has created financial benefits to those involved (most notably in cost reduction), has introduced a wide range of benefits and diminished much of the risk present that is present in an unbanked society.

Nick Hughes, one of the men behind the idea and creation of M-PESA, attributes the rise to two factors, "targeting the unbanked" and learning "to keep it simple" (Hughes and Lonie, 2007). Before M-PESA, formal financial institutions struggled with find appropriate means to connect with the general population in developing countries. Hughes and Lonie (2007) attributes this to every previous financial service platform being constructed for and by western banks, something M-PESA broke away from. Instead, M-PESA turned towards its potential customers and addressed their wants and needs. Jack and Suri (2011) describe the result as the facilitation of "the safe storage and transfer of money". The trade unlocked a new medium for transfers and created an easier way to receive credit.

In addition, it also improved and accelerated avenues for trade, created a higher demand for saving, as well as allowed risk to be spread out amongst a larger network (Jack and Suri, 2011). The transference of money has never been easier in the East African nation. In 2010, close to 50% of adopters of M-PESA claimed they saved upwards of three hours and $3 by using M-PESA rather than delivering the cash themselves (McKay and Pickens, 2010). Furthermore, the ease in which individuals can now save has improved the informal credit market. With an increase in transfers, individuals are now able to better smooth consumption, receiving aid from friends and family in times of hardship or economic shock (Jack and Suri, 2011). As empirical evidence has shown, the access to credit generally raises the wealth and consumption of individuals. Research by Jack and Suri (2014) showed that during times of economic shock, individuals using M-PESA saw negligible drops in consumption.

THE HIDDEN AND UNPLEASANT EFFECTS OF M-PESA IN THE KENYAN SOCIETY

Despite its documented success M-PESA has brought potential negative effects such as reinforcing patriarchy and gender inequalities, weakened social ties between migrants and their home communities, increasing women's workloads, increasing the risk of domestic violence and abuse, limiting women's access to formal financial services, and the disintegration of family and the clan system. In their study Morawczynski and Pickens (2009) established an interesting pattern of remittances between urban dwellers and rural relatives. First, they established that the urban migrants sent smaller but more frequent remittances, which resulted in overall larger remittances to rural areas. Second, they established that the urban migrants also visited their rural homes less frequently, potentially weakening the social ties between migrants and their home communities and that money flows between male urban senders and female rural recipients. Although M-PESA has empowered women to make financial decisions and run family budgets, there is a potential for husbands to keep a second wife in the urban centers given their less frequent visits in their rule homes and for wives to become unfaithful to their husbands. There is potential for spousal suspicion, tensions, family breakdown and the possibility of violence which has a negative impact on children as they grow up in a dysfunctional family.

Although, studies such as Jack & Zuri (2016), Cook & McKay (2017), Mbiti & Weil (2014), and Hughes & Lonie (2007) have demonstrated that M-PESA has helped to close the gender gap by empowering women who have traditionally had less bargaining power, it is also reinforcing patriarchy, toxic masculinity, and domestic violence. The only valid identification document required to open and maintain an M-PESA is the Kenyan National Identification Card (ID) and that is where the trouble begins. A married woman must lose her identity and adopt the name of her husband as a sign of leaving and cleaving to her husband and sticking together till death. Adopting the husband's name is viewed as security for property inheritance in the event of death of husband. Any women who fails to adopt her husband's name risks being chased by her husband's relatives in the event of a husband's death. Women are caught up in a dilemma whether to relinquish their maiden names and adopt their husband's names and lose their identity or maintain their identity and risk losing their husband's estate.

Summers, et all., (2020) argue that the use of M-PESA bring with it the potential of increasing the risk of domestic violence and abuse. This is particularly the case in societies where mobile phones are often seen as a symbol of empowerment for women. In such societies and Kenya is no exception such empowerment can challenge traditional gender roles and power dynamics within the household leading to physical, verbal, and psychological violence against women. According to a study by the Global Partnership for financial inclusion (2020) women who keep their money outside their home, in a bank or mobile money account, could be more vulnerable to violence from partners who are unable to seize their money. This is reinforced by Porter, et all (2019) who argue that any expansion of female autonomy in domestic incomes, business, education, or health decision, can pose a potent threat to husbands and boyfriends who have been conditioned by traditions of patriarchy to expect total control in the relationship. Male vulnerability in contexts of poverty reinforces patriarchy, domestic violence, and gender inequality.

M-PESA has also reinforced traditional gender roles, inequalities, and power imbalance. I recently facilitated a couples' church seminar in a rural setting and a female participant underscored the challenge of reinforcing traditional gender roles when she said, "My husband doesn't send me money directly. Whenever, I go shopping to buy grocery or to the saloon, he demands that I give him the stores Paybill number so that he can send money directly to the merchant. I am not foolish; I don't give him the exact amount for my shopping." She then added, "Women have become clever, we collude with the merchants and add a few more shillings on top of the actual cost of goods or service for our own use…after all, why do our husbands treat us like children…why don't they trust us to handle money." This is a serious matter that could easily lead to marital breakdown and plant seeds of discord, suspicion, and marital unfaithfulness.

M-PESA is aiding the disintegration of the family and clan system in Kenya. As argued earlier, urban migrants visit their rural homes less frequently potentially weakening the social ties between migrants and their home communities. In the past, families met together to strategise on how to raise funds for social events such as dowry payments, weddings, burials, and other communal events. However, nowadays people meet virtually and collect money via M-PESA denying families and clans the opportunity to meet and solidify relationships. Whenever there was a death, members of the clan and friends met in city restaurants in the evenings to plan for burials of their loved ones. Today, burial collections are done on M-PESA via WhatsApp without any physical meetings. This is contrary to the African spirit of community as epitomized by the words of Chinua Achebe (1958), When we gather in the moonlight village ground, it is not because of the moon. Every man can see it in his own compound. We come together because it is good for kinsmen to do so. Therefore, let us continue with the team spirit and enjoy the power of togetherness.

Julie Zollmann (2020), in a report for the Global Partnership for Financial Inclusion, discusses the ways in which digital money can create new responsibilities for women in rural areas. She notes that women are often expected to manage household finances and make transactions on behalf of family members, which can be time-consuming and limit their ability to engage in other activities, such as paid work or education. This is positive thing because it is giving women power in decision making and spending on behalf of the family, particularly where the husbands make limited visits to their rural homes. Jack & Suri (2016) in a paper published in Science, found that while the introduction of M-PESA in Kenya led to a reduction in poverty, it also had negative effects on women's social and economic empowerment. They argue that M-PESA can reinforce traditional gender roles by limiting women's ability to participate in the formal economy and increasing their dependence on male family members. Jack and Suri further argue that M-PESA has enabled more women to move out of subsistence agriculture into business, a position supported by a report of the World Bank (WB) and the International Labour Organization (ILO) (2021) on the percentage of women working in agriculture. According to this report the percentage of women working in agriculture has seen a steady decline from 68 percent in 2006 to 59 percent in 2019. The However, the reality is that subsistence agriculture has faced various challenges including high cost of inputs, climate change, and mushrooming of supermarkets in urban centers making it unprofitable for women. Therefore, unlike the argument of Jack and Suri, the reality is that women have become poor even though some have translated subsistence agriculture into digital business using mobile money, M-PESA .

The Global Partnership for financial inclusion in their report titled, Advancing Women's Financial Inclusion (2020), argues that women are less likely to have access to formal financial services than men due to limited property rights and particularly in rural areas. The report approximates that one billion women globally do not have formal financial services, due to persistent barriers in access to identification documents, mobile phones, digital skills, financial capability, as well as inappropriate products and more. The explosion of M-PESA in Kenya has given dishonest businessmen undue advantage over women's financial illiteracy to exploit them with expensive unregulated mobile loans. Moreover, due to low levels of literacy, particularly financial literacy, most women have their phone operated by their husbands, who will go to withdraw money giving them an opportunity to misuse the money through alcoholism, and other forms of irresponsibility.

According to a FAO (2020) report titled " Deconstructing the gender gap in rural financial inclusion – The cases of Mozambique and Tanzania", socio-cultural norms can significantly influence the extent of women's access to financial services, in several ways. The report cites time and mobility constraints in traditional cultures where rural women are responsible for a variety of household tasks which take up most of their day, making it harder for them to find the time to take care of their financial necessities. In their study on gender empowerment and access to financial services in Machakos County, Eastern Kenya, Wandibba, Nangendo, and Mulemi (2016) argue that the use of M-PESA has created a situation where husbands and sons in towns have neglected their roles as men because they rarely visited their rural homes as long as they could "M-PESA " (send) money to their families. As a result, some family matters, such as, family planning, decision making, attending to children's personal education needs, and child rearing concerns, which require the attention of both spouses have been neglected. Therefore, the introduction and use of M-PESA means that men are rapidly forgetting their family gender roles such as family leadership, important decision making, and child rearing roles (Wandibba, Nangendo, and Mulemi, 2007).

It is also worth mentioning that the expansion of M-PESA has led to the dwindling fortunate of the old money transfer method, Western Union due to increased competition, convenience and cost-effective alternative for money transfer offered by M-PESA, particularly for individuals in rural areas or those without access to traditional banking services. M-PESA's widespread availability through mobile phones has made it more accessible than Western Union's physical locations and as discussed earlier, in some cases western union has collaborated with M-PESA to facilitate remittances. In remote or underserved areas where traditional banking infrastructure is lacking, M-PESA provides a convenient way for individuals to send and receive money without the need to travel to a physical agent (Mas & Radcliffe, 2010). Secondly, M-PESA transactions often incur lower fees compared to Western Union transfers, particularly for smaller amounts. This cost advantage has made M-PESA a preferred option for many users, especially those sending remittances or conducting frequent transactions. Finally, M-PESA transactions are typically processed instantly, allowing recipients to access funds immediately. In contrast, Western Union transfers may take longer to process, especially if recipients need to visit a physical location to collect funds and has become deeply integrated into daily life in Kenya, used for a wide range of transactions beyond money transfers, including bill payments, purchases, and even salary disbursements (Mugambi, Njunge, & Yang, May 2014). This integration has reinforced M-PESA's convenience and utility, further diminishing the reliance on traditional money transfer services like Western Union.

THE ROLE OF M-PESA IN ECONOMIC GROWTH AND DEVELOPMENT

Jack and Suri (2011) noted that M-PESA transfers are immediate and less costly—for amounts large *and* small. This enables small emergency remittances in response to small-magnitude shocks. Moreover, the availability of M-PESA allow households to send members further, thus optimising migrants' revenues and increasing remittance flows. Morawczynski (2009:512), in her qualitative study, conducts two interrelated surveys (in an urban slum and in a rural village where some of the urban dwellers migrated from), and analyses the evolution in financial behaviour as a result of the adoption of M-PESA. Morawczynski (2009) observed an increase in total remittance flows, resulting in a decrease in the financial vulnerability of the receivers. Jack and Suri (2011) demonstrated quantitatively that M-PESA remittances effectively improve resilience to income shocks. Jack and Suri (2011) and Aker and Mbiti (2010) further found a reduction of per capita consumption of 7 per cent among non-user households facing shocks, while consumption of households with access to M-PESA is not affected. The qualitative study by Plyler et al. (2010:2-3) confirmed that users of M-PESA credited its services with boosting local consumption thanks to rescue money. This boosting was also remarkable in their agriculture industry where M-PESA services have been seen increasing agricultural productivity. Similarly, Kikulwe, Fischer and Qaim (2014) suggested that mobile money use is welfare-enhancing for smallholder households, who constituted most of the rural poor people. It not only results in higher remittances but also stimulates more commercially oriented farming (Kikulwe, Fischer & Qaim, 2014).

Beyond higher remittances, Jack and Suri (2011) found that M-PESA digital technology led to an increase in household savings because of the security provided by M-PESA services as well as higher investment in human and physical capital. However, Arestoff and Venet (2013:3) found that where saving is concerned the results are less clear-cut than for money transfers. Demombynes and Thegeya (2012) argued that M-PESA users are 20 per cent more likely to have "savings of any kind that boost their economic growth thus achieve sustainable income. However, in spite of the use of instrumental variables their regressions might suffer from an endogeneity problem, in that the causality could also run from savings to M-PESA adoption. It is argued that having any savings in any of the forms are undoubtedly poorer than average. Some of those people might not have access to a SIM card and are thus simply not able to register for M-PESA services (Ouma & Odongo, 2014).

Suri and Jack (2016) found that M-PESA services have tremendous effects on economic growth and sustainable livelihoods. Suri and Jack (2011) reported that three out of four M-Pesa users indicate that, they use it to save money. According to Anderson, Baland, & Moene (2008) previous studies have shown that such services provide a foundation for economic development and allow the poor to climb the banking ladder by facilitating access to the formal economy. Furthermore, it is asserted that through such participation the poor can increase their wealth, diversify their asset base, and become more resilient to shocks. Morawcyznski and Pickens (2009) observe that M-PESA users sent smaller but more frequent remittances, which resulted in overall larger remittances to rural areas. They also observe that urban migrants using M-Pesa visited their rural homes less frequently, potentially weakening the social ties between migrants and their home communities. Morawcyznski and Pickens (2009) observe that users often keep a balance on their Mpesa accounts, thereby using the system as a rudimentary bank account even though the system does not provide interest. Mbiti and Weil (2011) argued that the increases in employment due to M-PESA are driven by changes in farm employment. Gikunda, Abura, and Njeru (2014:350) noted that M-PESA services also "emerged that mobile phones allowed immediate money transfers from town to village and vice versa with consequences on time and money savings rapid solution to daily problems affecting vulnerable communities."

Grewal (2008) found that that Mpesa is a network in the sense of interconnected group of people linked in a way that makes them capable of beneficial cooperation which can take various forms including the exchange of goods and ideas. Gikunda, et al., (2014:354) indicated that the ease of sending and receiving rescue money had strengthened friendships and increased social interaction." Moreover, M-PESA services strengthened social bonds by enabling people to receive and send money to each other when in need and have made it easier for people to solicit cash from friends and relatives (Gikunda, et al., 2014:360). M-PESA has made access to credit through M-shwari services and thus have enhanced interpersonal transaction which made it possible for people to borrow from friends and relatives (Gikunda, et al., 2014:360). Furthermore, M-PESA has facilitated people to access credit from commercial banks through mobile phones and has led to increased money circulation and expansion of local market thus promoting economic growth and sustainable income (Gikunda, et al., 2014:360). M-PESA services have enabled immediate money transfers from town to village and vice versa, from national to international vice versa in less time. Therefore, M-PESA services have enhanced rapid transfers and savings thus initiating rapid solution to the eradication of poverty that affects vulnerable communities.

Suri and Jack (2016) estimated that access to M-PESA technology lifted at least 194,000 Kenyan households, or 2 per cent of the total, out of poverty. They point out that the higher observed consumption levels could be driven by multiple mechanisms and find some support for the savings and occupational choice channels. They further found that an increase in M-PESA agent density positively affects total financial savings of households, although not savings on M-PESA itself (Suri and Jack, 2016). Also, individuals who experienced larger increases in M-PESA technology access were less likely to be working in farming and more likely to be working in business or sales (Suri and Jack, 2016). Morawczynski and Pickens (2009), Klapper and Singer (2014) found that M-PESA technology is a financial stepping stone with positive effects on poverty reduction.

Moreover, Suri and Jack (2016) argued that M-PESA services in Kenya have led to women empowerment of women which is a foundation towards achieving sustainable development. Morawczynski and Pickens (2009) argued that M-PESA technology has led to a great improvement in women's independence because its services enable them to have more regular and immediate remittances. The explanation would seem to be that, in patriarchal households, the privacy offered by M-PESA as a store of value allows women to manage household liquidity with less control by the male head. Suri and Jack (2016) argued that M-PESA has led to great consumption growth with the ability reduce poverty and underdevelopment. Scholars have found that M-PESA services on economic transactions have influenced significant cost savings for users (Ivatury & Pickens, 2006; McKay & Pickens 2010). M-PESA technology services have decreased and reduced the amount of time needed to travel to the banks to do transactions (Williams & Torma, 2007), thus reducing the risks of theft and promoting savings that promote economic security (Morawczynski, 2008).

M-PESA technology services have equally enhanced economic income not only in urban but also in the rural areas as well (Morawczynski & Pickens, 2009; Morawczynski, 2008). These services have improved livelihood strategies, risk sharing to cope with shocks, improved perceptions of economic expansion, and capital accumulation (Morawczynski, 2009, Jack & Suri, 2011). M-PESA technology services have improved business environment, promoted the creation of new job opportunities that reduced unemployment and enhanced economic growth (World Bank, 2010; Plyler, et al., 2010; Gencer, 2011). The users of M-PESA technology have experienced security of their money and have also been enhanced, through better increasing the efficiency of existing social networks of support (Goodman & Walia, 2007). This technology has reinforced social capital, improved informal risk-sharing networks (Jack & Suri, 2011), and reduced vulnerability during financial crises (Morawczynski, 2009; Morawczynski, 2008).

CONCLUSION

The dawn and evolution of M-PESA in Kenya marks a significant chapter in Africa's contributions to digital technology. From its inception, M-PESA revolutionized financial transactions by bridging the gap between formal financial institutions and the unbanked population, fostering financial inclusion and socio-economic empowerment. Its success stems from its simplicity, accessibility, and transformative impact on economic development.

M-PESA's rapid adoption underscores its role as a game-changer in the digital financial services ecosystem. With its widespread use, M-PESA has become deeply ingrained in Kenyan society, altering traditional credit systems, and facilitating safer and more accessible money transfers. Moreover, M-PESA's functionality extends beyond money transfers to encompass savings, loans, bill payments, and international remittances, offering a comprehensive digital financial platform. However, alongside its success, M-PESA also brings forth challenges and unintended consequences. Its expansion may reinforce patriarchal structures, exacerbate gender inequalities, and weaken social ties within communities.

Furthermore, M-PESA has empowered women by providing them with greater control over their finances and fostering consumption growth. Its role in strengthening social bonds and reducing vulnerability during financial crises underscores its contribution to community resilience and cohesion. As Africa continues to innovate in the digital space, M-PESA stands as a testament to the continent's potential to drive technological advancement and socio-economic transformation on a global scale. Initiatives like M-PESA are beneficial in promoting financial inclusion, an important determinant for poverty alleviation, and in preventing social exclusion. Globally, a large proportion of the adult population is prevented from using financial services. Market failures are often responsible for many individuals relying on their own resources or informal instruments to deal with personal finances. M-PESA-like systems could help overcome many obstacles to financial inclusion. It is important for similar attempts to consider the specific needs of each society and the means to adapt supply to the new markets. Government involvement is also necessary in a regulatory and supervisory role, enabling new products to focus on the most vulnerable groups.

REFERENCES

Aker, J. C., & Mbiti, I. M. (2010). Mobile Phones and Economic Development in Africa. *The Journal of Economic Perspectives, 24*(3), 207–232. doi:10.1257/jep.24.3.207

Anderson, S., Baland, J. M., & Moene, K. (2008). *Sustainability and Organizational Design in Roscas: Some Evidence from Kenya.* University of Namur. https://www.fundp.ac.be/eco/recherche/cred/papers/enforce811.pdf

Arestoff, F., & Venet, B. (2013). *Learning to Walk Before You Run: Financial Behavior and Mobile Banking in Madagascar.* Working Paper No. DT/2013/09. Université Paris-Dauphine.

Center for Public Impact. (2016). *Mobile currency in Kenya: The M-PESA.* https://www.centreforpublicimpact.org/case-study/m-currency-in-kenya

Chib, A., Leon, L., & Rahim, F. (2015). *The Impact of mFinance Initiatives in the Global South: A Review of the Literature.* Available at: https://link.springer.com/chapter/10.1007/978-981-287-381-1_6

Chinua, A. (1958). *Things fall apart.* William Heinemann Ltd.

CISCO. (2013). *Safaricom M-Pesa.* Available: https://www.cisco.com/c/en/us/solutions/collateral/service-provider/vni-service-adoptionforecast/Cisco_Safaricom_CS.pdf

Colbourn, T., Seal, T., & Obadha, M. (2018). *Mobile money is increasing healthcare access.* https://www.scidev.net/sub-saharan-africa/opinions/mobile-money-increasing- healthcare-access/

Cook, W., & McKay, C. (2017). *Banking in the M-PESA Age.* https://www.cgap.org/research/publication/banking-in-M-PESA -age

Daily, B. (2021). *Traffic police gives up Sh26m bribes to State.* https://www.businessdailyafrica.com/bd/news/traffic-police-gives-sh26m-bribes-state-3605792

Daily, B. (2023). *Free mobile transfers push transactions above half of GDP.* Available from: https://www.businessdailyafrica.com/bd/economy/how-free- mobile-transfers-pushed-t-over-half-of-gdp--4079900#:~:text=Data%20from%20the%20Central%20Bank,to%20transact%20on% 20their%20handsets.

Danby, S., Fleer, M., Davidson, C., & Hatzigianni, M. (2018). Digital Childhoods across contexts and countries. In S. Danby, M. Fleer, C. Davidson, & M. Hatzigianni (Eds.), Digital childhoods: Technologies and children's everyday lives (pp. 1-14). Springer. doi:10.1007/978-981-10-6484-5_1

Dawson, S. (2017). *Why Does M-PESA Lift Kenyans Out of Poverty?* https://www.cgap.org/blog/why-does-M-PESA-lift-kenyans-out-of-poverty

Demombynes, G., & Thegeya, A. (2012). *Kenya's Mobile Revolution and the Promise of Mobile Savings*. Working Paper No. 5988; The World Bank.

di Castri, S. (2013). *A conversation with Professor Njuguna Ndung'u, Governor of the Central Bank of Kenya, on the critical policy issues around mobile money.* GSMA Blog. Available at https://www.gsma.com/mobilefordevelopment/a-conversation-with-professor-njuguna-ndungu-governor-of-the-central-bank-of-kenya-on-the-critical-policy-issues-around-mobile-money

Donovan, K. (2012). Mobile money and financial inclusion. In T. Kelly & M. Minges (Eds.), *Information and communication for development 2012*. World Bank. doi:10.1596/9780821389911_ch04

eCitizen. (2023). *All government services*. https://accounts.ecitizen.go.ke/

Edwards, S. (2014). Towards contemporary play: Sociocultural theory and the digital-consumerist context. *Journal of Early Childhood Research*, *12*(3), 219–233. doi:10.1177/1476718X14538596

Edwards, S. (2015). New concepts of play and the problem of technology, digital media and popular-culture integration with play-based learning in early childhood education. *Technology, Pedagogy and Education*, *25*(4), 1–20. doi:10.1080/1475939X.2015.1108929

FAO. (2020). *Deconstructing the gender gap in rural financial inclusion – The cases of Mozambique and Tanzania*. FAO. doi:10.4060/cb0569en

Gencer, M. (2011). The mobile money movement: Catalyst to jumpstart emerging markets. *Innovations: Technology, Governance, Globalization*, *6*(1), 101–117. doi:10.1162/INOV_a_00061

Gikunda, R. M., Abura, G. O., & Njeru, S. G. (2014). Socio-economic Effects of Mpesa Adoption on the Livelihoods of People in Bureti Sub County, Kenya. *International Journal of Academic Research in Business & Social Sciences*, *4*(12). doi:10.6007/IJARBSS/v4-i12/1369

Global Partnership for Financial Inclusion (GPFI). (2020). *Advancing Women's Digital Financial Inclusion.* Available at: https://www.gpfi.org/sites/gpfi/files/sites/default/files/saudig20_women.pdf

Goodman, J., & Walia, V. (2007). Airtime transfer services in Egypt. In *The transformational potential of m-transactions*. Vodafone.

Grewal, D. S. (2008). *Network Power: the Social Dynamics of Globalization*. Yale UP.

Haas, S. & Nagarajan, G. (2011). *MPESA and access to health in Kenya*. Academic Press.

Hughes, N., & Lonie, S. (2007). M-PESA: Mobile Money for the "Unbanked" Turning Cellphones into 24-Hour Tellers in Kenya. Innovations. *Innovations: Technology, Governance, Globalization*, *2*(1-2), 63–81. doi:10.1162/itgg.2007.2.1-2.63

Ivatury, G., & Pickens, M. (2006). *Mobile phone banking and low-income customers*. CGAP.

Jack, W., & Suri, T. (2011). *Mobile Money: The Economics of M-PESA*. Working Paper No. W16721; National Bureau of Economic Research.

Jacob, F. (2013). The Role of M-PESA in Kenya's Economic and Political Development. In Kenya after 50: Reconfiguring Education, Gender, and Policy. Academic Press.

Kikulwe, E. M., Fischer, E., & Qaim, M. (2014). Mobile Money, Smallholder Farmers, and Household Welfare in Kenya. *PLoS One*, *9*(10), 9. doi:10.1371/journal.pone.0109804 PMID:25286032

Kimeu, C. 2022). 'Traumatising': how rogue digital loan apps in Kenya intimidate borrowers. *The Guardian*. Available at: https://www.theguardian.com/global-

Klapper, L., & Singer, D. (2014). *The Opportunities of Digitizing Payments*; Working Paper; The World Bank: Washington, DC, USA.

KRA. (2023). *Lipa Ushuru na M-PESA*. https://kra.go.ke/individual/filing-paying/379-lipa-ushuru-na-M-PESA.

M-Changa Africa. (2023). *M-Changa*. https://www.mchanga.africa

M-Tiba. (n.d.). *Affordable and accessible healthcare for all*. Available at: https://mtiba.com/

Mas, I., & Radcliffe, D. (2010). *Mobile payments go viral: M-PESA in Kenya.* ResearchGate. https://www.researchgate.net/publication/227489474_Mobile_payments_go_viral_M-PESA_in_Kenya

Mbiti, I., & Weil David, N. (2011). *Mobile Banking: The Impact Of M-Pesa in Kenya.* Working Paper 17129 https://www.nber.org/papers/w17129

Mbiti, S., & Weil, D. (2014). *Mobile Banking: The Impact of M-PESA in Kenya.* https://www.nber.org/system/files/chapters/c13367/revisions/c13367.rev1.pdf

McKay, C., & Pickens, M. (2010). *Branchless banking 2010: Who's served? At what price? What's next?* CGAP.

Moin, R. (2009). *Growth building jobs and prosperity in developing countries.* Available at: https://www.eldis.org/document/A37900

Morawczynski, O. (2008). Surviving in the dual system: How M-PESA is fostering urban-to-rural remittances in a Kenyan slum. *Proceedings of the Eighth International Conference on Human Choice and Computers (HCC8).*

Morawczynski, O. (2009). Examining the usage and impact of transformational m-banking in Kenya. In N. Aykin (Ed.), *Internationalization, design and global development* (pp. 495–504). Springer. doi:10.1007/978-3-642-02767-3_55

Morawczynski, O. (2009). Exploring the Usage and Impact of 'Transformational' Mobile Financial Services: The Case of M-PESA in Kenya. *Journal of Eastern African Studies : the Journal of the British Institute in Eastern Africa, 3*(3), 509–525. doi:10.1080/17531050903273768

Morawczynski, O., & Mark, P. (2009). *Poor People Using Mobile Financial Services: Observations on Customer Usage and Impact from M-PESA*. https://openknowledge.worldbank.org/handle/10986/9492

Morawczynski, O., & Pickens, M. (2009). *Poor People Using Mobile Financial Services: Observations on Customer Usage and Impact from M-PESA*. CGAP Brief Online. https://www.cgap.org/gm/document1.9.36723/BR_Poor_People_Using_Mobile_Financial_Services.pdf

Morawczynski, O., & Pickens, M. (2009). *Poor people using mobile financial services: Observations on customer usage and impact from M-PESA*. CGAP.

Mugambi, A., Njunge, C., & Yang, S. (2014, May). Mobile-money benefits and usage: The case of M-PESA. *IT Professional, 16*(3), 16–21. doi:10.1109/MITP.2014.38

Munga, G. (2010). *The impact of mobile banking: A case study of M-PESA in the Kenyan society.* http://erepository.uonbi.ac.ke/bitstream/handle/11295/5647/Munga-The%20imapct%20of%20mobile%20banking.pdf?sequence=1

Ntara, C. (2015). An Analysis of M-Pesa Use in International Transactions. *European Journal of Business and Management*, *7*(17).

Nyaronga, S. (2015). *Social Health Insurance Contribution Payment Options Among Informal Sector Workers In Eastlands Area Of Nairobi County*. http://erepository.uonbi.ac.ke/bitstream/handle/11295/105100/Nyaronga_Social%20Health%20Insurance%20Contribution%20Payment%20Options%20Among%20Informal%20Sector%20Workers%20In%20Eastlands%20Area%20Of%20Nairobi%20County.pdf?sequence=2&isAllowed=y

OECD. (2022). *Leading Country Example: Kenya Digital Tax Payments.* Available at: https://www.oecd.org/tax/forum-on-tax-administration/publications-and products/taxadministration-3.0-kenya-country-example.pdf

Ouma, S. A., Odongo, T. M., & Were, M. (2017). Mobile Financial Services and Financial Inclusion: Is it a Boon for Savings Mobilization? *Review of Development Economics*, *7*, 29–35.

Pazarbasioglu, C., & Garcia Mora, A. (2020). *Digital Financial Services*. https://pubdocs.worldbank.org/en/230281588169110691/Digital-Financial- Services.pdf

Plyler, M. G., & Geetha, N. (2010). *Community-Level Economic Effects of M-PESA in Kenya*. Iris Center, University of Maryland.

Plyler, M. G., Haas, S., & Nagarajan, G. (2010). *Community-Level Economic Effects of M-PESA in Kenya*. Financial Services Assessment.

Poters, G. (2019). *Mobile phones, gender, and female empowerment in sub-Saharan Africa: studies with African youth.* Available at: https://www.tandfonline.com/doi/full/10.1080/02681102.2019.1622500

Safaricom. (2015). *M-Pesa Services & Tariffs*. Available: https://www.safaricom.co.ke/personal/m-pesa/m-pesaservices/international-money-transfer

Safaricom. (2023a). *M-PESA, Global*. Available at: https://www.safaricom.co.ke/personal/m- pesa/do-more- with-M-PESA /M-PESA-global

Safaricom. (2023b). *M-Shwari*. Available at: https://www.safaricom.co.ke/personal/m- pesa/credit-and-savings

Samantha, M. (2011). The Water Crisis in Kenya: Causes, Effects and Solutions. *Global Majority E-Journal, 2*(1), 31-45. https://www.american.edu/cas/economics/ejournal/upload/marshall_accessible.pdf

Shaikh, A., & Karjaluoto, H. (2014). Mobile banking adoption: A literature review. *Telematics and Informatics*, 1–14. doi:10.1016/j.tele.2014.05.003

Statista. (2023). *M-PESA customer numbers from 2017 to 2022*. https://www.statista.com/statistics/1139190/M-PESA -customer-numbers/

Suri, T., & Jack, W. (2011). *Mobile Money: The Economics of M-Pesa*. NBER Working Paper 16721.

Suri, T., & Jack, W. (2016). The Long-run Poverty and Gender Impacts of Mobile Money. *Science, 354*(6317), 1288–1292. doi:10.1126/science.aah5309 PMID:27940873

Triki, T., & Faye, I. (Eds.). (2013). *Financial inclusion in Africa.* Working papers of the African Development Bank (AfDB) https://www.afdb.org/fileadmin/uploads/afdb/Documents/Project-and- Operations/Financial_Inclusion_in_Africa.pdf

Van Hove, L. & Dubus, A. (2019). *M-PESA and Financial Inclusion in Kenya: Of Paying Comes Saving?* doi:10.3390/su11030568

Vodafone Foundation. (2023). *MPESA*. Available at: https://www.vodafone.com/about- vodafone/what-we- do/consumer-products-and-services/M-PESA

Wandibba, S., Nangendo, S., & Mulemi, B. (2007). *Gender Empowerment and access to financial services in Machakos County, Eastern Kenya*. available at: https://www.imtfi.uci.edu/files/docs/2014/wandibba_imtfi_final_report.pdf

Waweru, N., & Kalani, V. (2009). Commercial banking crises in Kenya: Causes and remedies. *African Journal of Accounting, Economics, Finance and Banking Research, 4*(4). Available at: https://citeseerx.ist.psu.edu/document?repid=rep1&type=pdf&doi=ce4a3019b540f1c4e29741d19beb51e8653126d8

Wikipedia. (2023). *M-PESA*. Available from: https://en.m.wikipedia.org/wiki/M-PESA

Williams, H., & Torma, M. (2007). Trust and fidelity: From "under the mattress" to the mobile phone. In *The transformational potential of m-transactions* (Vol. 6, pp. 10–20). Vodafone.

World Bank. (2003). *Births attended by skilled health staff (% of total)*. Available at: https://data.worldbank.org/indicator/SH.STA.BRTC.ZS

World Bank. (2010). *Kenya economic update: Kenya at the tipping point?* The World Bank.

World Bank. (2021). *Employment in agriculture, female (% of female employment) (modeled ILO estimate) – Kenya*. Available at: https://data.worldbank.org/indicator/SL.AGR.EMPL.FE.ZS?locations=KE

INTRODUCTION

As the contributions of this volume show, Africa's contributions to the wave of digital technology are still hindered by the effects of the trans-Atlantic slave trade, colonialism, and contemporary coloniality forces, which have led to the creation of colonized minds resulting from colonized education that is ravaging the continent. Pre-trans-Atlantic slave trade history literature shows that science and technology started in Alkebulan (the original name of Africa). For example, historical literature presents that Africa is the origin of metallurgy, which later became the origin of digital technology that is shaping the modern world. Per definition, metallurgy is a technology for transforming metallic ores into usable metals (Becerra & Pieroni, 2018). Scholars have argued that this technology for transforming metallic occurs in a sequence of steps starting with mining, ore processing, fuel preparation, smelting, purification, alloying, casting, and/or hammering, and components like raw materials, means of production, skills, knowledge (Lechtman, 1976; Shimada & Craig 2013; Serneels 2017). This shows that the history of metallurgy helps us understand the origins of digital technology as well as the contributions to the African continent to the waves of digital technology in the modern world. This is because metallurgy is the branch of science and technology concerned with the properties of metals and their production and purification (Becerra & Pieroni, 2018).

It is a process used to extract metals in their pure form. The compounds of metals mixed with soil, limestone, sand, and rocks are known as minerals. Ores are rocks, minerals, or alluvial sediments that contain metals that are considered valuable and that can be mined and processed with the technologies available at a particular time (Killick 2014). Gold, and sometimes silver, copper, lead, mercury, and more rarely iron, can be found as native metals (Childs, 1986; Becerra & Pieroni, 2018). Metals are commercially extracted from minerals at low cost and minimum effort. Metallurgy involves scientists examining the microstructure of a metal, allowing then the mechanical properties of the metal to be determined. It empowers engineers to utilize that metal for various purposes such as welding or metalworking, thus leading to the wave of digital technology.

From antiquity, Egypt and adjacent regions closely mimicked the metallurgical trajectories of the nearby Middle East, which was, by then, part of Africa. This is because the name Middle East existed during colonialism when British colonizers called the region to demarcate their boundaries and prevent interruption from other colonizers in the region. Ancient Egyptian metallurgy started with the working of copper around 4000 BC, and by 3000 BC, the Bronze Age was fully established, with iron appearing much later in the last millennium BC (Scheel, 1989). Ancient Egypt had cultural interactions with regions south of the Nile, so metallurgy was established in Nubia by 2600 BC (Emery, 1963). It has been noted that Iron smelting appeared much later in Egypt (600 BC; Miller, 2002) when compared to the rest of the Middle East and was established even later (500 BC) at places such as Meroe in the Sudan (Rehren & Pernicka, 2008). In North Africa, by 600 BC or shortly after, Carthaginians (now Tunisia) worked iron, copper, and bronze (Alpern, 2005).

In the sub-Sahara, metallurgy began with the working of iron and, in some cases, iron and copper (Holl, 2009). This is the same case in other regions of the continent, like Southern Africa, Central Africa, West Africa, and East Africa (Craddock, 2010). Metallurgy in West, East, and Central Africa began sometime between 800 and 400 BC in the radiocarbon black hole created by fluctuations in atmospheric radiocarbon concentration (Killick, 2004). However, in Southern Africa, metallurgy only appeared with the advent of agriculturalists early in the first millennium AD (Phillipson, 2005; Alpern, 2005). This shows that the workings of metallurgy in Africa's different regions provide important lessons for the origins and spread of digital technology. It also confirms that Africa is at the center of technology and has a lot of contributions to the wave of digital technology. It serves us lessons such as technological innovation, technology transfer, and cultural interactions that spread to the rest of the world. For example, it has been noted that the workings of metallurgy in Africa were neither static nor homogenous throughout antiquity (Alpern, 2005). It developed locally and regionally, creating a rich history of local innovation and cross-cultural borrowing (Alpern, 2005). For example, it was argued that metallurgy spread to Aksum which witnessed the height of its power from the early first millennium AD and minted its own coinage in gold and silver (Phillipson 2005, p. 230). The iron work spread to Ethiopia and the neighbourhood regions around 300 BC (Childs,1996). The technology spread to the rest of the world starting with Mediterranean regions on the other via the Red Sea trade.

Metallurgy, which became the origin of digital technology in the modern world that brought about iron and other technological workings, was informed by Africa's indigenous knowledge. However, events such as the Arab invasions, the trans-Atlantic slave trade, and colonialism phased out this evidence and destroyed these truths about the origins of science and technology in Africa and the contributions of the continent to the wave of digital technology. While the trans-Atlantic slave trade destroyed the local ideology that informed technological advancement in Africa and took away approximately 12 million African people, energetic and talented in science and technology, to go build the West, colonialism promoted the stealing of raw materials on which metallurgy and other technology depend on. Colonialism also became worse as it introduced a colonial education that affected Africa's indigenous knowledge and promoted the production of colonial minds in Africa that opposed Africa's reality in science and technology. However, scholars have found that African indigenous knowledge plays a vital social ordering role in many postcolonial African states (Abidogun & Falola, 2020; Bob-Milliar, 2009). Moreover, Omolewa (2007) argued that several educational initiatives in different parts of the continent highlight the value of existing African Indigenous knowledge and practices. However, the historical realities about the way Africa's indigenous knowledge and system invented metallurgy, which became the foundation of modern digital technology, are mutedly spoken about. African Indigenous epistemology and systems have a footing in the global enterprise of knowledge construction and the wave of digital technology.

In post-colonial Africa, some alienating circumstances that resulted from colonialism and its legacies are still present and remain problematic and need to be discussed and addressed (Onwuegbuchulam, 2023). Boahen (1980:144) noted that colonialism's impact on indigenous knowledge and practices is huge and has prevented industrialization in Africa. The creation of colonial minds resulting from colonial education in Africa has led to many Africans becoming gatekeepers and thus continue spreading the colonial agenda in Africa. This is done through colonial forces, which was found to be horrible than colonialism. Many scholars credit African elites for ensuring that colonial attitudes, ideas, and knowledge have enduring appeal and a negative impact on African Indigenous knowledge and epistemology. Nyerere (1968) asserts that colonial education influenced beneficiaries' attitudes, ideas, and knowledge. These elites act as the diffusion agents of Western ways of being and knowing by privileging Western and Islamic education and political systems (Boahen, 1980). Their lifestyle and growing status in the postcolonial world also appeal to young people, who desire those kinds of lifestyles (Mandela, 1995; Nkrumah, 1965b; Williams, 1980). Yale and Gestrin (1998) argued that many African elites view the colonial elites lavish and ostentatious lifestyles as the standard of a good life and try to mimic them.

Coloniality resulting from unaddressed colonial legacies has affected Africa's contributions to the wave of digital technology and as well as digital technology growth in Africa. coloniality has affected the search for a tenable political thesis that would tackle the impacts of colonial legacies on Africa's indigenous knowledge and the contributions of the African continent to the wave of digital technology in the modern world. Addressing coloniality forces would also position the continent on a sound and surer footing to challenge the existing discrimination within science and technology. For example, the effects of coloniality, such as the exploitation of Africa's minerals, corruption, nepotism, economic slowness, and ineffective political leadership, have hindered the continent's contributions to the wave of digital technology. Modern African leaders and scholars lack the political will needed to address the trans-Atlantic slave trade, colonialism, and coloniality effects on the science and technology in Africa. On the contrary, those effects have become multi-faceted problems that are dragging on African society in the 21st century, preventing the continent from benefitting from the role of digital technology in achieving SDGs.

Those legacies have empowered coloniality and enabled it to create numerous puppet and irresponsible governments in Africa. It also stands for governments that are used by the Western forces in furtherance of neo-colonial interests by undermining the sovereignty of independent African states. Coloniality can also be described as a government represented by stooges, either through fabricated elections or an anti-people system, compradors, and corrupt civil and military functionaries. However, African leaders who took over from colonialists did not have a proper vision and failed to understand the ugly faces of colonialism. For this reason, their political governance failed to address the legacies of trans-Atlantic slavery and colonialism to prevent the growth and spread of coloniality. This had repercussions on their political leadership as they found being stooges of Western colonialists and served as tools of coloniality. They thus became colonizers of their fellow Africans, an issue that deprived the continent of total political liberation and total socio-economic liberation as well. This lack of visionary leaders who understand the effects of colonialism and the trans-Atlantic slave trade on Africa has remained one of the major setbacks for the continental development of digitalization and continued to hinder Africa's contributions to the wave of digitalization.

In the absence of visionary and courageous leadership, purposeful articulation and execution of people-oriented programs have continued to remain a mirage. A cryptic investigation into the leadership challenge in Africa shows that African leaders are an endangered species, and, in many cases, they are being replaced with rulers rather than leaders (Iliffe, 2007, p. 133). This may account for because personalized leadership, as opposed to collective leadership, is very common in Africa. Ineffective leadership in Africa has, in turn, given rise to multiple other problems, such as discrimination, corruption, social injustice, hunger, insecurity, threats of wars, and general backwardness in the contemporary world of progress and civilization. The major effects of coloniality that hindered Africa's contributions to the wave of digital technology have been that of political leadership. Modern African political leadership exists not for Africans but for the colonizers and their interests. Contemporary political leadership in Africa is characterized by mediocrity, irresponsible leaders, colonized scholars, high-level corruption, ethnic fragmentation, careless nepotism, and myopic pedestrianism. Any leader who gains a minimum level of visibility and respectability at any level tends to be intoxicated by selfish woolliness (Iliffe, 2007, p. 188). Hence, affects Africa's contributions to the wave of digital technology because what is left behind is a group of intellectually sterile descendants or disciples, full of malicious ego and irredeemable illogicality in all its ramifications (Iliffe, 2007, pp. 188-189).

THE WAY FORWARD

There is a need to understand the role of Indigenous knowledge in promoting digital technology governance. Indigenous knowledge is the traditional, cultural, and community knowledge produced and owned by local people in their specific communities and passed on from one generation to the next generation through practice and oral channels (Govender, Mudaly & James, 2013). It is mainly derived from local experiments, innovations, creativity, and experiences, embedded in the minds and activities of communities with long histories of close interactions (Ngulube, 2002; Govender, Mudaly & James, 2016). This knowledge serves as the basis for problem-solving, communication, teaching, decision-making, and science and technology in Africa's society where it is embedded (Cámara-Leret, Fortuna & Bascompte, 2019). It is the basis of technological advancement in agriculture, education, healthcare sectors, and the wide range of other activities that have sustained a society and its environment for many centuries (Senanayake, 2006). Indigenist thinkers have advocated for recovering and promoting Africa's indigenous knowledge systems (Battiste, 2002) as influential in decolonizing digital technology. This knowledge, therefore, needs to be safeguarded, protected and decolonized (Ndinda, 2011) to advance Africa's contributions to the wave of digital technology. Decolonization would thus mean the recovery from the effects of the trans-Atlantic slave trade, colonialism, and coloniality in the digitalization and restoration of indigenous people's identities, languages, and experiences (Datta, 2018). In this manner, indigenous communities can disentangle themselves from the oppressive control of colonizing state government through decolonized strategies (Simpson, 2004). Denzin et al. (2008) further described decolonization as a continuous process of anti-colonial struggle that honors indigenous approaches to knowing the world, recognizing indigenous land, indigenous peoples, and indigenous sovereignty.

Decolonial thinking has been used to critique the colonial-modern function of assumptions and knowledge forms that are deeply embedded in the discipline and the broader field of Western social science (Seth, 2013). Smith (1999) sees decolonization as a process for conducting research that places indigenous voices and epistemologies at the center of the research process. Smith (1999) suggests that the decolonization of indigenous research would help regain control over indigenous ways of knowing and being and how research can be used for social justice (Heleta, 2018). There has also been a general disregard for the indigenous knowledge system amongst academics and scientists, and as a result, the value of primary knowledge is still strategically rejected among African academics (Mji, 2019).

Even indigenous researchers who are aware of the benefits and superiority of indigenous techniques are afraid to admit Africa's contributions to the wave of digitalization, fearing to be ridiculed either by their fellow African scholars or Western peers (Mji, 2019). Researchers thus require contextualized research processes relevant to indigenous communities' challenges and contribute to their development, using acceptable indigenous research methods and theories. Indigenous decolonizing methods and theories suitable to our African contexts include decolonizing research methodologies (Briggs, 2005). Decolonizing research methodology is an approach that is used to challenge the Eurocentric research methods that undermine the local knowledge and experiences of marginalized population groups (Smith, 1999). Furthermore, Goduka, et al. (2013) argued that for research to be relevant and improve the quality of life of indigenous people, it needs to be rooted in indigenous worldviews, cultural values, and languages.

Much literature review, studies, and research on indigenous knowledge has been carried out by researchers without decolonizing the research (Datta, 2018; Keane, Khupe & Seehawer, 2017). It has been argued that many Africans who have benefitted from Western research without decolonization are referred to as oppressors towards indigenous knowledge (Kovach, 2010). Scholars rarely think about sharing their indigenous research with the indigenous knowledge owners (Datta, 2018; Keane, et al., 2017). Colonial education and Western perspective domination in the research constitute a rape model of research because a Western researcher comes in, takes what they need, and leaves when they feel like it, as well as failing to acknowledge Africa as the origin. On the other hand, African scholars write to please and applaud their Western counterparts but fail to critique and interrogate what Western academia talks about in Africa and digitalization in the global world. This has had the repercussions of hindering Africa's contributions to the wave of digital technology. Indigenous knowledge researchers, particularly Africans should be decolonized to engage in research not only to produce knowledge but also to make positive change in the lives of Africans.

Louis (2007) noted that African studies must emphasize indigenous knowledge, and the African origins of metallurgy, science, and technology must be put forward. Knowledge outcomes from these African studies must be shared in ways that benefit the community through consultations with research participants (Keane & Malcolm, 2003). However, indigenous researchers in African studies have still not done enough to redress this travesty of conducting research that is not benefitting the indigenous knowledge owners (Battiste, 2002). Thus, Indigenous knowledge must be decolonized and shared to benefit indigenous owners (Sithole, 2007). The protection of indigenous knowledge systems is also important in the field of previously suppressed indigenous knowledge systems. The indigenous knowledge system policy needs to be well understood by indigenous communities.

It has been noted that African indigenous knowledge is poorly managed (Adeniyi & Mrs, 2013), and yet it is the origin of metallurgy, which later became the basis of digitization. For the sake of promoting contributions of the African continent in the wave of digital technology, there is a need to preserve Africa's indigenous knowledge. A concern over the loss of indigenous knowledge due to colonization and that indigenous communities may lose control and rights over their knowledge has also raised a need to preserve this knowledge in digital formats (Stevens, 2008). It has been argued that if indigenous knowledge makes the origins of science and technology that transformed into digitization and coloniality continue to make it unavailable for generations to come (Biyela, 2016). Therefore, scholars in African studies must preserve indigenous knowledge, share it with others, and be passed on to upcoming generations of African scholars.

Documentation and preservation of indigenous knowledge must thus be part of a legislative system that recognizes rights over this knowledge (Battiste, 2002). Proper knowledge management policies and procedures thus need to be implemented to preserve indigenous knowledge (Kaniki & Mphahlele, 2002). Sithole (2007) described preservation as an acceptable way to validate and grant indigenous knowledge protection from colonial legacies. Kumar (2010) and Capel (2014) argued that preserving Africa's indigenous knowledge means decolonizing it first and maintaining and conserving it in a way that people must not lose it but use it for the benefit of the African continent. Yadav (2013) argued that preserving Africa's indigenous knowledge means assisting in keeping Africa's contributions to the modern world. This means that for Africa's contributions to the wave of digitalization to be heard and understood, Africans must embrace the indigenous knowledge and decolonize it to put on board what history holds on

Africa in terms of science and technology. The aim and purpose of this volume were to first argue that Africa was never dark nor primitive and that the metallurgy that gave birth to science and technology, which grew to become digitalization, originated in Africa before it spread to the rest of the world. Africa is rich in oil, copper, diamonds, gold, and many other resources that have been converted into metals that are accelerating the wave of digitalization. Yet, Africa is nowhere to be found as one of the key contributors to digitalization that is shaping the modern world. The volume appeals to African scholars and policymakers to understand the role of decolonizing Africa's indigenous knowledge to bring on board Africa's contributions to the wave of digital technology.

REFERENCES

Abidogun, J. M., & Falola, T. (Eds.). (2020). *Palgrave Handbook of African Education and Indigenous Knowledge*. Palgrave Macmillan. doi:10.1007/978-3-030-38277-3

Adeniyi & Mrs. (2013). Accessing Indigenous Knowledge Resources in Libraries and the Problems Encountered by Librarians Managing IK in Oyo State, Nigeria. *Library Philosophy and Practice (e-journal)*. 988.

Alpern, S. (2005). Did They or Didn't They Invent It? Iron in Sub-Saharan Africa. *History in Africa*, *32*, 41–94. doi:10.1353/hia.2005.0003

Battiste, M. (2002). *Protecting Indigenous knowledge and heritage*. Purich Publisher.

Becerra, M. F., & Pieroni, M. J. P. (2018). Metallurgy. Springer Nature Switzerland AG. doi:10.1007/978-3-319-51726-1_2715-1

Biyela, A. N. (2016). *The management and preservation of indigenous knowledge in Dlangubo village*. Academic Press.

Boahen, A. A. (1980). *Topics in West African history*. Longman.

Bob-Milliar, G. (2009). Chieftaincy, diaspora, and development: The institution of Nksuohene in Ghana. *African Affairs*, *108*(433), 541–558. doi:10.1093/afraf/adp045

Briggs, J. (2005). The use of indigenous knowledge in development: Problems and challenges. *Progress in Development Studies*, *5*(2), 99–114. doi:10.1191/1464993405ps105oa

Cámara-Leret, R., Fortuna, M. A., & Bascompte, J. (2019). Indigenous knowledge networks in the face of global change. *Proceedings of the National Academy of Sciences of the United States of America*, *116*(20), 9913–9918. doi:10.1073/pnas.1821843116 PMID:31043567

Capel, C. (2014). Mindfulness, indigenous knowledge, indigenous innovations and entrepreneurship. *Journal of Research in Marketing and Entrepreneurship*, *16*(1), 63–83. doi:10.1108/JRME-10-2013-0031

Childs, S. T. (1986). *Style in technology: A view of African Early Iron Age iron smelting through its refractory ceramics* [Ph.D. dissertation]. Boston University.

Childs, S. T. (1996). Technological history and culture in western Tanzania. In P. R. Schmidt (Ed.), *The culture and technology of African iron production* (pp. 277–317). University Press Florida.

Craddock, P. (2010). New Paradigms for Old Iron. Thoughts on E. Zangato and AFC Holl's 'On the Iron Front.'. *Journal of African Archaeology*, *8*(1), 29–36. doi:10.3213/1612-1651-10157

Datta, R. (2018). Decolonizing both researcher and research and its effectiveness in Indigenous research. *Research Ethics*, *14*(2), 1–24. doi:10.1177/1747016117733296

Denzin, N. K., Lincoln, Y. S., & Smith, L. T. (2008). *Handbook of Critical and Indigenous Methodologies*. SAGE. doi:10.4135/9781483385686

Goduka, N., Madolo, Y., Rozani, C., Notsi, L., & Talen, V. (2013). Creating spaces for eZiko Sipheka Sisophula theoretical framework for teaching and researching in higher education: A philosophical exposition. *Indilinga*, *12*(1), 1–12.

Govender, N., Mudaly, R., & James, A. (2013. *Indigenous Knowledge of Custodians of Zulu Culture – Implications for Multilogical Dialogue in the Academy*. Available at https://www.academia.edu/31738684/Indigenous_Knowledge_of_Custodians_of_Zulu_Culture_Implications_for_Multilogical_Dialogue_in_the_Academy

Govender, N., Mudaly, R., & James, A. (2016). Engaging Indigenous Knowledge Holders in Teaching Preservice Teachers in IKS Food Production and Practices: Implications for Higher Education. *Alternation, 23*(1), 180 – 207.

Heleta, S. (2018). Decolonizing Knowledge in South Africa: Dismantling the pedagogy of big lies. Ufahamu. *Journal of African Studies*, *40*(2), 47–65.

Iliffe, J. (2007). *Africans: The History of A Continent*. Cambridge University Press. doi:10.1017/CBO9780511800375

Kaniki, A. M., & Mphahlele, M. E. K. (2002). Indigenous knowledge for the benefit of all: Can knowledge management principles be used effectively? *South African Journal of Library and Information Science*, *68*(1), 1–16.

Keane, M., Khupe, C., & Seehawer, M. (2017). Decolonising methodology: Who benefits from indigenous knowledge research? *Educational Research for Social Science, 6*(1), 12–24. doi:10.17159/2221-4070/2017/v6i1a2

Keane, M., & Malcolm, C. (2003). Relevant science: "but relevant for what?" *Labtalk, 47*(2), 36–38.

Killick, D. (2004). Review Essay: What Do We Know about African Iron Working? *Journal of African Archaeology, 2*(1), 97–112. doi:10.3213/1612-1651-10021

Killick, D. (2014). From ores to metals. In B. Roberts & C. P. Thornton (Eds.), *Archaeometallurgy in global perspective. Methods and syntheses* (pp. 11–45). Springer. doi:10.1007/978-1-4614-9017-3_2

Kovach, M. (2010). *Indigenous methodologies: Characteristics, conversations, and contexts*. University of Toronto Press.

Kumar, S. A. (2010). Knowledge management and new generation of libraries information services: A concept. *International Journal of Library and Information Science, 1*(2), 24–30.

Lechtman, H. (1976). A Metallurgical Site Survey in the Peruvian Andes. *Journal of Field Archaeology, 3*(1), 1–42. doi:10.1179/009346976791547995

Louis, R. P. (2007). Can you hear us now? Voices from the margin: Using indigenous methodologies in geographic research. *Geographical Research, 45*(2), 130–139. doi:10.1111/j.1745-5871.2007.00443.x

Mandela, N. (1995). *Long walk to freedom: The autobiography of Nelson Mandela*. Little, Brown and Company.

Miller, D. (2002). Smelter and Smith: Iron Age Metal Fabrication Technology in Southern Africa. *Journal of Archaeological Science, 29*(10), 1083–1131. doi:10.1006/jasc.2001.0758

Mji, G. (2019). *Opting for a walk without limbs: the avoidance of Stellenbosch University to be an African University*. Stellenbosch University Library Auditorium.

Ndinda, C. (2011). *Indigenous knowledge to preserve and protect. Human science Research Council*. HSRC.

Ngulube, P. (2002). Managing and Preserving Indigenous Knowledge in the Knowledge Management Era: Challenges and opportunities for information professionals. *Information Development, 18*(2), 95–101. doi:10.1177/026666602400842486

Nkrumah, K. (1965b). *Neo-colonialism: The last stage of imperialism*. International Publishers.

Nyerere, J. (1968). Education for self-reliance. *Cross Currents, 18*(4), 415–434.

Omolewa, M. (2007). Traditional African modes of education: Their relevance in the modern world. *International Review of Education, 53*(5–6), 593–612. doi:10.1007/s11159-007-9060-1

Onwuegbuchulam, S. P. C. (2023). Decolonization, knowledge construction, and legitimation at African universities in the 21st century: Relevance of François Lyotard. *Journal of Education*, 94–110.

Phillipson, D. W. (2005). *African Archaeology*. Cambridge University Press. doi:10.1017/CBO9780511800313

Rehren, T., & Pernicka, E. (2008). Coins, Artefacts and Isotopes: Archaeometallurgy and Archaeometry. *Archaeometry, 50*(2), 232–248. doi:10.1111/j.1475-4754.2008.00389.x

Scheel, B. (1989). *Egyptian Metalworking and Tools*. Shire Publications.

Senanayake, S. G. J. N. (2006). Indigenous knowledge as a key to sustainable development. *The Journal of Agricultural Sciences, 2*. Available at: https://www.researchgate.net/publication/265197993

Serneels, V. (2017). Metals. In A. S. Gilbert (Ed.), *Encyclopedia of geoarchaeology* (pp. 521–532). Springer. doi:10.1007/978-1-4020-4409-0_35

Seth, S. (2013). Once Was Blind but Now Can See': Modernity and the Social Sciences. *International Political Sociology, 7*(2), 136–151. doi:10.1111/ips.12014

Shimada, I., & Craig, A. (2013). The style, technology, and organization of Sicán mining and metallurgy, Northern Peru: Insights from holistic study. *Chungara (Arica), 45*(1), 3–31. doi:10.4067/S0717-73562013000100001

Simpson, L. R. (2004). Anticolonial Strategies for the Recovery and Maintenance of Indigenous Knowledge. *American Indian Quarterly, 28*(3/4), 373–384. doi:10.1353/aiq.2004.0107

Sithole, J. (2007). The challenges faced by African libraries and information centres in documenting and preserving indigenous knowledge. *IFLA Journal, 33*(2), 117–122. doi:10.1177/0340035207080304

Smith, L. (1999). *Decolonizing methodologies: Research and indigenous peoples*. Zed Books.

Stevens, A. (2008). A Different Way of Knowing: Tools and Strategies for Managing Indigenous Knowledge. *Libri*, *58*(1), 25–33. doi:10.1515/libr.2008.003

Williams, J. (1980). *The educated and professional elite in the Gold Coast and Sierra Leone, 1885-1914*. University of California.

Yadav, R. H. (2013). Medicinal plants in the folk medicine system of Ethiopia. *Journal of Poisonous and Medicinal Plants Research*, *1*(1), 7–11.

Yale, R., & Gestrin, P. (1998). *Into Africa: Intercultural insights*. Intercultural Press.

Compilation of References

Abidogun, J. M., & Falola, T. (Eds.). (2020). *Palgrave Handbook of African Education and Indigenous Knowledge*. Palgrave Macmillan. doi:10.1007/978-3-030-38277-3

Abrahams, L. (2015). *Framing the digital complexity economy*. Keynote paper delivered at CyberAbuja 2015, the IEEE International Conference on Cyberspace Governance: The Imperative for National and Economic Security, Abuja. 10.1109/CYBER-Abuja.2015.7360506

Abrahams, L. (2016). *ICT regulation for fostering the digital complexity economy in the SADC region 2016-2030*. Unpublished conference paper delivered at the ACER Conference, Livingstone, Zambia. Retrieved from https://static1.squarespace.com/

Acemoglu, D., Johnson, S., & Robinson, J. (2001). The Colonial Origins of Comparative Development: An Empirical Investigation. *The American Economic Review*, *91*(5), 1369–1401. doi:10.1257/aer.91.5.1369

Acemoglu, D., & Robinson, J. (2010). Why is Africa Poor. *Economic History of Developing Regions*, *25*(1), 21–50. doi:10.1080/20780389.2010.505010

Acemoglu, D., & Robinson, J. (2012). *Why Nations Fail: The Origins of Power, Prosperity and Poverty*. Crown.

Acton, R., Salter, P., Lenoy, M., & Stevenson, R. (2017). Conversations on cultural sustainability: Stimuli for embedding indigenous knowledges and ways of being into curriculum. *Higher Education Research & Development, 36*(7), 1311-1325. . doi:10.1080/07294360.2017.1325852

Adadi, M., Hilson, G., & Garvin, T. (2020). Indigenous knowledge systems and non-indigenous actors in Ghana's artisanal and small-scale gold mining sector. *Journal of Cleaner Production*, *244*, 118849.

Adams, H. (1983). African Observers of the Universe: The Sirius Question. *Blacks in Science: Ancient and Modern Science*.

Adams, H. (1983). African Observers of the Universe: The Sirius Question. *Blacks in Science: Ancient and Modern*, 27–46.

Adamu, M. S. (2021). Rethinking technology design and deployment in Africa: Lessons from an African standpoint. AfriCHI 2021.

Adebanwi, W., & Obadare, E. (2013). *Democracy and Prebendalism in Nigeria: Critical Perspectives*. Palgrave Macmillan. doi:10.1057/9781137280770

Adebayo, A. G. (1990). The Production & Export of Hides and Skins in Colonial Northern Nigeria, 1900-1945. The Journal of African History, 33(2), 273 - 300.

Adedeji, A. A., & George, G. P. (2016). The impact of mineral resources on the environment and community development: A critical analysis. *International Journal of Developmental Research*, *6*(11), 10326–10332.

Adeniji G. (2012, Dec. 19). Slavery, colonialism haunting Africa – Soyinka. *Punch.*

Adeniyi & Mrs. (2013). Accessing Indigenous Knowledge Resources in Libraries and the Problems Encountered by Librarians Managing IK in Oyo State, Nigeria. *Library Philosophy and Practice (e-journal).* 988.

Adepoju, A. (2008). *Migration in sub-Saharan Africa*. Nordiska Afrikainstitutet.

Adeyemi, I. O., Uzamot, W. O., & Modupe, F. (2022). Knowledge Transfer and Use as Predictors of Law Firm Performance: Nigerian Lawyer's Perspectives. *International Journal of Knowledge Management*, *18*(1), 1–17. doi:10.4018/IJKM.291097

Adigbuo, E. R. (2017). African indigenous knowledge: The challenges of industrialization. *International Journal of Humanities Social Sciences and Education*, *4*(5), 76-84. www.arcjournals.org doi:10.20431/2349-0381.0405009

Adigbuo, E. R. (2017). African Indigenous Knowledge. *The Challenges of Industrialization International Journal of Humanities Social Sciences and Education*, *4*(5), 76–84.

Afisi. (2009). Tracing contemporary Africa's conflict situation to colonialism: A breakdown of communication among natives. *Philosophical Papers and Reviews, 1*(4), 59-66.

African Bank. (2023, May). *Digital technologies key to inclusive growth in Africa - African Union Commissioner*. https://www.afdb.org/en/news-and-events/digital-technologies-key-inclusive-growth-africa-african-union-commissioner-60818

African Development Bank Group. (2023, May 3). *Digital technologies key to inclusive growth in Africa.* https://www.afdb.org/en/news-and-events/digital-technologies-key-inclusive-growth-africa-african-union-commissioner-60818

African Development Bank. (2016). *Catalyzing growth and development through effective natural resources management.* African Development Bank (AfDB).

African Development Bank. (2017). African *Economic Outlook*. Entrepreneurship and Industrialization. African Development Bank.

African Union. (2015). *Agenda 2063: The Africa We Want*. Retrieved from https://au.int/sites/default/files/documents/36453-doc-africa_2063_eng.pdf

African Union. (2023, May 3). *Digital technologies key to inclusive growth in Africa - African Union Commissioner*. https://www.afdb.org/en/news-and-events/digital-technologies-key-inclusive-growth-africa-african-union-commissioner-60818

Ahmed, S. (2022, January 25). *Leveraging the potential of digitalisation, disruptive technologies, and cross-border data flows to facilitate aspirations of a Pan-African Digital Single Market*. https://researchictafrica.net/2022/01/25/leveraging-the-potential-of-digitalisation-disruptive-technologies-and-cross-border-data-flows-to-facilitate-aspirations-of-a-pan-african-digital--single-market/v

Aikenhead, G. S., & Ogawa, M. (2007). Indigenous knowledge and science revisited. *Cultural Studies of Science Education, 2*(3), 539n620. https://doi.org// doi:10.1007

Aikenhead, G. (2006a). *Science education for everyday life: Evidence-based practice*. The Althouse Press.

Aikenhead, G. (2006b). Towards decolonizing the Pan-Canadian Science Framework. *Canadian Journal of Science, Mathematics, &. Technology Education*, *6*(4), 387–399.

Ajulu, R. (1997). *Kenya's 1992 Election and its implications for Democratisation in Sub-Saharan Africa.* Foundation for Global Dialogue Occasional Paper No.9., Johannesburg.

Akeredolu-Ale, E. O. (1975). *The Underdevelopment of Indigenous Entrepreneurship in Nigeria*.

Aker, J. C., & Mbiti, I. M. (2010). Mobile Phones and Economic Development in Africa. *The Journal of Economic Perspectives*, *24*(3), 207–232. doi:10.1257/jep.24.3.207

Akileswaran, K., & Hutchinson, G. (2019). *Adapting to the 4IR: Africa's development in the age of automation*. Tony Blair Institute for Global Change.

Akpor, O. B., Ohiembor, M. O., & Olaolu, T. D. (2018). Environmental degradation in the Niger Delta: A human security challenge. *Journal of Human Security*, *14*(2), 11–22.

Akrich, M. (1997). The de-scription of technical objects. In W. E. Bijker & J. Law (Eds.), *Shaping Technology/Building Society* (pp. 205–224). The MIT Press.

Akuamoah, K.D. (2017). *History of Technological innovations and the changing trends in African societies: Pottery as a case study*. Academic Press.

Alcott, W. (2015). *The Underdevelopment of Africa by Europe*. Revealing Histories. Retrieved at http://www.revealinghistories.org.uk/africa

Alemayehu, D. A., & Sun, Y. (2017). E-government service delivery in Ethiopia: Achievements and challenges. *International Journal of Information Management*, *37*(5), 378–386.

Alemazung, J. A. (2010). Post-Colonial Colonialism: An Analysis of International Factors and Actors Marring African Socio-Economic and Political Development. *The Journal of Pan African Studies*, *3*(10).

Alessa, L., Kliskey, A., Gamble, J., Fidel, M., Beaujean, G., & Gosz, J. (2016). The role of Indigenous science and local knowledge in integrated observing systems: moving toward adaptive capacity indices and early warning systems. *Sustainability Science, 11*(1), 91-102. doi:10.1007/s11625-015-0295-7

Alessandrini, M. (2017). Colonialism and energy policies in Africa. *Journal of Energy in Southern Africa, 28*(2), 1–10.

Al-Khatib, S. A. (2011). *General psychology*. Bright Horizon.

Allain, J. (2008). Nineteenth Century Law of the Sea and the British Abolition of the Slave Trade. *British Yearbook of International Law, 78*, 342–88.

Allain, J. (2017). Contemporary slavery and its definition in law. In A. Bunting & J. Quirk (Eds.), *Contemporary Slavery: Popular Rhetoric and Political Practice* (1st ed., pp. 36–66). University of British Columbia Press. doi:10.59962/9780774832458-004

Allen, N. (2021, March 12). *The promises and perils of Africa's digital revolution*. https://www.brookings.edu/articles/the-promises-and-perils-of-africas-digital-revolution/

Almeida, M. B., & Barbosa, R. R. (2009). Ontologies in Knowledge Management Support: A Case Study. *Journal of the American Society for Information Science and Technology, 60*(10), 2032–2047. doi:10.1002/asi.21120

Almond, G. A., & Verba, S. (1965). *The Civic Culture — Political Attitudes and Democracy in Five Nations*. Little, Brown and Company.

Alpern, S. (2005). Did They or Didn't They Invent It? Iron in Sub-Saharan Africa. *History in Africa, 32*, 41–94. doi:10.1353/hia.2005.0003

Amara, J. (1987). Indigenous technology of Sierra Leone and the science education of girls. *International Journal of Science Education, 9*(3), 317–324. doi:10.1080/0950069870090308

Amaury de Feydeau, A., Menski, M., & Perry, S. (2022, May 26). *Africa's digital infrastructure transformation*. https://www.whitecase.com/insight-our-thinking/africas-digital-infrastructure-transformation

Andersen, T. B. (2009). E-government as an anti-corruption strategy. *Information Economics and Policy, 21*(3), 201–210. doi:10.1016/j.infoecopol.2008.11.003

Anderson, S., Baland, J. M., & Moene, K. (2008). *Sustainability and Organizational Design in Roscas: Some Evidence from Kenya*. University of Namur. https://www.fundp.ac.be/eco/recherche/cred/papers/enforce811.pdf

Aniakor, C. C. (2011). Strategies for human capital development in Nigeria: A perspective. In The humanities and human capital development. The Humanities, Delta State University.

Annarelli, A., Battistella, C., Nonino, F., Parida, V., & Pessot, E. (2021). Literature review on digitalization capabilities: Co-citation analysis of antecedents, conceptualization and consequences. *Technological Forecasting and Social Change*, *166*(3), 120635. doi:10.1016/j.techfore.2021.120635

Apffel-Marglin, F. (2011). *Subversive spiritualities: How rituals enact the world.* Oxford University Press.

Arestoff, F., & Venet, B. (2013). *Learning to Walk Before You Run: Financial Behavior and Mobile Banking in Madagascar.* Working Paper No. DT/2013/09. Université Paris-Dauphine.

Armstrong, M. (2009). *Handbook of Human Resource Management Practice* (11th ed.). Kogan Page Ltd.

Arunachallam, S. (2007). *Global Village: A View from the Periphery.* Background paper commissioned by International Development Research Network (IDRC). Available at: www.idrc.ca/uploads/user-S/10583649160Arun_paper.pdf

Asante, M., & Asante K. (1983). Great Zimbabwe: An Ancient African City-State. *Blacks in Science: Ancient and Modern,* 84-91.

Asante, M. K. (2013). *Afrocentricity: The theory of social change.* CRC Press.

Attwood, B., & Markus, A. (1999). *The struggle for aboriginal rights: A documentary history.* Allen & Unwin.

Augustine, S. J. (1997). *Traditional aboriginal knowledge and science versus accidental science.* Paper prepared for the Biodiversity Convention Office of Environment Canada. http://www.nativemaps.org/?q=node/1399

Austen, R. A., & Headrick, D. (1983). The role of technology in the African past. *African Studies Review*, *26*(3-4), 163–184. doi:10.2307/524168

Awolowo, O. (1970). *The Strategy and Tactics of the People's Republic of NIGERIA.* Macmillan.

Ayentimi, D. T. (2020.) The 4IR and the challenges for developing economies. Developing the Workforce in an Emerging Economy, 18-30.

Ayentimi, D. T., & Burgess, J. (2019). Is the fourth industrial revolution relevant to sub-Sahara Africa? *Technology Analysis and Strategic Management*, *31*(6), 641–652. doi:10.1080/09537325.2018.1542129

Ayittey, G. B. N. (2002). *Sustainable Development: Promoting Progress or Perpetuating Poverty?* Profile Books.

Azi, J.I. (2012). Appraising the role of Afrimation (African-Animation) in promoting Africa's rich cultural heritage in a digital age. *International Journal of computer graphics and animation (IJCGA),* *2*(2/3).

Azikiwe, N. (1968). The Future of Pan-Africanism. *Presence Africaine*, *12*, 7.

Babatola, J. T. (2014). *Neo-Colonialism In Africa: A Perpetuation Of Western Interest And Subjugation Of Africa*. Available at: https://www.researchgate.net/publication/313666367_NEO-COLONIALISM_IN_AFRICA_A_PERPETUATION_OF_WESTERN_INTEREST_AND_SUBJUGATION_OF_AFRICA

Bah & Scott. (2022). *Evaluation of the UNESCO's Local and Indigenous Knowledge Systems (LINKS) programme*. Academic Press.

Bai, D.B., Hu, J., Irfan, M., & Hu, M.J. (2023). Unleashing the Impact of Ecological Civilization Pilot Policies on Green Technology Innovation: Evidence from a Novel SC-DID Model. *Energy Economics, 106813*. https://doi.org/.eneco.2023.106813. doi:10.1016/j

Baku, A. A. (2022). Digitalisation and new public management in Africa. *New Public Management in Africa: Contemporary Issues*, 299-316.

Baldwin, K. (2016). *The paradox of traditional chiefs in democratic Africa*. Cambridge University Press.

Bandama, F. (2020). *Preindustrial mining and metallurgy in Africa*. Available at: https://www.researchgate.net/publication/345918784

Baniaman, H. M. (2015). *Controlling corruption through e-governance: Case evidence from Bangladesh U4 Brief.* Bergen, Norway: U4 Anti-Corruption Resource Centre.

Barnhardt, R., & Kawagley, A. O. (1999). Education Indigenous to Place: Western Science Meets Indigenous Reality. In *Ecological Education In Action* (pp. 117 – 140). State University Of New York Press. http://www.ankn.uaf.edu/Curriculum/Articles/BarnhardtKawagley/EIP.html

Barnhardt, R., & Oscar Kawagley, A. (2005). Indigenous Knowledge Systems And Alaska Native Ways Of Knowing. *Anthropology & Education Quarterly*, *36*(1), 8–23. doi:10.1525/aeq.2005.36.1.008

Basil, D. (1961). *The African Slave Trade: Pre-Colonial History, 1450- 1850.* Atlantic Monthly Press.

Battiste, M. (2002). *Indigenous knowledge and pedagogy in first nations education a literature review with recommendations.* Retrieved from https://www.afn. ca/uploads/files/education/24._2002_oct_marie_battiste_indigenousknowledge andpedagogy_lit_review_for_min_working_group.pdf

Battiste, M. (2017). Protecting indigenous knowledge and heritage. In *African indigenous knowledge: An underutilized and neglected resource for development. Library Philosophy and Practice (e-journal).* 1663. https://digitalcommons.unl.edu/libphilprac/1663

Battiste, M. (2002). *Protecting Indigenous knowledge and heritage*. Purich Publisher.

Bayeh, E. (2015). The Political And Economic Legacy Of Colonialism In The Post-Independence African States. International Journal in Commerce, IT & Social Sciences, 2(2).

Bebbington, A. (1991). Indigenous Agricultural Knowledge Systems, Human Interests, and Critical Analysis: Reflections on Farmer Organizations in Ecuador. *Agriculture and Human Values*, *8*(1-2), 14–24. doi:10.1007/BF01579652

Beccerra-Fernandez, I., Gonzalez, A., & Sabherwa, R. (2004). *Knowledge management: challenges, solutions and technologies*. Pearson Prentice Hall.

Becerra, M. F., & Pieroni, M. J. P. (2018). Metallurgy. Springer Nature Switzerland AG. doi:10.1007/978-3-319-51726-1_2715-1

Bekele, G. Z., & Arnqvist, P. (2015). Indigenous knowledge for renewable energy production: Insights from small-scale farmers in Ethiopia. *Journal of Cleaner Production*, *87*, 870–879.

Bell, P. M. H. (2001). *The world since 1945, and international history*. Oxford University Press Inc.

Ben-Jochannan, Y. (1997). *Africa: Mother of Western Civilization. African-American Heritage Series*. Black Classic Press.

Benyera, E. (2022). The Fourth Industrial Revolution and Africa: A Cure Which Kills the Patient. Africa and the Fourth Industrial Revolution, 145-158. doi:10.1007/978-3-030-87524-4_8

Berman, M. (2021, October 28). *What is Digital Technology? Programming Insider*. Programming Insider. https://programminginsider.com/what-is-digital-technology/

Bettina, L. (2011). A Critical Review On The Concept Of Social Technology. *Socialinės Technologijos Social Technologies*, *1*(1), 7–24.

Beyette, A. K. (2021). *Environmental Legacies of Colonialism: How Past Colonization Affects Current Social and Environmental Outcomes*. Academic Press.

Bidwell, N. J. (2020). Wireless in the weather-world and community networks made to last. In PDC'20: Vol.1, June 15-20, Manizales, Colombia (pp. 126–136). doi:10.1145/3385010.3385014

Biko, S. (1978). *I Write What I Like: Selected Writings*. University of Chicago Press.

Bimberg & Resnick. (1977, June). Colonial Development' An Econometric Study. *The Journal of Economic History*, *37*(2), 489–491.

Bi, S., Shao, L., Tu, C., Lai, W., Cao, Y., & Hu, J. (2023). Achieving carbon neutrality: The effect of China pilot Free Trade Zone policy on green technology innovation. *Environmental Science and Pollution Research International*, *30*(17), 50234–50247. doi:10.1007/s11356-023-25803-1 PMID:36790713

Biyela, A. N. (2016). *The management and preservation of indigenous knowledge in Dlangubo village*. Academic Press.

Blakemore, E. (2019, February 19). Colonialism facts and information. *National Geographic*. Available at: https://www.nationalgeographic.com/culture/article/colonialism

Blatch, S. (2013). *Great achievements in science and technology in ancient Africa*. ASMB. https://www.asbmb.org

Blichfeldt, H., & Faullant, R. (2021). Performance effects of digital technology adoption and product & service innovation: A process-industry perspective. *Technovation, 105*, 102275. doi:10.1016/j.technovation.2021.102275

Blyden, E. W. (1967). *Christianity, Islam and the Negro Race*. Edinburgh University Press.

Boadi, L. N., Markwei, C. M., & Osei-Bonsu, A. (2021). Rethinking the concept of sustainable mining practices in Ghana: The need for a 'mining development agenda'. *Journal of Cleaner Production, 280*, 124485.

Boahen, K.A. (2012). The Impact of European Presence on Slavery in the Sixteenth to Eighteenth Century Gold Cost. *Transactions of the Historical Society of Ghana, 14*.

Boahen, A. A. (1980). *Topics in West African history*. Longman.

Bob-Milliar, G. (2009). Chieftaincy, diaspora, and development: The institution of Nksuohene in Ghana. *African Affairs, 108*(433), 541–558. doi:10.1093/afraf/adp045

Bourreau, M. & Valletti, T. (2015). *Enabling digital financial inclusion through improvements in competition and interoperability: What works and what doesn't?* Academic Press.

Bouzar, K. (2015). *No Peace, No Sustainable Development: A Vicious Cycle that We Can Break*. Academic Press.

Bratianu, C. (2015). *Organizational Knowledge Dynamics: Managing Knowledge Creation, Acquisition, Sharing, and Transformation*. IGI Global. doi:10.4018/978-1-4666-8318-1

Brier, B., & Hobbs, H. (2013). *Ancient Egypt: Everyday Life in the Land of the Nile*. Sterling.

Briggs, J. (2005). The Use of Indigenous Knowledge in Development: Problems and Challenges. *Progress in Development Studies, 5*(2), 99–114. doi:10.1191/1464993405ps105oa

Briggs, J., & Sharp, J. (2004). Indigenous knowledges and development: A postcolonial caution. *Third World Quarterly, 25*(4), 661–676. doi:10.1080/01436590410001678915

Brock-Utne, B. (2000). Whose education for all: The recolonization of the African mind. In *African indigenous knowledge: An underutilized and neglected resource for development. Library Philosophy and Practice (e-journal)*. 1663. https://digitalcommons.unl.edu/libphilprac/1663

Brooks, L. (1971). *African Achievements: Leaders*. Civilizations and Cultures of Ancient Africa.

Brooks, L. (2013). African achievements: Leaders, civilisations and culture of ancient. *Africa*.

Bruchac, M. M. (2014). Indigenous Knowledge and Traditional Knowledge. In C. Smith (Ed.), *Encyclopedia of Global Archaeology* (pp. 3814–3824). Springer Science and Business Media. doi:10.1007/978-1-4419-0465-2_10

Brush, S., & Stabinsky, D. (1996). *Valuing Local Knowledge: Indigenous Peoples and Intellectual Property Rights*. Island Press.

Bühnen. (1993). Ethnic Origins of Peruvian Slaves (1548-1650): Figures for Upper Guinea. *Paiduema*, 39.

Bunson, M. (1991). *The Encyclopedia of Ancient Egypt*. Gramercy Books.

Burnard, T. (2008). Review of Flight to Freedom: African Runaways and Maroons in the Americas. *The American Historical Review*, *113*(1).

Büschges, G. (2002). *Wörterbuch der Soziologie. 2. völlig neubearbeitete und erweiterte Auflage*. Lucius & Lucius.

Cai, D., Hu, J., Jiang, H., Ai, F., & Bai, T. (2023). Research on the relationship between defense technology innovation and high-quality economic development: Gray correlation analysis based on panel data. *MDE. Managerial and Decision Economics*, *44*(7), 3867–3877. Advance online publication. doi:10.1002/mde.3925

Calhoun, C (2002). Cold War. *Dictionary of the Social Sciences*.

Cámara-Leret, R., Fortuna, M. A., & Bascompte, J. (2019). Indigenous knowledge networks in the face of global change. *Proceedings of the National Academy of Sciences of the United States of America*, *116*(20), 9913–9918. doi:10.1073/pnas.1821843116 PMID:31043567

Capel, C. (2014). Mindfulness, indigenous knowledge, indigenous innovations and entrepreneurship. *Journal of Research in Marketing and Entrepreneurship*, *16*(1), 63–83. doi:10.1108/JRME-10-2013-0031

Carvalho, N., & Gomes, I. (2017). Knowledge Sharing between Enterprises of the Same Group. *International Journal of Knowledge Management*, *13*(1), 34–52. doi:10.4018/IJKM.2017010103

Castro, C., & Lopes, C. (2022). Digital government and sustainable development. *Journal of the Knowledge Economy*, *13*(2), 880–903. doi:10.1007/s13132-021-00749-2

Center for Public Impact. (2016). *Mobile currency in Kenya: The M-PESA*. https://www.centreforpublicimpact.org/case-study/m-currency-in-kenya

Chahine, I., & Kinuthia, W. (2013). Juxtaposing form, function and social symbolism: An ethnomathematical analysis of indigenous technologies in the Zulu culture. *Journal of Mathematics & Culture*, *7*(1), 1–30.

Chakrabarty, D. (2000). *Provincialising Europe: Postcolonial Thought and Historical Difference*. Princeton University Press.

Chandra, D. V. (2014). *Re-examining the importance of indigenous perspectives in the Western environmental education for sustainability: iFrom tribal to mainstream*. Academic Press.

Chang, A. (2018). Analysis on corporate governance compliance standards in New Zealand–a qualitative study on disclosures using content analysis and interviews. *Journal of Financial Regulation and Compliance*.

Chanias, S., Myers, M. D., & Hess, T. (2019). Digital transformation strategy making in pre-digital organizations: The case of a financial services provider. *The Journal of Strategic Information Systems*, *28*(1), 17–33. doi:10.1016/j.jsis.2018.11.003

Chengu, G. (2015). *Africa's "Second Liberation" against Today's Neo-Colonialism*. https://www.globalresearch.ca/africas-second-liberation-against-todays-neo-colonialism/5451647

Chib, A., Leon, L., & Rahim, F. (2015). *The Impact of mFinance Initiatives in the Global South: A Review of the Literature*. Available at: https://link.springer.com/chapter/10.1007/978-981-287-381-1_6

Chigbu, U. E. (2019). Digital technologies, youth entrepreneurship, and sustainable development in sub-Saharan Africa: Prospects, challenges, and policies. *Journal on Sustainability*, *11*(4), 1032–1062.

Chihambakwe, N. (2021). Covid-19, Africa and indigenous knowledge systems. *African Journal of Science, Technology, Innovation and Development*, *13*(1), 59–66.

Chikwata, C., Mangara, N., & Tsamba, A. (2020). The Use of Indigenous Plant Species for Mine Rehabilitation in the Copperbelt Province of Zambia. *Journal of African Earth Sciences*, *163*, 103739.

Childs, S. T. (1986). *Style in technology: A view of African Early Iron Age iron smelting through its refractory ceramics* [Ph.D. dissertation]. Boston University.

Childs, S. T. (1996). Technological history and culture in western Tanzania. In P. R. Schmidt (Ed.), *The culture and technology of African iron production* (pp. 277–317). University Press Florida.

Chinua, A. (1958). *Things fall apart*. William Heinemann Ltd.

Chisholm, R. M. (1982). *Foundations of Knowing*. University of Minnesota Press.

Chiwanza, K., Musingafi, M., & Mupa, P. (2013). Challenges in preserving indigenous knowledge systems: Learning from past experiences. *Information and Knowledge Management*, *3*(2), 19–25.

Chocian, E. (2017). *Wherever it rains, we go there: Afar nomadic architecture: tradition and challenge* (Doctoral dissertation).

Choi, T., & Chandler, S. M. (2020). Knowledge vacuum: An organisational learning dynamic of how e-government innovations fail. *Government Information Quarterly*, *37*(1), 101416. doi:10.1016/j.giq.2019.101416

Chove, B. E., Mwita, E., Mlyuka, H., & Mrema, G. E. (2015). Assessment of the potential of bioenergy for electricity generation in Tanzania. *International Journal of Renewable Energy Research*, *5*(4), 977–986.

Chowdhury, T. (2019). *Wind Energy in Africa: A Case Study*. Springer.

Christensen, C. (2003). *The innovator's dilemma. The revolutionary book that will change the new way you do business*. HarperBusiness Essentials.

Christopher, D. R. (2001). *An Archaeology of El-Mina: Africana and Europeans on the Gold-Cost, 1400-1900*. Smithsonian Institution Press.

Churchill, W. (1996). I Am Indigenist: Notes on the Ideology World. In *A Native Son: Selected Essays on Indigenism, 1985-1995*. South End Press.

Çınar, M. A. (2020). Historical Perspective on Epidemics and New Beginning after *COVID-19*. Karabuk University.

Ciriello, R. F., Richter, A., & Schwabe, G. (2018). Digital Innovation. *Business and Information Systems Engineering*. https://link.springer.com/article/10.1007%2Fs12599-018-0559-8

CISCO. (2013). *Safaricom M-Pesa*. Available: https://www.cisco.com/c/en/us/solutions/collateral/service-provider/vni-service-adoptionforecast/Cisco_Safaricom_CS.pdf

Clancey, W. (1993). The knowledge level reinterpreted: Modeling socio-technical systems. *International Journal of Intelligent Systems*, *8*(1), 33–50. doi:10.1002/int.4550080104

Clarke, R. V. (1995). Situational Crime Prevention. In M. Tonry & D. Farrington (Eds.), *Building a Safer Society: Strategic Approaches to Crime Prevention* (pp. 91–150). University of Chicago Press.

Cockburn, A. (2000). Selecting a project's methodology. *IEEE Software*, *17*(4), 64–71. doi:10.1109/52.854070

Colbourn, T., Seal, T., & Obadha, M. (2018). *Mobile money is increasing healthcare access*. https://www.scidev.net/sub-saharan-africa/opinions/mobile-money-increasing- healthcare-access/

Congressional Research Service (CRS). (2023). *Report on, African Continental Free Trade Area (AfCFTA): overview and issues for congress*. CRS.

Connelly, M. (2000). Taking Off the Cold War Lens: Visions of North — South Conflict during the Algerian War for Independence. *The American Historical Review*, *105*(3), 739–769. doi:10.2307/2651808

Cook, W., & McKay, C. (2017). *Banking in the M-PESA Age*. https://www.cgap.org/research/publication/banking-in-M-PESA -age

Cook, C., Koenigstorfer, J., & Biswas, A. (2017). Comparing hydro-power resource and needs of India, Nepal, Bhutan, and Rwanda. *Renewable & Sustainable Energy Reviews*, *77*, 12–21.

Cook, M. (1974). *The African origin of civilization: Myth or reality*. Lawrence Hill & Company.

Cooper, F. (1940). Africa since 1940: The past of the present. Cambridge University Press.

Copson, R. W. (2001). *Africa Backgrounder: History, U.S. Policy, Principal Congressional Actions.* CRS Report for Congress.

Craddock, P. (2010). New Paradigms for Old Iron. Thoughts on E. Zangato and AFC Holl's 'On the Iron Front.'. *Journal of African Archaeology, 8*(1), 29–36. doi:10.3213/1612-1651-10157

Cronje, A., de Beer, J., & Ankiewicz, P. (2015). The development and use of an instrument to investigate science teachersi views on indigenous knowledge. *African Journal of Research in Mathematics, Science and Technology Education, 19*(3), 319-332. . doi:10.1080/10288457.2015.1108567

Curtin, P. D. (1971). The Atlantic Slave Trade, 1600-1800. In History of West Africa (Vol. 1). Longman.

Curtin, P. D. (1967). *Africa Remembered: Narratives of West Africans from the Era of the Slave Trade.* University of Wisconsin Press.

Custer, R. L. (1995). Examining the dimensions of technology. *International Journal of Technology and Design Education, 5*(3), 219–244. doi:10.1007/BF00769905

Cuthbertson, L. M., Robb, Y. A., & Blair, S. (2020). Theory and application of research principles and philosophical underpinning for a study utilising interpretative phenomenological analysis. *Radiography, 26*(2), e94–e102. doi:10.1016/j.radi.2019.11.092 PMID:32052765

Daily, B. (2021). *Traffic police gives up Sh26m bribes to State.* https://www.businessdailyafrica.com/bd/news/traffic-police-gives-sh26m-bribes-state-3605792

Daily, B. (2023). *Free mobile transfers push transactions above half of GDP.* Available from: https://www.businessdailyafrica.com/bd/economy/how-free- mobile-transfers-pushed-t-over-half-of-gdp-- 4079900#:~:text=Data%20from%20the%20Central%20Bank,to%20transact%20on% 20their%20handsets.

Dalkir, K. (2005). *Knowledge management in theory and practice.* Elsevier.

Dalleo, R. (2012). Ngũgĩ wa Thiong'o's *Wizard of the Crow* and Postcolonial Pedagogy. *Research in African Literatures, 43*(2), 138–154. doi:10.2979/reseafrilite.43.2.138

Danby, S., Fleer, M., Davidson, C., & Hatzigianni, M. (2018). Digital Childhoods across contexts and countries. In S. Danby, M. Fleer, C. Davidson, & M. Hatzigianni (Eds.), Digital childhoods: Technologies and children's everyday lives (pp. 1-14). Springer. doi:10.1007/978-981-10-6484-5_1

Danielle, C. (2019). Digital colonialism: The 21st century scramble for Africa through the extraction and control of user data and the limitations of data protection laws. *MICH. J. RACE & L, 24*(2), 417. https://repository.law.umich.edu/mjrl/vol24/iss2/6

Datta, R. (2018). Decolonizing both researcher and research and its effectiveness in Indigenous research. *Research Ethics, 14*(2), 1–24. doi:10.1177/1747016117733296

David, M. (2001). Truth as epistemic goal. In M. Steup (Ed.), *knowledge, truth, and duty: Essays on epistemic justification, responsibility, and virtue*. Oxford University Press. doi:10.1093/0195128923.003.0010

David, R. (2007). *Handbook to Life in Ancient Egypt Revised*. Oxford University Press.

Davies, T., & Fumega. S. (2014). *Mixed incentives: Adopting innovations for transparency, accountability, and anti-corruption. U4 Issue*. Bergen, Norway: U4 Anti-Corruption Resource Centre.

Dawson, S. (2017). *Why Does M-PESA Lift Kenyans Out of Poverty?* https://www.cgap.org/blog/why-does-M-PESA-lift-kenyans-out-of-poverty

De Beer, J., & Whitlock, E. (2009). Indigenous knowledge in the life sciences classroom: Put on your de Bono hats! *The American Biology Teacher*, *71*(4), 209–216. doi:10.2307/27669414

de Reuver, M., Sørensen, C., & Basole, R. C. (2018). The digital platform: A research agenda. *Journal of Information Technology*, *33*(2), 124–135. doi:10.1057/s41265-016-0033-3

De Vries, M. J. (1996). Technology Education: Beyond the "Technology is applied science" paradigm. *Journal of Technology Education*, *8*(1), 7–15. doi:10.21061/jte.v8i1.a.1

Delwar H. & Abu Md. (2016). *Africar Itihas* [History of Africa]. Dhaka: Bishobidiyaloy Prokashoni.

Demombynes, G., & Thegeya, A. (2012). *Kenya's Mobile Revolution and the Promise of Mobile Savings*. Working Paper No. 5988; The World Bank.

Denzin, N. K., Lincoln, Y. S., & Smith, L. T. (2008). *Handbook of Critical and Indigenous Methodologies*. SAGE. doi:10.4135/9781483385686

Department of Basic Education. (2011). *Curriculum and Assessment Policy Statement Grades 7-9. Technology*. Government Printers.

Department Of Telecommunications and Postal Services. (2018). *Invitation To Nominate Candidates for The Presidential Commission On Fourth Industrial Revolution*. https://www.gov.za/sites/default/files/gcis_document/201812/42078gen764.pdf

Development Bank of Southern Africa. (2024). *How digital transformation will secure a sustainable future for Africa*. https://www.dbsa.org/article/how-digital-transformation-will-secure-sustainable-future-africa#:~:text=This%20is%20because%20Africa%20is,a%20state%20of%20cataclysmic%20pause

DeWalt, B. R. (1994). Using indigenous knowledge to improve agriculture and natural resource management. *Human Organization*, *53*(2), 123–131. doi:10.17730/humo.53.2.ku60563817m03n73

di Castri, S. (2013). *A conversation with Professor Njuguna Ndung'u, Governor of the Central Bank of Kenya, on the critical policy issues around mobile money*. GSMA Blog. Available at https://www.gsma.com/mobilefordevelopment/a-conversation-with-professor-njuguna-ndungu-governor-of-the-central-bank-of-kenya-on-the-critical-policy-issues-around-mobile-money

Diop, S. (2012). African elites and their post-colonial legacy: Cultural, political and economic discontent–by way of literature. *Alter Native*, *19*(1), 71-79. https://digitalcommons.unl.edu/libphilprac/1663

Diop, A. (1988). *Precolonial Black Africa*. Lawrence Hill.

Diop, S. (2012). African elites and their Post-Colonial legacy: Cultural, political and economic discontent–by way of literature. *Africa Development. Afrique et Developpement*, *37*(4), 221–235.

Dodge, M. A. (2013). The Search for Resistance: A Layperson's Reflection on the Historiography of Slavery in the African Atlantic. *The History Teacher*, *47*(1).

Doloitte. (2021). *Future of Cyber Survey: Complexity is the new normal*. Accessed: 17 February 2024. https://www.deloitte.com/za/en/services/risk-advisory/services/2021-future-of-cyber-survey.html

Domínguez de Olazábal, I. (2022). *On Indigenous Refusal against Externally Imposed Frameworks in Historic Palestine*. Academic Press.

Donovan, K. (2012). Mobile money and financial inclusion. In T. Kelly & M. Minges (Eds.), *Information and communication for development 2012*. World Bank. doi:10.1596/9780821389911_ch04

Doubleday & Co. Inc. (1976). *The overseers also forced the slaves to watch as they whipped the headless body of the rebellious slave*. Author.

Dourish, P., & Mainwaring, S. D. (2012). Ubicomp's colonial impulse. UbiComp'12. doi:10.1145/2370216.2370238

Du Preez, J. (2019, June 24). *The 4th industrial revolution in Africa: The next great frontier*. https://www.inonafrica.com/2019/06/24/the-4th-industrial-revolution-in-africa-the-next-great-frontier/

Du, H., Bradbury, T. N., Lavner, J. A., Meltser, A. L., McNulty, J. K., Neff, L. A., & Karney, B. R. (2020). A comparison of Bayesian synthesis approaches for studies comparing two means: A tutorial. *Research Synthesis Methods*, *11*(1), 36–65. doi:10.1002/jrsm.1365 PMID:31782621

Dwayne, S. J. (1996). The Impact of Colonialism on African Economic Development. *University of Tennessee Honors Thesis Projects*.

eCitizen. (2023). *All government services*. https://accounts.ecitizen.go.ke/

Ecofin Agency. (2021, January 14). *Mali inaugurates its first solar power plant in Kita*. Retrieved August 3, 2021, from https://www.ecofinagency.com/energy/1401-42941-mali-inaugurates-its-first-solar-power-plant-in-kita

Economic Development in Africa Debt Sustainability. (2004). *Oasis or Mirage?* United Nations.

Edwards, S. (2014). Towards contemporary play: Sociocultural theory and the digital-consumerist context. *Journal of Early Childhood Research*, *12*(3), 219–233. doi:10.1177/1476718X14538596

Edwards, S. (2015). New concepts of play and the problem of technology, digital media and popular-culture integration with play-based learning in early childhood education. *Technology, Pedagogy and Education*, *25*(4), 1–20. doi:10.1080/1475939X.2015.1108929

Egbon, M. (2021). Impact of Mining on Sustainable Development in Africa: A Review. *Journal of Sustainable Mining*, *20*(1), 1–10.

Eglash, R. (1999). *African fractals: Modern computing and indigenous design*. Rutgers University Press.

Ehret, C. (2002). *The Civilizations of Africa*. University of Virginia Press.

Eling, M., & Lehmann, M. (2018). The impact of digitalization on the insurance value chain and the insurability of risks. *Geneva Pap. R. I.*, *43*(3), 359–396.

Ellen, R., & Harris, H. (1996). Concepts of indigenous environmental knowledge in scientific and development studies literature: A critical assessment. East-West Environmental Linkages Network Workshop.

Ellen, R., Parker, P., & Bicker, A. (2000). *Indigenous environmental knowledge and its transformations: Critical anthropological perspectives*. Harwood Academic.

Emeagwali, G., & Shizha, E. (Eds.). (2016). *African Indigenous knowledge and the sciences*. Sense Publishers. doi:10.1007/978-94-6300-515-9

Emery, A. (1996). *The Participation of Indigenous Peoples and Their Knowledge in Environmental Assessment and Development Planning (draft)*. Centre for Traditional Knowledge.

Enaifoghe, A. (2021). Digitalisation of African Economies in the Fourth Industrial Revolution: Opportunities for Growth and Industrialisation. *African Journal of Development Studies*, *11*(2), 31.

Englebert, P. (2000). Pre-Colonial Institutions, Post-Colonial States, and Economic Development in Tropical Africa. *Political Research Quarterly*, *53*(1), 7–36. doi:10.1177/106591290005300101

Etzemüller, T. (2009). Social engineering als Verhaltenslehre des kühlen Kopfes. In Die Ordnung der Moderne. Social Engineering im 20. Jahrhundert. Transcript Verlag.

Eyong, C. T. (2007). Indigenous knowledge and sustainable development in Africa: Case study on central Africa. *Tribes and Tribals. Special Volume*, (1), 121–139.

Ezeanya, C. (2015). Research, Innovation and Indigenous Knowledge in Africa. In *African indigenous knowledge: An underutilized and neglected resource for development. Library Philosophy and Practice (ejournal)*. 1663. https://digitalcommons.unl.edu/libphilprac/1663

Ezinwa Nwagwu, W. (2007). Africa's indigenous knowledge Creating science and technology information databases for developing and sustaining sub-Saharan Citations Creating science and technology information databases for developing and sustaining sub-Saharan Africa's indigenous knowledge. *Journal of Information Science*, *33*(6), 737–751. doi:10.1177/0165551506077374

Fadle, H.Y. (2005). *Studies in Sudanese History.* Khartoum: SUDATek Limited.

Falola, T. (2016). *Nigerian Political Modernity and Postcolonial Predicaments.* Pan-African University Press.

Fanon, F. (1952). *Black Skin, White Masks.* Grove Press.

Fanon, F. (1967). *Black skin, white masks.* Grove Press.

FAO. (2020). *Deconstructing the gender gap in rural financial inclusion – The cases of Mozambique and Tanzania.* FAO. doi:10.4060/cb0569en

Farah, I., Kiamba, S., & Mazongo, K. (2011). *Major challenges facing Africa in the 21st century: A few provocative remarks At the International Symposium on Cultural Diplomacy in Africa - Strategies to confront the Challenges of the 21st Century: Does Africa have what is required?* International Symposium on Cultural Diplomacy in Africa 14th – 17th July 2011, Berlin.

Farah, I., Kiamba, S., & Mazongo, K. (2011). Major challenges facing Africa in the 21st century: A few provocative remarks At the International Symposium on Cultural Diplomacy in Africa - Strategies to confront the Challenges of the 21st Century: Does Africa have what is required? *International Symposium on Cultural Diplomacy in Africa.*

Fayola, T., & Fleming, T. (n.d.). Afican civilisations: From the pre-colonial to the modern day. *Encyclopedia of Life Support Systems (EOLSS).* Available at: https://www.eolss.net/ebooklib/sc

Fedosejeva, J., BoEe, A., Romanova, M., Ilisko, D., & Ivanova, O. (2018). Education for sustainable development: The choice of pedagogical approaches and methods for the implementation of pedagogical tasks in the anthropocene age. *Journal of Teacher Education for Sustainability, 20*(1), 157-179. . doi:10.2478/jtes-2018-0010

Feenberg, A. (2005). Critical Theory of Technology: An Overview. *Tailoring Biotechnologies*, *1*(1), 47–64.

Fields, L. (1991). Tools for Indigenous Agricultural Development in Latin America: An Anthropologist's Perspective. *Agriculture and Human Values*, *8*(1-2), 85–92. doi:10.1007/BF01579659

Fitzgerald, M., Kruschwitz, N., Bonnet, D., & Welch, M. (2014). Embracing digital technology-a new strategic imperative. *MIT Sloan Management Review*, *55*(2), 1–12.

Fletcher, M. S., Hamilton, R., Dressler, W., & Palmer, L. (2021). Indigenous knowledge and the shackles of wilderness. *Proceedings of the National Academy of Sciences of the United States of America*, *118*(40), e2022218118. doi:10.1073/pnas.2022218118 PMID:34580210

Food and Agriculture Organization. (2020). Overcoming water challenges in agriculture. Food and Agriculture Organization. *The State of Food and Agriculture.*

Food and Agriculture Organization. (2021). *Digital technologies in agriculture and rural areas: Status report.* Food and Agriculture Organization (FAO) of the United Nations.

Foucault, M. (2013). The right of death and power over life. In Biopolitics: A Reader. Duke University Press.

Fox, L., & Signe, L. (2022, September 23). *Inclusion, inequality, and the Fourth Industrial Revolution (4IR) in Africa.* https://www.brookings.edu/articles/inclusion-inequality-and-the-fourth-industrial-revolution-4ir-in-africa/#:~:text=Adoption%20of%20Fourth%2DIndustrial%2DRevolution,discussed%20in%20our%20recent%20report

Franklin, S. (2022). *Embodied progress: A cultural account of assisted conception.* Taylor & Francis. doi:10.4324/9781003284499

Fricker, M. (2007). *Epistemic Injustice: Power and the Ethics of Knowing.* doi:10.1093/acprof:oso/9780198237907.001.0001

Friedman, L., & Schipper, L. (2011). Mining colonial legacies: Extractive regimes and indigenous movements in the Andes. *Development and Change, 42*(4), 1017–1035.

Fuchs, C. (2008). The implications of new information and communication technologies for sustainability. *Environment, Development and Sustainability, 10*(3), 291–309. doi:10.1007/s10668-006-9065-0

Fukuyama, F. (2019). *Dimension of development. Global Empowerment Conference.* Harvard University. Available from https://www.youtube.com/watch?v=5R-dty5V9DM&t=458s

Gaines, B. (1993). Modeling practical reasoning. *International Journal of Intelligent Systems, 8*(1), 51–70. doi:10.1002/int.4550080105

Gallagher, K. P., Matsuura, M., & Tang, X. (2019). Rediscovering African sustainability: Integrating indigenous knowledge and social capital into development. *Ecology and Society, 24*(3), 21.

Gaura, A. S., Hongjia, M. A., & Baoshan, G. E. (2019). MNC strategy, knowledge transfer context, and knowledge flow in MNEs. *Journal of Knowledge Management, 23*(9), 1885–1900. doi:10.1108/JKM-08-2018-0476

Gawer, A. (2014). Bridging differing perspectives on technological platforms: Toward an integrative framework. *Research Policy, 43*(7), 1239–1249. doi:10.1016/j.respol.2014.03.006

Gemma, R. B. (2018). *Introduction to positivism, interpretivism and critical theory.* https://www.researchgate.net/publication/323811451

Gencer, M. (2011). The mobile money movement: Catalyst to jumpstart emerging markets. *Innovations: Technology, Governance, Globalization, 6*(1), 101–117. doi:10.1162/INOV_a_00061

GennaioliN.RainerI. (2005). The Modern Impact of Precolonial Centralization in Africa. doi:10.2139/ssrn.848164

Gennaioli, N., & Rainer, L. (2007). The Modern Impact of Precolonial Centralization in Africa. *Journal of Economic Growth*, *12*(3), 185–234. doi:10.1007/s10887-007-9017-z

Ghanem, H. (2020). *Shooting for the moon: An agenda to bridge Africa's digital divide*. https://www.brookings.edu/articles/shooting-for-the-moon-an-agenda-to-bridge-africas-digital-divide/

Ghosh, K., & Dohan, M., & Veldandi, H. (2018). Digital Transformation Strategies for Healthcare Providers : Perspectives from Senior Leadership. *Americas Conference on Information Systems*, 1–5.

Gikunda, R. M., Abura, G. O., & Njeru, S. G. (2014). Socio-economic Effects of Mpesa Adoption on the Livelihoods of People in Bureti Sub County, Kenya. *International Journal of Academic Research in Business & Social Sciences*, *4*(12). doi:10.6007/IJARBSS/v4-i12/1369

Gil-Garcia, J.R., & Flores-Zúniga, M.A. (2020). Towards a comprehensive understanding of digital government success: integrating implementation and adoption factors. *Govern. Inf. Q.*, *37*(4), 101518. . doi:10.1016/j.giq.2020.101518

Global Partnership for Financial Inclusion (GPFI). (2020). *Advancing Women's Digital Financial Inclusion*. Available at: https://www.gpfi.org/sites/gpfi/files/sites/default/files/saudig20_women.pdf

Global System for Mobile Communications Association. (2019). *The Mobile Economy Sub-Saharan Africa 2019*. GSMA Headquarters.

Gobble, M. (2018). Digitalization, digitization, and innovation. *Research Technology Management*, *61*(4), 56–59. doi:10.1080/08956308.2018.1471280

Goduka, N., Madolo, Y., Rozani, C., Notsi, L., & Talen, V. (2013). Creating spaces for eZiko Sipheka Sisophula theoretical framework for teaching and researching in higher education: A philosophical exposition. *Indilinga*, *12*(1), 1–12.

Gómez-Baggethun, E. (2022). Is there a future for Indigenous and local knowledge? *The Journal of Peasant Studies*, *49*(6), 1139–1157. doi:10.1080/03066150.2021.1926994

Goodman, J., & Walia, V. (2007). Airtime transfer services in Egypt. In *The transformational potential of m-transactions*. Vodafone.

Gordon, A. A., & Gordon, D. L. (Eds.). (2013). *Understanding Contemporary Africa* (5th ed.). Lynne Rienner Publishers.

Gorman, R. F. (1984). Soviet Perspectives on the Prospects for Socialist Development in Africa. *African Affairs*, *83*(331), 163–187. doi:10.1093/oxfordjournals.afraf.a097604

Govender, N., Mudaly, R., & James, A. (2013. *Indigenous Knowledge of Custodians of Zulu Culture – Implications for Multilogical Dialogue in the Academy*. Available at https://www.academia.edu/31738684/Indigenous_Knowledge_of_Custodians_of_Zulu_Culture_Implications_for_Multilogical_Dialogue_in_the_Academy

Govender, N., Mudaly, R., & James, A. (2016). Engaging Indigenous Knowledge Holders in Teaching Preservice Teachers in IKS Food Production and Practices: Implications for Higher Education. *Alternation, 23*(1), 180 – 207.

Graham, G. (2010). Behaviorism. *Stanford Encyclopedia of Philosophy*. Retrieved February 26, 2011, from https://plato.stanford.edu/entries/behaviorism/

Greenleaf, V. (2011). *'Lion of Judah' Animated Movie Premieres in June*. Charisma News Online. http://www.charismamag.com/index.php/news/31147-lion-of-judah-animated-movie-premieres-injune

Greig, A., Hulme, D., & Turner, M. (2007). *Challenging Global Inequality*. Palgrave Macmillan. doi:10.1007/978-0-230-20840-7

Grenier, L. (1998). *Working With Indigenous Knowledge: A Guide For Researchers*. IDRC.

Grewal, D. S. (2008). *Network Power: the Social Dynamics of Globalization*. Yale UP.

Grimsdottir, E., & Edvardsson, I. R. (2018). Knowledge Management, Knowledge Creation, and Open Innovation in Icelandic SMEs. *International Journal of Knowledge Management, 14*(4), 1–13.

Grosfoguel, R. (2011). Decolonizing Post-Colonial Studies and Paradigms of Political-Economy: Transmodernity, Decolonial Thinking, and Global Coloniality. *TRANSMODERNITY: Journal of Peripheral Cultural Production of the Luso-Hispanic World, 1*(1).

Grosfoguel, R. (2011). Decolonizing Post-Colonial Studies and Paradigms of Political-Economy: Trans-modernity, Decolonial Thinking, and Global Coloniality. *Transmodernity: Journal of Peripheral Cultural Production of the Luso-Hispanic World, 1*(1). Retrieved from https://escholarship.org/uc/item/21k6t3fq doi:10.5070/T411000004

Grosfoguel, R. (2007). The Epistemic Decolonial Turn: Beyond Political Economy Paradigms 1. *Cultural Studies, 21*(2-3), 2–3. doi:10.1080/09502380601162514

Groves, C. P. (1948). *The Planting of Christianity in Africa* (Vol. 1). Lutterworth Press.

Gumbo Mishack, T. (2020). Indigenising Technological Pedagogical Content Knowledge in Open Distance Learning, Africa. *Education Review, 17*(4), 72–86. doi:10.1080/18146627.2020.1868073

Gumbo, M. T. (2003). *Indigenous technologies: Implications for a Technology Education curriculum* [Unpublished PhD thesis]. Vista University.

Gumbo, M. T., & Msila, V. (Eds.). (2017). African Voices on Indigenisation of the Curriculum: Insights from Practice. Wandsbeck: Reach.

Gumbo, M. T. (2015). Indigenous technology in Technology Education curricula and teaching. In P. J. Williams, A. Jones, & C. Buntting (Eds.), *Contemporary issues in Technology Education: The future of Technology Education* (pp. 57–75). Springer.

Gumbo, M. T. (2017). An indigenous perspective on Technology Education. In P. Ngulube (Ed.), *Handbook of research on Indigenous Knowledge Systems in developing countries* (pp. 137–160). IGI.

Gumbo, M. T., & Williams, J. P. (2014). Investigating Grade 8 Technology teachers' pedagogical content knowledge in the Tshwane District of Gauteng Province of South Africa. *International Journal of Educational Sciences*, *6*(3), 479–488. doi:10.1080/09751122.2014.11890159

Haas, S. & Nagarajan, G. (2011). *MPESA and access to health in Kenya*. Academic Press.

Habibi, A., Sofyan, S., & Mukminin, A. (2023). Factors affecting digital technology access in vocational education. *Scientific Reports*, *13*(1), 5682. doi:10.1038/s41598-023-32755-6 PMID:37029180

Handayani, I.D., Wilujeng, I., & Prasetyo, Z.K. (2018). *Elaborating Indigenous Knowledge*. Academic Press.

Harding, S. (1998). *Is science multicultural? Postcolonialisms, feminisms, and epistemologies*. Indiana University Press.

Haron, H., & Hamiz, M. (2014). An Ontological Framework to Preserve Malay Indigenous Health Knowledge. *Journal of Computational and Theoretical Nanoscience*, *20*(1), 226–23.

Hart, M. A. (2010). Indigenous worldviews, knowledge, and research. *The development*.

Hawley, S. (2000). *Exporting Corruption: Privatization, Multinationals and Bribery*. http://www.psrast.org/corrupt.htm

Heeks, R. (2006). E-governance for development: A focus on rural areas. *Journal of Urban Technology*, *13*(2), 123–137.

Heleta, S. (2018). Decolonizing Knowledge in South Africa: Dismantling the pedagogy of big lies. Ufahamu. *Journal of African Studies*, *40*(2), 47–65.

Hendrickson, J., Burnett, N., & Haruvy, E. (2019). Bitcoin and the charity of digital giving. *Journal of Political Economy*, *127*(6), 234–261.

Herlehy, T. J., & Wickins, P. (1983). An Economic History of Africa from the Earliest Times to the Partition. New York: Oxford University Press, 1981. 323 pp. Maps, tables, biblio., index. $29.95, hardcover. *African Studies Review*, *26*(1), 157–158. doi:10.2307/524615

Herman, M., Vizina, Y., Augustus, C., & Sawyer, J. (2008). *Learning indigenous science from the place*. Retrieved 23 October, 2015 from https://iportal.usask.ca/docs/Learningindigenousscience.pdf

Herusatoto, B. (2012). *Mitologi Jawa* [The Javanese mythology]. Oncor Semesta.

Higginbotham, Jr. & Jacobs. (1992). The Law Only as an Enemy: The Legitimization of Racial Powerlessness through the Colonial and Antebellum Criminal Laws of Virginia. *N.C. L. Rev. 70*(969), 1063.

Hildebrandt, J. (1981). *History of the Church in Africa. Achimota.* Christian Press.

Hilson, G., & Maconachie, R. (2008). "Good Governance" and the Extractive Industries in Sub-Saharan Africa. *Mineral Processing and Extractive Metallurgy Review: An International Journal, 30*(1).

Hong, Y. X. (2000). *Research on the Transformation of Economic Growth Mode.* Nanjing University Press.

Honoré. (2012). Antony 2012. "Slavery: From Ancient to Modern. In J. Allain (Ed.), *The Legal Understanding of Slavery: From the Historical to the Contemporary* (pp. 9–39). Oxford University Press.

Hoppers, C. A. (2001). *Decolonising the curriculum, indigenous knowledge systems and globalisation.* HSRC.

Horvath, R. J. (1972). A definition of colonialism. *Current Anthropology, 13*(1), 45–57. doi:10.1086/201248

Hossain, M. M., & Raihan, A. (2020). An enquiry into knowledge management practices of indigenous communities: A case study of Bagdi tribe in Bangladesh. *African Journal of Science, Technology, Innovation and Development, 12*(1), 1–11.

Hountondji, P. J. (2017.) Knowledge appropriation in a post-colonial context. In*African indigenous knowledge: An underutilized and neglected resource for development. Library Philosophy and Practice (e-journal).* 1663. https://digitalcommons.unl.edu/libphilprac/1663

Hountondji, P. (2019). *African modes of thought: The challenges of indigenous knowledge.* Routledge.

Howe, A. C. (2002). *Engaging children in science* (3rd ed.). Prentice Hall.

Huang, Q. H., Yu, Y. Z., & Zhang, S. L. (2019). Internet development and manufacturing productivity improvement: Internal mechanism and China's experience. *China's Industrial Economy*, (8), 5–23.

Hughes, L., & Rogei, D. (2020). Feeling the heat: Responses to geothermal development in Kenya's Rift Valley. *Journal of Eastern African Studies : the Journal of the British Institute in Eastern Africa, 14*(2), 165–184. doi:10.1080/17531055.2020.1716292

Hughes, N., & Lonie, S. (2007). M-PESA: Mobile Money for the "Unbanked" Turning Cellphones into 24-Hour Tellers in Kenya. Innovations. *Innovations: Technology, Governance, Globalization, 2*(1-2), 63–81. doi:10.1162/itgg.2007.2.1-2.63

Hu, J., Hu, M., & Zhang, H. (2023). Has the construction of ecological civilization promoted green technology innovation? *Environmental Technology & Innovation*, *29*, 102960. doi:10.1016/j.eti.2022.102960

Huntington, S. P. (1991). *The Third Wave: Democratization in the Late Twentieth Century*. University of Oklahoma Press.

IEA. (2019). *Africa Energy Outlook 2019*. International Energy Agency.

IIRR (International Institute of Rural Reconstruction). (1996a). *Recording and Using Indigenous Knowledge A Manual. IIRR: Silang, Philippines. IUCN/UNEP/WWF 1991 Summary - Caring for the Earth: A Strategy for Sustainable Living*. IUCN/UNEP/WWF.

Iliffe, J. (2007). *Africans: The History of A Continent*. Cambridge University Press. doi:10.1017/CBO9780511800375

IMF. (2021, March 1). *Africa Goes Digital – IMF F&D*. https://www.imf.org/en/Publications/fandd/issues/2021/03/africas-digital-future-after-COVID19-duarte

Inikori, J. E. (2003). *The Struggle against the Trans-Atlantic Slave Trade," in Fighting the Slave Trade: West African Strategies* (A. Diouf, Ed.). Ohio University Press.

International Organization for Standardization, ISO Survey of Management System Standard Certifications. (2018). *International Telecommunications Union, Measuring the Information Society Report 2018* (Vol. 1). International Telecommunications Union.

International Telecommunication Union (ITU). (2020). *The State of Broadband Report 2020*. International Telecommunication Union HQ.

Ivatury, G., & Pickens, M. (2006). *Mobile phone banking and low-income customers*. CGAP.

Jack, W., & Suri, T. (2011). *Mobile Money: The Economics of M-PESA*. Working Paper No. W16721; National Bureau of Economic Research.

Jacob, F. (2013). The Role of M-PESA in Kenya's Economic and Political Development. In Kenya after 50: Reconfiguring Education, Gender, and Policy. Academic Press.

James GM. (2009). Stolen Legacy: Greek Philosophy is Stolen Egyptian Philosophy. *The Journal of Pan African Studies*.

James, W. (1890). *The Principles of Psychology*. Holt.

Jana, P. (2001). African Renaissance and the millennium action plan. *Quest*, *15*, 41–46.

Jeater, D. (2018). Academic standards or academic imperialism? Zimbabwean perceptions of a hegemonic power in the global construction of knowledge. *African Studies Review*, *61*(2), 8–27. doi:10.1017/asr.2017.132

Johnson, G., Whittington, R., & Scholes, K. (2011). *Exploring strategy. Text and case* (9th ed.). Financial Times/ Prentice Hall.

Johnson, M. (1992). *Lore: Capturing Traditional Environmental Knowledge.* IDRC.

Johnston, R. J. (2000). The Dictionary of Human Geography (4th ed.). Wiley-Blackwell.

Jordan, C. (2018, May 17). *Incorporating indigenous knowledge into 21st-century innovation.* Retrieved August 3, 2021, from https://www.un.org/africarenewal/magazine/may-2018/incorporating-indigenous-knowledge-21st-century-innovation

Jost, J. T. (2006). The end of the end of ideology. *The American Psychologist, 61*(7), 651–670. doi:10.1037/0003-066X.61.7.651 PMID:17032067

Kaba, A., & Ramaiah, K. C (2020). Measuring Knowledge Acquisition and Knowledge Creation: A Review of the Literature. *Library Philosophy and Practice (e-journal).* 4723.

Kallinikos, J., Aaltonen, A., & Marton, A. (2013). The Ambivalent Ontology of Digital Artifacts. *Management Information Systems Quarterly, 37*(2), 357–370. doi:10.25300/MISQ/2013/37.2.02

Kalu, K. (2017). State–Society Relations, Institutional Transformation and Economic Development in Sub-Saharan Africa. *Development Policy Review, 35*(S2), O234–O245. doi:10.1111/dpr.12320

Kalu, K. (2018). Africa and the Cold War. In M. S. Shanguhyia & T. Falola (Eds.), *The Palgrave Handbook of African Colonial and Postcolonial History* (pp. 661–680). Palgrave Macmillan. doi:10.1057/978-1-137-59426-6_27

Kalu, K. (2020). The Cold War and Africa's Political Culture. Vestnik RUDN. *International Relations, 20*(1), 11–21. doi:10.22363/2313-0660-2020-20-1-11-21

Kaniki, A. M., & Mphahlele, M. E. K. (2002). Indigenous knowledge for the benefit of all: Can knowledge management principles be used effectively? *South African Journal of Library and Information Science, 68*(1), 1–16.

Kasanda, Lubben, F., Gaoseb, N., Kandjeo-Marenga, U., Kapenda, H., & Campbell, B. (1823). The role of everyday contexts in learner centred teaching: The practice in Namibian secondary schools. *International Journal of Science Education, 27*(15), 1805. doi:10.1080/09500690500277854

Kaya, H. O., & Seleti, Y. N. (2013). African indigenous knowledge systems and relevance of higher education in South Africa. *The International Education Journal: Comparative Perspectives, 12*(1), 30–44.

Kaya, H., & Seleti, Y. (2014). African Indigenous knowledge systems and relevance of higher education in South Africa. *International Education Journal: Comparative Perspectives, 12*(1), 30–44.

Keane, M., Khupe, C., & Seehawer, M. (2017). Decolonising methodology: Who benefits from indigenous knowledge research? *Educational Research for Social Science, 6*(1), 12–24. doi:10.17159/2221-4070/2017/v6i1a2

Keane, M., & Malcolm, C. (2003). Relevant science: "but relevant for what?" *Labtalk, 47*(2), 36–38.

Keane, M., Malcolm, C., & Rollnick, M. (2004). What is science education doing for rural South Africans, and what would they like it to do. In *Proceedings of the 12th Annual Southern African Association for Research in Mathematics, Science and Technology Education Conference* (pp. 453-458). Cape Town: University of Cape Town.

Kende, M., & Gillwald, A. (2017). The impact of mobile broadband in Sub-Saharan Africa: Policy considerations for digital diversity. *Journal of Global Information Technology Management, 20*(2), 63–87.

Kettl, D. F. (2015). The job of government: Interweaving public functions and private hands. *Public Administration Review, 75*(2), 219–229. doi:10.1111/puar.12336

Khan, S. N. (2014). Qualitative research method. *International Journal of Business and Management, 9*(11), 224–233. doi:10.5539/ijbm.v9n11p224

Khumalo, N. B., & Baloyi, C. (2017). African indigenous knowledge: An underutilized and neglected resource for development. *Library Philosophy and Practice (e-journal).* 1663. https://digitalcommons.unl.edu/libphilprac/1663

Kienon-Kabore, T. H. (2017). History of traditional science and technology in Sub-Saharan Africa: Problematic and methodology of approach. *International Journal of Academic Research and Reflection, 5*(5), 47–59.

Kiggundu, J. (2007). Intellectual property law and the protection of indigenous knowledge. In I. Mazonde & P. Thomas (Eds.), *Indigenous knowledge systems and intellectual property in the twenty-first century; perspectives from Southern Africa* (pp. 26–47). Dakar Codesria.

Kikulwe, E. M., Fischer, E., & Qaim, M. (2014). Mobile Money, Smallholder Farmers, and Household Welfare in Kenya. *PLoS One, 9*(10), 9. doi:10.1371/journal.pone.0109804 PMID:25286032

Killick, D. (2004). Review Essay: What Do We Know about African Iron Working? *Journal of African Archaeology, 2*(1), 97–112. doi:10.3213/1612-1651-10021

Killick, D. (2014). From ores to metals. In B. Roberts & C. P. Thornton (Eds.), *Archaeometallurgy in global perspective. Methods and syntheses* (pp. 11–45). Springer. doi:10.1007/978-1-4614-9017-3_2

Kimeu, C. 2022). 'Traumatising': how rogue digital loan apps in Kenya intimidate borrowers. *The Guardian.* Available at: https://www.theguardian.com/global-

Kimwaga, S. (2010). *African indigenous psychology and eurocentricism.* Unpublished Manuscript. Dar es Salaam, Tanzania: College of Business Education.

Kinyua, E. A., Maina, M., & Ngaruiya, G. M. (2020). Energy poverty in developing countries: A review of literature on the history, causes, and solutions of energy poverty. *International Journal of Energy Economics and Policy, 10*(1), 468–481.

Klapper, L., & Singer, D. (2014). *The Opportunities of Digitizing Payments*; Working Paper; The World Bank: Washington, DC, USA.

Knobloch, H. (2006). *Sozialtechnologie, Soziologie und Rethorik.* http://www. as.tu-berlin.de/v-menue/mitarbeiter/prof_ hubert_knoblauch/publikationen/

Knudston, P., & Suzuki, D. (2017.) The Wisdom of the Elders. Toronto, Ontario: StoddartLePena. In*African indigenous knowledge: An underutilized and neglected resource for development. Library Philosophy and Practice (e-journal).* 1663. https://digitalcommons.unl.edu/libphilprac/1663

Knudston, P., & Suzuki, D. (1992). *Native and Scientific Ways of Knowing about Nature.* Mountain Man Graphics.

Kochalumchuvattil, T. (2010). The Crisis of Identity in Africa: A Call for Subjectivity. Academic Press.

Kohsaka, R., & Rogel, M. (2021). Traditional and local knowledge for sustainable development: Empowering the Indigenous and local communities of the world. In *Partnerships for the Goals.* Academic Press.

Kovach, M. (2010). *Indigenous methodologies: Characteristics, conversations, and contexts.* University of Toronto Press.

Kozhakhmet, S., & Nazri, M. (2017). Governing Knowledge Sharing Behavior in Post-Soviet Kazakhstan. *Journal of Workplace Learning*, *29*(3), 1–18. doi:10.1108/JWL-06-2016-0053

KRA. (2023). *Lipa Ushuru na M-PESA.* https://kra.go.ke/individual/filing-paying/379-lipa-ushuru-na-M-PESA.

Kramlinger, T., & Huberty, T. (1990). Behaviorism versus humanism. *Training and Development Journal*, *44*(12), 41–45.

Kresge, N. (2011). *A history of black scientists.* ASBMB Today.

Kumar, S. A. (2010). Knowledge management and new generation of libraries information services: A concept. *International Journal of Library and Information Science*, *1*(2), 24–30.

Kusimba, C. M. (2004). Archaeology of Slavery in East Africa. *African Archaeological Review*, *21*(2), 59–88. doi:10.1023/B:AARR.0000030785.72144.4a

Lambert, L. (2017). *The Four Challenges of the Fourth Industrial Revolution.* Market Mogul. Retrieved from https://themarketmogul.com/industry-4-0- challenges/?hvid=2Gt2CE

Langill, S. (1999). *Indigenous knowledge: A resource kit for sustainable development researchers in dryland Africa.* IDRC.

Lashitew, A. (2022, November). *Digital technologies open vast business opportunities in Africa.* Academic Press.

Laufenberg, V. (2009). *About Technology*. Retrieved May 15, 2015 from mhtml:file://B:\TEKIT\AboutTechnology.mht.

Law, R. (1980). Wheeled transport in pre-colonial West Africa. *Africa*, *50*(3), 249–262. doi:10.2307/1159117

Leakey, R. E. (2002). *Exploration of Africa: The Emerging Nations: Central and East Africa. 1880 to the Present: From Colonialism to the Civil War*. Chelsea House Publishers.

Lechtman, H. (1976). A Metallurgical Site Survey in the Peruvian Andes. *Journal of Field Archaeology*, *3*(1), 1–42. doi:10.1179/009346976791547995

Lee, K. (2019). *The Art of Economic Catch-Up. Barriers, Detours and Leapfrogging in Innovation Systems*. Cambridge University Press. doi:10.1017/9781108588232

Lee, K., & Lim, C. (2001). Technological Regimes, Catching-up and Leapfrogging: Findings from the Korean Industries. *Research Policy*, *30*(3), 459–483. doi:10.1016/S0048-7333(00)00088-3

Le, T. H. (2022). Are digital business and digital public services a driver for better energy security? Evidence from a European sample. *Environmental Science and Pollution Research International*, *29*(18), 27232–27256. doi:10.1007/s11356-021-17843-2 PMID:34978034

Likalu, M., Abdullah, R., Selamat, M. H., Ibrahim, H., & Nor, M. Z. M. (2010). A framework of collaborative knowledge management system in open-source software development environment. *Computer and Information Science*, *3*(1), 81–90. doi:10.5539/cis.v3n1p81

Li, M., Hu, J., Liu, P., & Chen, J. (2023). How can digital finance boost enterprises' highquality development?: Evidence from China. Environ. Sci. Pollut. *Control Ser. 1–*, *15*. Advance online publication. doi:10.1007/s11356-023-28519-4 PMID:37440136

Lindsey, L. (2004). *Democratization in the Third Wave: Political Violence as a Transition Tactic*. Centre for the Study of Democracy Department of Political Science University of California, Irvine; Presentation at the Midwestern Political Science Association Annual Meeting Chicago, IL.

Liu. (2020). Digital innovation management: Theoretical framework and future research. *Management World*, *36*(7), 198–217.

Liu, N., Teng, L., Tian, W., & Li, Y. (2023). Does digitalization enhance fossil fuels resources efficiency? *Resources Policy*, *85*, 103878. doi:10.1016/j.resourpol.2023.103878

Liu, X. X., Yang, Y. Q., & Sun, Z. J. (2022). *The construction and evolutionary development of enterprise digital capabilities: A multi-case exploratory study based on leading digital*. Enterprises Reform.

Li, X. Y. (2022). *Research on the impact and mechanism of digital technology driving enterprise innovation*. East China Normal University.

Li, X., Wu, T., Zhang, H., & Yang, D. (2022). Digital Technology Adoption and Sustainable Development Performance of Strategic Emerging Industries: The Mediating Role of Digital Technology Capability and the Moderating Role of Digital Strategy. *Journal of Organizational and End User Computing*, *34*(8), 1–18. doi:10.4018/JOEUC.315645

Ljutic, A., & McPhee, C. (2017). Editorial: Blockchain. *Technology Innovation Management Review*, *7*(10).

Locke, J. (2001). *An essay concerning human understanding*. Batoche Books.

López Peláez, A., Erro-Garcés, A., & Gómez-Ciriano, E. J. (2020). Young people, social workers and social work education: The role of digital skills. *Social Work Education*, *39*(6), 825–842. doi:10.1080/02615479.2020.1795110

Louis, R. P. (2007). Can you hear us now? Voices from the margin: Using indigenous methodologies in geographic research. *Geographical Research*, *45*(2), 130–139. doi:10.1111/j.1745-5871.2007.00443.x

Lovejoy, P. E., Mahadi, A., & Mukhtar, M. I. (1993). C.L. Temple's Notes on the History of Kano (1909): A Lost Chronicle on Political Office. Sudanic Africa, 4(40).

Lovejoy, P. E. (1974). *International Slave Trade: Causes and Consequences*. York University Press.

Lusch, R. F., & Nambisan, S. (2015). Service innovation: A service-dominant logic perspective. *Management Information Systems Quarterly*, *39*(1), 155–175. doi:10.25300/MISQ/2015/39.1.07

Lynch, B. M., & Robbins, L. H. (1978). Namoratunga: The First Archeoastronomical Evidence in Sub-Saharan Africa. *Science*, *4343*(4343), 766–768. doi:10.1126/science.200.4343.766 PMID:17743241

Mack, E., & Meadowcroft, J. (Eds.). (2009). *Major Conservative and Libertarian Thinkers: John Locke*. Continuum.

Ma, D., & Zhu, Q. (2022). Innovation in emerging economies: Research on the digital economy driving high-quality green development. *Journal of Business Research*, *145*, 801–813. doi:10.1016/j.jbusres.2022.03.041

Madiba, R. (2023, July 19). *Connective collaboration key to the 4IR agenda in Africa*. https://www.bcx.co.za/insights/connective-collaboration-key-to-the-4ir-agenda-in-africa/

Maiese, M. (2003). *Dehumanization. In Beyond Intractability*. University of Colorado.

Maksymiv, Y. V. (2016). Reporting as an important tool in ensuring interaction between stakeholders. *Actual Problems of Economics*, *178*(4), 304–310.

Maldonado-Torres, N. (2007). On The Coloniality Of Being. *Cultural Studies*, *21*(2-3), 2–3, 240–270. doi:10.1080/09502380601162548

Mamdani, M. (2012). *Define and rule: Native as political identity (Kindle)*. Harvard University Press. doi:10.4159/harvard.9780674067356

Mandela, N. (1995). *Long walk to freedom: The autobiography of Nelson Mandela*. Little, Brown and Company.

Manning, P. (2006). Slavery and Slave Trade in West Africa, 1430- 1930. In E. K. Akyeampong (Ed.), *Themes in West Africa's History*. Ohio University Press. doi:10.2307/j.ctv136c1xk.11

Manny, L., Duygan, M., Fischer, M., & Rieckermann, J. (2021). Barriers to the digital transformation of infrastructure sectors. *Policy Sciences*, *54*(4), 943–983. doi:10.1007/s11077-021-09438-y PMID:34751195

Mantzou, P., Bitsikas, X., & Floros, A. (2023). Enriching Cultural Heritage through the Integration of Art and Digital Technologies. *Social Sciences (Basel, Switzerland)*, *12*(11), 594. doi:10.3390/socsci12110594

Markowitz, C. (2019). Harnessing the 4IR in SADC: Roles for Policymakers. *Occasional paper, 303*. https://saiia.org.za/research/harnessing-the-4ir-in-sadc-roles-for-policymakers/

MartinaO. 2015. The Trans-Atlantic Slave Trade and the long-term effects on African Economies. Case study: Angola and Namibia. DOI: doi:10.13140/RG.2.1.3798.4808

Martin, K., & Mirraboopa, B. (2003). Ways of knowing, being and doing: A theoretical framework and methods for indigenous and indigenist research. *Journal of Australian Studies*, *27*(76), 203–214. doi:10.1080/14443050309387838

Maru, Y., & Masu, S. (2020). *Navigating Digital Disruptions and Opportunities for Inclusive Development: Proceedings of the 9th International Conference on ICT for Development, Education and Training*. Springer.

Maruta, R. (2014). The creation and management of organizational knowledge. *Knowledge-Based Systems*, *67*, 26–34. doi:10.1016/j.knosys.2014.06.012

Mas, I., & Radcliffe, D. (2010). *Mobile payments go viral: M-PESA in Kenya*. ResearchGate. https://www.researchgate.net/publication/227489474_Mobile_payments_go_viral_M-PESA_in_Kenya

Materer, S. M., Valdivia, C., & Gilles, J. L. (2002). *Indigenous Knowledge Systems: Characteristics and Importance to Climatic Uncertainty*. Working or Discussion Paper.

Mathew, R. M. (1985). Social analysis of information producation and consumption: the new challenges and tasks of Third World Countries. In A. I. Mikhalov (Ed.), *Theoretical problems of informatics: social aspects of modern informatics* (pp. 37–47). All Union Institute for Scientific and Technical Information.

Matike, E. (2008). *Knowledge and Perceptions of Educators and Learners in the Incorporation of IKS into School Curriculum*. Unpublished BA (Honours) thesis. North-West University, South Africa.

Matimu, V. F., Tawodzera, R., & Chikodzi, D. (2021). Sustainable mining, enhancing community livelihoods and socio-economic development through artisanal small-scale mining in Zimbabwe. *Journal of Cleaner Production*, *278*, 123938.

Matyana, M., & Thusi, X. (2023). Unemployment and poverty in South Africa: Assessing the National Development Plan 2030 predictions. *International Journal of Development and Sustainability, 12*(6), 212–226.

Mavhunga, C. (2014). *Transient workspaces: Technologies of everyday innovation in Zimbabwe*. MIT Press. doi:10.7551/mitpress/9780262027243.001.0001

Mavhunga, C. (2017). *What do science, technology, and innovation mean from Africa?* MIT Press. doi:10.7551/mitpress/10769.001.0001

Mawere, J. (202)0. Decolonising legal education in South Africa: A review of African indigenous law in the curriculum. *Pretoria Student Law Review, 14*(1).

Mawere, M., & Awuor, F. O. (2019). Local knowledge systems and sustainable development in Africa: A discursive view. *African Journal of Science, Technology, Innovation and Development, 11*(6), 699–707. doi:10.1080/20421338.2018.1560587

Mazimpaka, J. D., Nzeyimana, I., & Harelimana, F. (2018). Colonialism and economic development in Rwanda: Can history explain the post-colonial state of affairs? *Journal of International Development, 30*(3), 497–518.

Mazrui A. (1986). *The Africans: A Triple Heritage*. Academic Press.

Mazrui, A. A. (1994). *The bondage and boundaries*. IBRU Boundary and Security Bulletin.

Mazrui, A., & Tidy, M. (1984). *Nationalism and New States in Africa*. Heinemann Educational Books Inc.

Mbeki, T. (1996). *I am an African.* Statement on Behalf of the African National Congress on the occasion of the adoption by the Constitutional Assembly of the Republic of South Africa Constitution Bill.

Mbembe, A. (2017). Strategic Narratives of Technologies and Africa. Academic Press.

Mbembe, A. (2019). Bodies as borders. *From the European South, 4*, 5–18.

Mbembe, A. (2017). *Critique of Black Reason*. Duke University Press.

Mbiti, I., & Weil David, N. (2011). *Mobile Banking: The Impact Of M-Pesa in Kenya*. Working Paper 17129 https://www.nber.org/papers/w17129

Mbiti, S., & Weil, D. (2014). *Mobile Banking: The Impact of M-PESA in Kenya*. https://www.nber.org/system/files/chapters/c13367/revisions/c13367.rev1.pdf

Mbiza, M. (2023, June 21). *The Ethics of AI in the Fourth Industrial Revolution: Ensuring Responsible and Fair Use*. https://www.linkedin.com/pulse/ethics-ai-fourth-industrial-revolution-ensuring-fair-use-mbiza/

McCarthy, S. (2017). In globalization and education. In *African indigenous knowledge: An underutilized and neglected resource for development. Library Philosophy and Practice (e-journal).* 1663. https://digitalcommons.unl.edu/libphilprac/1663

M-Changa Africa. (2023). *M-Changa.* https://www.mchanga.africa

McIlwain, C. D. (2020). Black software: The internet and racial justice. Oxford University Press.

McInnes, B. D. (2017). Preparing teachers as allies in indigenous education: Benefits of an American Indian content and pedagogy course. *Teaching Education, 28*(2), 145-161. doi:10.1080/10476210.2016.1224831

McKay, C., & Pickens, M. (2010). *Branchless banking 2010: Who's served? At what price? What's next?* CGAP.

Melo, D.J., & Solleder, M.J. (2022). *Challenges of digitalisation in MENA and sub-Saharan Africa.* Academic Press.

Mengesha, T., & Hassen, N. (2019). Bridging the gap between indigenous knowledge and scientific knowledge for improved land use management in Ethiopia. *African Journal of Science, Technology, Innovation and Development, 11*(5), 601–610.

Mengisteab, K. (2011). Critical Factors in the Horn of Africa's Raging Conflicts. In *Discussion papers 67.* Nordiska Afrikainstitutet.

Meng, Q. S., & Yu, J. (2021). Impact mechanism of digital technology innovation on new generation of information technology industry upgrading. *R & D Management, 33*(1), 90–100.

Mensah, A. D. J. T., & Mensah, A. D. J. T. (2023, October 2). *Ten Facts About Digital Technology Adoption in Ghana.* World Bank Blogs. https://blogs.worldbank.org/africacan/ten-facts-about-digital-technology-adoption-ghana

Mestre, J., & Touger, J. (1989). Cognitive research — What's in it for physics teachers? *The Physics Teacher, 27*(6), 447–456. doi:10.1119/1.2342824

Meyer, X., & Crawford, B. A. (2011). Teaching science as a cultural way of knowing: Merging authentic inquiry, nature of science, and multicultural strategies. *Cultural Studies of Science Education, 6*(3), 525–547. Advance online publication. doi:10.1007/s11422-011-9318-6

Mhlanga, D., Ndhlovu, E., & Hofisi, C. (2021). Assessment of the 4IR Challenges of Radical Innovation in Service Delivery in Africa. *Journal of Public Administration, 56*(4.1), 1002-1017.

Michalopoulos, S., & Papailoannou, E. (2010). *Divide and Rule or the Rule of the Divided? Evidence from Africa.* Discussion Paper Series No. 8088. Center for Economic Policy Research. doi:10.1016/j.jbusres.2019.06.019

Mignolo, W. D. (2011). *Local Histories/Global Designs; Mignolo, The Darker Side of Renaissance.* Mignolo, The Darker Side of Western Modernity.

Mignolo, W. D., & Walsh, C. E. (2018). *On Decoloniality: Concepts, Analytics, Praxis*. Duke University Press.

Miller, S. T. (2016). *Geothermal Electricity Compared To Electricity From Other Sources In Kenya* (Master's thesis).

Miller, D. (2002). Smelter and Smith: Iron Age Metal Fabrication Technology in Southern Africa. *Journal of Archaeological Science*, *29*(10), 1083–1131. doi:10.1006/jasc.2001.0758

Minter W. (2008). *Apartheid's Contras: An Inquiry into the Roots of War in Angola and Mozambique*. Academic Press.

Mises Institute. (2022, August 12.). *Yes, Precolonial Africa Had Technology and Economic Life before*. https://mises.org/wire/yes-precolonial-africa-had-technology-and-economic-life-colonialism

Mji, G. (2019). *Opting for a walk without limbs: the avoidance of Stellenbosch University to be an African University*. Stellenbosch University Library Auditorium.

Mji, G. (2019). *Opting for a walk without limbs: the avoidance of Stellenbosch University to be an African University*. Stellenbosch University Library Auditorium.

Mkabela, Q. (2005). Using the Afrocentric Method in Researching Indigenous African Culture. *Qualitative Report*, *10*(1), 178–189.

Moin, R. (2009). *Growth building jobs and prosperity in developing countries*. Available at: https://www.eldis.org/document/A37900

Moodie, T. (2017). Alternative ways of knowing–doing justice to non-Western intellectual traditions in a postmodern era. In *African indigenous knowledge: An underutilized and neglected resource for development. Library Philosophy and Practice (e-journal)*. 1663. https://digitalcommons.unl.edu/libphilprac/1663

Moore, N., & Czerwinska, K. (2019). *Understanding the use of digital technology in the career development sector*. The University of Derby.

Morawczynski, O. (2008). Surviving in the dual system: How M-PESA is fostering urban-to-rural remittances in a Kenyan slum. *Proceedings of the Eighth International Conference on Human Choice and Computers (HCC8)*.

Morawczynski, O., & Mark, P. (2009). *Poor People Using Mobile Financial Services: Observations on Customer Usage and Impact from M-PESA*. https://openknowledge.worldbank.org/handle/10986/9492

Morawczynski, O., & Pickens, M. (2009). *Poor People Using Mobile Financial Services: Observations on Customer Usage and Impact from M-PESA*. CGAP Brief Online. https://www.cgap.org/gm/document1.9.36723/BR_Poor_People_Using_Mobile_Financial_Services.pdf

Morawczynski, O. (2009). Examining the usage and impact of transformational m-banking in Kenya. In N. Aykin (Ed.), *Internationalization, design and global development* (pp. 495–504). Springer. doi:10.1007/978-3-642-02767-3_55

Morawczynski, O. (2009). Exploring the Usage and Impact of 'Transformational' Mobile Financial Services: The Case of M-PESA in Kenya. *Journal of Eastern African Studies : the Journal of the British Institute in Eastern Africa, 3*(3), 509–525. doi:10.1080/17531050903273768

Morawczynski, O., & Pickens, M. (2009). *Poor people using mobile financial services: Observations on customer usage and impact from M-PESA*. CGAP.

Morgan, D. L. (2003). Appropriation, Appreciation, Accommodation: Indigenous Wisdoms and Knowledges in Higher Education. *International Review of Education, 49*(1/2), 35–49. doi:10.1023/A:1022937003545

Moroka, V. (2021). Sustainable mining practices in Africa with particular reference to Botswana. *Journal of Sustainable Mining, 20*(1), 29–37.

Morsy, H. (2020, August 18). *How Africa Can Harness the Fourth Industrial Revolution*. https://www.project-syndicate.org/commentary/how-africa-can-close-education-skills-gap-by-hanan-morsy-2020-08

Moshi, A. P., Msangi, E. S., Njau, K. N., & Khalid, R. (2019). The impact of biogas technologies on household air pollution in Babati District, Tanzania. *Energy for Sustainable Development, 46*, 89–99.

Moyo, P. V., & Kizito, R. (2014). Prospects and challenges of using the argumentation instructional method to indigenise school science teaching. *African Journal of Research in Mathematics, Science and Technology Education, 18*(2). https://www.tandfonline.com/doi/abs/10.1080/10288457.2014.912831

Moyo, S., & Yeros, P. (2019). *Africa in the age of the fourth industrial revolution*. Pluto Press.

Mpofu, P. (2013). *Multilingualism, localism and the nation: identity politics in the Zimbabwe Broadcasting Corporation*. University of South Africa. http://hdl.handle.net/10500/18663

Mpofu, W. J. (2013). Coloniality in the Scramble for African Knowledge: A Decolonial Political Perspective. *Africanus: Journal of Development Studies, 43*(2), 105–117. doi:10.25159/0304-615X/2305

M-Tiba. (n.d.). *Affordable and accessible healthcare for all*. Available at: https://mtiba.com/

Mugambi, A., Njunge, C., & Yang, S. (2014, May). Mobile-money benefits and usage: The case of M-PESA. *IT Professional, 16*(3), 16–21. doi:10.1109/MITP.2014.38

Muhammad, E., & Patricia, M. (2003). The Trans-Atlantic Slave Trade: A Forgotten Crime Against Humanity as Defined by International Law. *American University International Law Review, 19*(4), 883–947.

Munga, G. (2010). *The impact of mobile banking: A case study of M-PESA in the Kenyan society.* http://erepository.uonbi.ac.ke/bitstream/handle/11295/5647/Munga- The%20imapct%20of%20 mobile%20banking.pdf?sequence=1

Munyoki, J. M. (2019). A Paradigm Shift for Innovation and Creativity in Africa in the 21st Century. *European Journal of Business and Management*, *11*(21). Advance online publication. doi:10.7176/EJBM

Murithi, T. (2007). A local response to the global human rights standard: The ubuntu perspective on human dignity. *Globalisation, Societies and Education*, *5*(3), 277–286. doi:10.1080/14767720701661966

Murombedzi, J. C. (2016). *Inequality and natural resources in Africa. World Social Science Report 2016*. UNESCO.

Murunga Godwin, R. (2004). The State, Its Reform and the Question of Legitimacy in Kenya, Identity. *Cultural Politics*, *5*(1 & 2), 179–206.

Musoni, M. (2023, September 18). *Looking into the crystal ball: Artificial intelligence policy and regulation in Africa.* https://ecdpm.org/work/looking-crystal-ball-artificial-intelligence-policy-regulation-africa

Mustonen, T. (2014). *2021 Towards Inclusion of Indigenous Knowledge and Local Knowledge in Global Reports on Climate Change*. Available from: https://www.researchgate.net/publication/352440686_2021_Towards_Inclusion_of_Indigenous_Knowledge_and_Local_Knowledge_in_Global_Reports_on_Climate_Change

Mustonen, T., Harper, S. L., Rivera Ferre, M., Postigo, J., Ayanlade, A., Benjaminsen, T., Morgan, R., & Okem, A. (Eds.). (2021). 2021 Compendium of Indigenous Knowledge and Local Knowledge: Towards Inclusion of Indigenous Knowledge and Local Knowledge in Global Reports on Climate Change. Snowchange Cooperative.

Mwirigi, J. N. (2020). Sustainability assessment of mini hydro power projects in Rwanda: A comparative analysis. *Energy Sources. Part B, Economics, Planning, and Policy*, *15*(8), 675–687.

Naamwintome, B. A., & Millar, D. (2015). Indigenous knowledge and the african way forward: Challenges and opportunities. *OAlib*, *2*(3), e1295. doi:10.4236/oalib.1101295

Nadaraj, Ronicka, & Angela. (2016). *Engaging Indigenous Knowledge Holders in Teaching Preservice Teachers in IKS Food Production and Practices: Implications for Higher Education.* Available at: https://www.semanticscholar.org/paper/Engaging-Indigenous-Knowledge-Holders-in-Teaching-Govender-Mudaly/f99a4ebcd3839b60b4344699a80c56c573caa2e8

Naimah, I. (2022). The Practice of Imperialism and Colonialism in Development and Their Impact on the Community Economy: An Overview. *Heritage, 3*(1), 39-48.

Nalubega, T., & Uwizeyimana, D. E. (2019). *Public sector monitoring and evaluation in the Fourth Industrial Revolution: Implications for Africa* (Vol. 7). Africa's Public Service Delivery and Performance Review.

Nambisan, S., Wright, M., & Feldman, M. (2019). The digital transformation of innovation and entrepreneurship: Progress challenges and key themes. *Research Policy*, *48*(8), 103773. doi:10.1016/j.respol.2019.03.018

Nasidi N.A. (2022). Slavery And Its Characteristics. *A Journal of Contemporary Research 2022, 19*(2), 135-146.

Nasiri, M., Saunila, M., & Ukko, J. (2020). Shaping digital innovation via digital-related capabilities. *Information Systems Frontiers*, 1–18.

Nathan, N., & Leonard, W. (2011, December). The Slave Trade and the Origins of Mistrust in Africa. *The American Economic Review*, *101*(7), 3221–3252. doi:10.1257/aer.101.7.3221

Ndemo, B., & Weiss, T. (2017). Making sense of Africa's emerging digital transformation and its many futures. *Africa Journal of Management*, *3*(3-4), 328–347. doi:10.1080/23322373.2017.1400260

Ndinda, C. (2011). *Indigenous knowledge to preserve and protect. Human science Research Council*. HSRC.

Ndiribe, M. O., & Aboh, S. C. (2022). Multilingualism and marginalisation: A Nigeria diversity approach. *International Journal of Multilingualism*, *19*(1), 1–15. doi:10.1080/14790718.2020.1818752

Ndlela, L. L. (2019). Green energy solutions in Africa: Rediscovering indigenous knowledge for fourth industrial revolution. In F. Van Wyk (Ed.), *African Studies: Current and Future Challenges*. Harvard Citation Style.

Ndlovu-Gatsheni, S. J. (2013). *Empire, Global Coloniality and African Subjectivity*. Berghahn Books.

Ndlovu-Gatsheni, S. J. (2015). Decoloniality as the Future of Africa. *History Compass*, *13*(10), 485–496. doi:10.1111/hic3.12264

Ndlovu-Gatsheni, S. J. (2018). *Epistemic Freedom in Africa: Deprovincialization and Decolonization*. Routledge. doi:10.4324/9780429492204

Ndlovu-Gatsheni, S. J. (2020). *Decolonization, development, and knowledge in Africa: Turning over a new leaf*. Routledge. doi:10.4324/9781003030423

Ndung'u, N., & Signé, L. (2020). *The Fourth Industrial Revolution and digitization will transform Africa into a global powerhouse*. Foresight Africa Report.

Ngiasike, J. T. (2011). Turkana children's rights to education and indigenous knowledge in science teaching in Kenya. *New Zealand Journal of Teacher's Work, 8*(1), 55-67. Retrieved from https://ir-library.ku.ac.ke/handle/123456789/9530

Nguendi, F. (2012). Africa's international borders as potential sources of conflict and future threats to peace and security. *Institute for Security Studies Paper*, *233*, 1–14.

Ngũgĩ wa Thiong'o. (1986). *Decolonising the Mind: The Politics of Language in African Literature*. James Currey Publishers.

Ngugi, T. (1986). Decolonising the mind. The International Education Journal: Comparative Perspectives, 12(1), 30–44.

Ngulube, P. (2002). Managing and Preserving Indigenous Knowledge in the Knowledge Management Era: Challenges and opportunities for information professionals. *Information Development, 18*(2), 95–101. doi:10.1177/026666602400842486

Ninnes, P. (2000). Representation of Indigenous Knowledge in Secondary School Science Textbooks in Australia and Canada. *International Journal of Science Education, 22*(6), 603–617. doi:10.1080/095006900289697

Niyitunga, E. B. (2017). *Exploring the Challenges of the African Union (Au)-Led Mediation in Peace Processes: The Cases of Burundi, Libya, and Zimbabwe*. Thesis Submitted to the University of Johannesburg, Auckland Park.

Niyitunga, E. B. (2021). Colonial Legacies: Africa's International Relations and Agenda 2063. In Indigenous Systems and Africa's Development. Africa Institute of South Africa (AISA). The National Institute for the Humanities and Social Science Pretoria.

Niyitunga, E.B. (2014). Who owns mediation at the African Union (AU)? Understanding the causes of the AU mediation ineffectiveness in the Libya crisis in 2011. *African Insight, 43*(4).

Niyitunga, E. B. (2022). The Current and Potential Role of Blockchain-based Technology in Managing Medical Records in Africa. *Digital Policy Studies Journal, 1*(1), 52–68. Retrieved December 01, 2023, from. doi:10.36615/dps.v1i1.1253

Niyitunga, E. B. (2023). Economic Insecurity Threatens Public Safety and Social Order in South Africa. *Can Blockchain Technology Provide a Sustainable Solution*.

Njoku, U. J. (2005). Colonial Political Re-Engineering and the Genesis of Modern Corruption in African Public Service: The Issue of the Warrant Chiefs of South Eastern Nigeria as a Case in Point. *Nordic Journal of African Studies, 14*(1), 99–116.

Njuguna, N., & Signe, L. (2020, January 8). *The Fourth Industrial Revolution and digitization will transform Africa into a global powerhouse*. https://www.brookings.edu/articles/the-fourth-industrial-revolution-and-digitization-will-transform-africa-into-a-global-powerhouse/

Nkondo, M. (2012). *Indigenous African knowledge systems in a polyepistemic world: The capabilities approach and the translatability of knowledge systems*. Paper presented at the Southern African Regional Colloquium on Indigenous African knowledge systems: methodologies and epistemologies for research, teaching, learning and community engagement in higher education. Howard College Campus: University of KwaZulu-Natal.

Nkrumah, K. (1963). *Africa Must Unite*. Frederick Praeger. A. Publisher.

Nkrumah, K. (1965). *Neo-colonialism: the Last Stage of Imperialism*. International Publishers.

Nkrumah, K. (1965b). *Neo-colonialism: The last stage of imperialism.* International Publishers.

Nnoli, O. (2003). *Introduction to Politics* (Revised 2nd Edition). PACREP.

Nonaka, I., & Takeuchi, H. (1995). *The knowledge-creating company. How Japanese companies create the dynamics of innovation.* Oxford University Press. doi:10.1093/oso/9780195092691.001.0001

Norris, P. (2004). *Democratic phoenix: Reinventing political activism.* Cambridge University Press.

Northrop, D. (2012). *A Companion to World History.* Willi Blackwell. doi:10.1002/9781118305492

Nozick, R. (1981). *Philosophical explanations.* Harvard University Press.

Ntara, C. (2015). An Analysis of M-Pesa Use in International Transactions. *European Journal of Business and Management, 7*(17).

Nunn, N. (2008). Shackled to the Past: The Causes and Consequences of Africa's Slave Trades. In Natural Experiments of History (pp. 142–184). Harvard University Press.

Nunn, N. (2007). Historical Legacies: A Model caineLinking Africa's Past to its Current Underdevelopment. *Journal of Development Economics, 83*(1), 145. doi:10.1016/j.jdeveco.2005.12.003

Nunn, N. (2008). The long-term effects of Africa's slave trades. *The Quarterly Journal of Economics, 123*(1), 139–176. doi:10.1162/qjec.2008.123.1.139

Nyagadza, B., Pashapa, R., Chare, A., Mazuruse, G., & Hove, P. K. (2022). Digital technologies, Fourth Industrial Revolution (4IR) & Global Value Chains (GVCs) nexus with emerging economies' future industrial innovation dynamics. *Cogent Economics & Finance, 10*(1), 2014654. doi:10.1080/23322039.2021.2014654

Nyaronga, S. (2015). *Social Health Insurance Contribution Payment Options Among Informal Sector Workers In Eastlands Area Of Nairobi County.* http://erepository.uonbi.ac.ke/bitstream/handle/11295/105100/Nyaronga_Social%20Health%20Insurance%20Contribution%20Payment%20Options%20Among%20Informal%20Sector%20Workers%20In%20Eastlands%20Area%20Of%20Nairobi%20County.pdf?sequence=2&isAllowed=y

Nyerere, J. (1968). Education for self-reliance. *Cross Currents, 18*(4), 415–434.

Nyssens, M., & Lensink, R. (2015). Explaining the success of social enterprise in the renewable energy sector: The case of Tanzania. *Journal of Cleaner Production, 98,* 75–83.

O'Hara, K. (2001). *Plato and the Internet.* Icon Books.

O'Riley, P. (2001). *A different storytelling of technology education curriculum revisions: A storytelling of difference.* Retrieved May 31, 2015 from https://scholar.lib.vt.edu/ejournals/JTE/v7n2/pdf/oriley.pdf

Oba, N. F., & Onyije, E. L. (2011). Colonialism and education. *Proceedings of the International Conference on Teaching, Learning and Change, International Association for Teaching and Learning (IATEL).* Accessed 28 September 2019, https://hrmars.com/admin/pics/170.pdf

Oba, N. F., & Onyije, E. L. (2011). Colonialism and education. *Mediterranean Journal of Social Sciences, 2*(4), 41. Retrieved September 28, 2019, from https://www.richtmann.org/journal/index.php/mjss/article/view/10879

Obikeze, D. S. (2011). *Indigenous knowledge systems and the transformation of the academy in Africa: The CULPIP model.* Retrieved February 11, 2015 from https://citeseerx.ist.psu.edu/viewdoc/download?doi=10.1.1.164.9357&rep=rep1&type= pdf.

Ocheni, S. B., & Nwankwo, B. C. (2012). Analysis of colonialism and its impact in Africa. *Cross-Cultural Communication, 8*(3), 46–54.

Ocheni, S., & Nwankwo, B. C. (2012). Analysis of colonialism and Its impact in Africa. *Cross-Cultural Communication, 8*(3), 46–54. doi:10.3968/j.ccc.1923670020120803.1189

Ocholla, D. N., & Dlamini, P. (2017). *Exploring the use of ICT tools for capturing, preserving, and disseminating Indigenous knowledge among IK owners and users: A preliminary study of IK users in the KwaZulu Natal Province, South Africa.* Paper Presented at the 3rd Moi University Annual International Conference, Moi University, Kenya. https://digitalcommons.unl.edu/libphilprac/1663

Ocholla, D. N., & Onyancha, O. B. (2017). The marginalized knowledge: An Afrocentric analysis of indigenous knowledge publications 1990-2004. In *African indigenous knowledge: An underutilized and neglected resource for development. Library Philosophy and Practice (e-journal).* 1663. https://digitalcommons.unl.edu/libphilprac/1663

OECD. (1997). *National Innovation Systems.* Organisation for Economic Co-operation and Development.

OECD. (2022). *Leading Country Example: Kenya Digital Tax Payments.* Available at: https://www.oecd.org/tax/forum-on-tax-administration/publications-and products/taxadministration-3.0-kenya-country-example.pdf

Offiong, D. (1980). *Imperialism and dependency.* Fourth Dimension.

Ogawa, M. (1995). Science education in a multiscience perspective. *Science Education, 79*(5), 583–593. doi:10.1002/sce.3730790507

Ogola, P. F. A., Davidsdottir, B., & Fridleifsson, I. B. (2012). Potential contribution of geothermal energy to climate change adaptation: A case study of the arid and semi-arid eastern Baringo lowlands, Kenya. *Renewable & Sustainable Energy Reviews, 16*(6), 4222–4246. doi:10.1016/j.rser.2012.01.081

Ogunbure, A. A. (2011). The possibilities of technological development in Africa: An evaluation of the role of culture. *The Journal of Pan African Studies, 4*(3), 86–100.

Ojo, E. O. (2015). The Atlantic Slave Trade and Colonialism: Reasons for Africa's Underdevelopment? European Scientific Journal, 11(17).

Okolo, A. (1986). Dependency: The Highest Stage of Capitalist Domination in Africa. In The Future of Africa and the New International Economic Order. Macmillan.

Okoth, A. (2006). *A history of Africa: African nationalism and the de-colonisation process*. East African Educational Publishers.

Olaopa, O. R., & Ayodele, O. A. (2022). Building on the strengths of African Indigenous knowledge and innovation (AIK&I) for sustainable development in Africa. *African Journal of Science, Technology, Innovation and Development, 14*(5), 1313–1326. doi:10.1080/20421338.2021.1950111

Olashore, O. (2021). African-centered development: An Afrocentric critique of development theories and practices. In J. C. Okeibunor & M. M. Kone (Eds.), *African Development* (pp. 91–113). Springer.

Olatunji, S.A. (2010). Thematic changes in postcolonial African literature: From colonialism to neocolonialism. *SinoUS English Teaching, 7*(10).

Oliver, R., & Fage, J. D. (1969). Streams of Trade in Guinea States. In Africa from Early Times to 1800. Oxford University Press.

Oloidi, J. (2014). African traditional technology: The Ekiti of southwestern Nigeria experience. *International Journal of Social Science and Humanities Research, 2*(3), 310–315.

Oloruntade, G., & Omoniyi, F. (2023, May 26). *Where is Africa in the global conversation on regulating AI?* https://techcabal.com/2023/05/26/where-is-africa-in-the-global-conversation-on-regulating-ai/

Olukayode, F. A. (2014). Africa and international relations theory: Acquiescence and responses. *Globalistics and Globalization Studies,* 154–163. Accessed 14 May 2019, https://www.sociostudies.org/almanac/articles/africa_and_international_relations_theory-_acquiescence_and_responses/

Olukoshi, A., & Odhiambo, I. (2016). Indigenous Knowledge Systems (IKS) and Sustainable Development in Africa. In Indigenous Knowledge Systems and Development in Africa (pp. 33-48). Springer.

Omoju, O. E., & Faloye, F. A. (2018). The role of indigenous knowledge in renewable energy development in Africa: A review. *Renewable & Sustainable Energy Reviews, 81*, 1856–1868.

Omolewa, M. (2007). Traditional African modes of education: Their relevance in the modern world. *International Review of Education, 53*(5–6), 593–612. doi:10.1007/s11159-007-9060-1

Omonov, Q., & Karimov, N. (2020). Importance Of Ancestral Heritage. *The American Journal of Social Science and Education Innovations, 2*(09), 196–202. doi:10.37547/tajssei/Volume02Issue09-29

Onwu, G., & Mosimege, M. (2004). Indigenous knowledge systems and science and technology education: A dialogue. *African Journal of Research in Mathematics, Science and Technology Education, 8*(1), 1-12. doi:10.1080/10288457.2004.10740556

Onwuegbuchulam, S. P. C. (2023). Decolonization, knowledge construction, and legitimation at African universities in the 21st century: Relevance of François Lyotard. *Journal of Education*, 94–110.

Osaghae, E. (2006). *Ethnicity and the State in Africa*. Afrasian Center for Peace and Development Studies.

Osha, S. (2020). African perspectives on energy transitions and sustainable development. *Energy Policy*, *140*, 111337.

Ouma, S. A., Odongo, T. M., & Were, M. (2017). Mobile Financial Services and Financial Inclusion: Is it a Boon for Savings Mobilization? *Review of Development Economics*, *7*, 29–35.

Ou, Y. Y. (2021). *Research on the impact of digital maturity of construction enterprises on environmental performance*. Chang'an University.

Ovadia, J. S. (2019). Resource extraction and security in Africa: What role for the private sector? *African Security Review*, *28*(2), 147–163.

Owusu-Ansah, F. E., & Mji, G. (2013). African indigenous knowledge and research. *African Journal of Disability*, *2*(1), 1–5. doi:10.4102/ajod.v2i1.30 PMID:28729984

Owusu, P. A. (2017). Renewable energy in Ghana: Status, prospects, and policies. *Renewable & Sustainable Energy Reviews*, *68*, 14–24.

Pacharapha, T., & Ractham, V. (2012). Knowledge acquisition: The roles of perceived value of knowledge content and source. *Journal of Knowledge Management*, *16*(5), 724–739. Advance online publication. doi:10.1108/13673271211262772

Pardi, P. (2012). *What is knowledge?* Philosophy News. Retrieved March 18, 2012 from: https://www.philosophynews.com/post/2011/09/22/What-is-Knowledge.aspx

Parker, K. I. (2004). *Biblical Politics of John Locke*. Wilfrid Laurier University Press.

Parviainen, P., Tihinen, M., Kääriäinen, J., & Teppola, S. (2017). Tackling the digitalization challenge: How to benefit from digitalization in practice. *IJISPM*, *5*(1), 63–77. doi:10.12821/ijispm050104

Pazarbasioglu, C., & Garcia Mora, A. (2020). *Digital Financial Services*. https://pubdocs.worldbank.org/en/230281588169110691/Digital-Financial- Services.pdf

Perez, C., & Soete, L. (1988). Catching-up in technology: entry barriers and windows of opportunity. In G. Dosi, C. Freeman, R. Nelson, G. Silverberg, & L. Soete (Eds.), *Technical Change and Economic Theory*. Francis Printer.

Perifanou, M., Economides, A. A., & Tzafilkou, K. (2021). *Teachers' digital skills readiness during COVID-19 pandemic*. Academic Press.

Petter, G. (2007). *Knowledge Management Systems: Value Shop Creation*. Idea Group Publisher.

Pheko, M. (2001). *How Africans can Regain their Lost Power*. Tokloho Development Association.

Phillipson, D. W. (2005). *African Archaeology*. Cambridge University Press. doi:10.1017/CBO9780511800313

Pierson, P. (2000). Increasing Returns to Scale, Path Dependence and the Study of Politics. *The American Political Science Review*, *94*(2), 251–267. doi:10.2307/2586011

Plyler, M. G., & Geetha, N. (2010). *Community-Level Economic Effects of M-PESA in Kenya*. Iris Center, University of Maryland.

Poters, G. (2019). *Mobile phones, gender, and female empowerment in sub-Saharan Africa: studies with African youth*. Available at: https://www.tandfonline.com/doi/full/10.1080/02681102.2019.1622500

Progler, Y. (1999). *Towards defining Western research on Indigenous Knowledge*. Muslimedia.

Pryshchepa, O., Kardash, O., Yakymchuk, A., Shvec, M., Pavlov, K., Pavlova, O., & Kramarenko, I. (2020). Optimization of multi-channel queuing systems with a single retail attempt: Economic approach. *Decision Science Letters*, *9*(4), 559–564. doi:10.5267/j.dsl.2020.8.002

Qi, Y. D., & Xiao, X. (2020). Enterprise management reform in the digital economy era. *Management World*, (6), 135–152.

Quijano, A. (1992). Colonialidad y modernidad/racionalidad. *Perú Indígena*, *13*(29), 11–20.

Quijano, A. (2000a). Coloniality of Power, Eurocentrism, and Latin America. *International Sociology*, *1*(3), 533–580. doi:10.1177/0268580900015002005

Rahaman, S., Yeazdani, R., & Rashed, M. (2017). The Untold History of Neocolonialism in Africa (1960-2011). *Historical Research*, *5*(1), 9–16. doi:10.11648/j.history.20170501.12

Rakometsi, M. S. (2008). *The transformation of black school education in South Africa, 1950-1990: A historical perspective* [Unpublished PhD thesis]. Bloemfontein: University of Free State.

Rawat, G., & Mishra, S. (2021). Spirituality and environment: significance of Indigenous knowledge systems in Africa and challenges. *Literary Endeavour, 12*(1).

Rawat, S., Kala, C. P., & Bhatt, D. (2013). Traditional water management system in Western Himalayas, India: Need for revival. *Journal of Mountain Science*, *10*(3), 399–413.

Regmi, J., & Fleming, M. (2012). Indigenous knowledge and science in a globalized age. *Cultural Studies of Science Education, 7*(2), 479-484. doi:10.1007/s11422-012-9389-z

Rehren, T., & Pernicka, E. (2008). Coins, Artefacts and Isotopes: Archaeometallurgy and Archaeometry. *Archaeometry*, *50*(2), 232–248. doi:10.1111/j.1475-4754.2008.00389.x

Remmling, G. W., & Campbell, R. C. (1970). *Basic Sociology: An Introduction to the Study of Society*. Littlefield Adams & Co.

Ren, Z. Z., & Deng, F. (2022). Digital technology, factor structure transformation, and high-quality economic development. *Soft Science*, (9), 1–10.

Rice, M. F. (2003). Information and communication technologies and the global digital divide: Technology transfer, development, and least developing countries. *Comparative Technology Transfer and Society*, *1*(1), 72–88. doi:10.1353/ctt.2003.0009

Richards, S., & Gezu, B. (2011). *Conflict in the Somali Region of Ethiopia: Can Education Promote Peace-Building?* Feinstein International Center.

Richter, M. (2017). Discussion paper prepared for the Treatment Action Campaign and AIDS Law Project. In *African indigenous knowledge: An underutilized and neglected resource for development. Library Philosophy and Practice (e-journal).* 1663. https://digitalcommons.unl.edu/libphilprac/1663

Riggs, F. W. (2000). *Globalization key concepts*. COCTA Panel on Globalization, World Sociology Congress, Montreal, Canada.

Rocheleau, D. E. (1991). Gender, Ecology, and the Science of Survival: Stories and Lessons from Kenya. *Agriculture and Human Values*, *8*(Winter-Spring), 156–165. doi:10.1007/BF01579669

Rock, I., & Plamer, S. (1990). The legacy of gestal psychology. *Scientific American*, *263*(6), 48–61. doi:10.1038/scientificamerican1290-84 PMID:2270461

Rodil, K., Winschiers-Theophilus, H., Bidwell, N. J., Eskildsen, S., Rehm, M., & Kapuire, G. K. (2011). *A new visualization approach to re-contextualize indigenous knowledge in rural Africa.* Paper presented at INTERACT2011, Lisbon, Portugal. 10.1007/978-3-642-23771-3_23

Rodney W. (1972). *How Europe Underdeveloped Africa*. Bogle-L'Ouverture Publications.

Rodney, W. (1972). How Europe underdeveloped Africa. *Beyond borders: Thinking critically about global issues*, 107-125.

Rodney, W. (1972). *How Europe Underdeveloped Africa*. Bogle-L'Ouverture Publications.

Rodney, W. (1982). How Europe Underdeveloped Africa. In African indigenous knowledge and the sciences. Sense Publishers.

Rothenberg, W. A., Sternberg, A., Blake, A., Waddell, J., Chassin, L., & Hussong, A. (2020). Identifying adolescent protective factors that disrupt the intergenerational transmission of cannabis use and disorder. *Psychology of Addictive Behaviors*, *34*(8), 864–876. doi:10.1037/adb0000511 PMID:31524418

Runde, D., & Bryja, B. (2023, November 7). *The Role of AGOA in Accelerating Africa's Digital Transformation.* https://www.csis.org/analysis/role-agoa-accelerating-africas-digital-transformation

Russel, B. (1972). *A history of Western philosophy*. Simon and Schuster.

Russel, B. (2009). *Human knowledge. Its scope and limits. With an introduction bu John G. Slater*. Routledge. doi:10.4324/9780203875353

Sabai, K. (2019). Digital technology, digital capability, and organizational performance: A mediating role of digital innovation. *International Journal of Innovation Science*, *11*(2), 177–195. doi:10.1108/IJIS-08-2018-0083

Safaricom. (2015). *M-Pesa Services & Tariffs*. Available: https://www.safaricom.co.ke/personal/m-pesa/m-pesaservices/international-money-transfer

Safaricom. (2023a). *M-PESA, Global*. Available at: https://www.safaricom.co.ke/personal/m-pesa/do-more- with-M-PESA /M-PESA-global

Safaricom. (2023b). *M-Shwari*. Available at: https://www.safaricom.co.ke/personal/m- pesa/credit-and-savings

Sagsan, M. (2009). Knowledge management discipline: Test for an undergraduate program in Turkey. *Electronic Journal of Knowledge Management*, *7*(5), 627–636.

Said, E. (1978). *Orientalism*. Vintage Books.

Salih, M. A. M. (2007). Political Parties in Africa: Challenges for Sustained Multiparty Democracy. In *Africa Regional Report Based on research and dialogue with political parties*. International Institute for Democracy and Electoral Assistance.

Samantha, M. (2011). The Water Crisis in Kenya: Causes, Effects and Solutions. *Global Majority E-Journal, 2*(1), 31-45. https://www.american.edu/cas/economics/ejournal/upload/marshall_accessible.pdf

Sánchez, D. L., & Botelho, T. (2019). Renewable Energy Transition in Africa: The Role of Community-Based Energy Development. *World Development*, *119*, 42–52.

Sanderson, S. K. (2005). World-systems analysis after thirty years. *International Journal of Comparative Sociology*, *46*(3), 179–213. doi:10.1177/0020715205058606

Sassson, J.R., & Douglas, I. (2006). A conceptual integration of performance analysis. *Knowledge Management and Technology: From concept to prototype*, 81-99.

Scheel, B. (1989). *Egyptian Metalworking and Tools*. Shire Publications.

Schlesinger Arthur, M., Jr. (2004). The Imperial Presidency. Houghton Mifflin.

Schmidt, E. (2013). *Foreign Interventions in Africa: from the Cold War to the War on Terror*. Cambridge University Press. doi:10.1017/CBO9781139021371

Scholl, H. J., & Scholl, M. C. (2014). Smart governance: A roadmap for research and practice. *iConference 2014 Proceedings*. http://www.effra.eu/attachments/article/129/Factories%20of%20the%20Fut ure%202020%20Roadmap.pdf

Schumpeter, J. A. (1912). The Theory of Economic Development. Oxford University Press.

Schumpeter, J. A. (1912). *The Theory of Economic Development.* Oxford University Press.

Schutte, G. (1999). Colonialism and Eurocentrism. *Revista de Antropologia, 22*(2), 120–134.

Schwab, K. (2015). *The Fourth Industrial Revolution: What It Means and How to Respond.* Retrieved from https://www.foreignaffairs.com/articles/2015-12-12/fourth-industrial-revolution

Schwab, K. (2016). *The Fourth Industrial Revolution: What It Means and How to Respond.* World Economic Forum. Available online: https://www.weforum.org/agenda/2016/01/the-fourth-industrial-revolution-what-it-means-and-how-to-respond/

Scoones, I. (2017). Sustainable rural livelihoods and the links between agriculture and HIV/AIDS in Africa: A case study from Zimbabwe. *The Journal of Development Studies, 53*(7), 923–936.

Sebastian, I. M., Ross, J. W., Beath, C., Mocker, M., Moloney, K. G., & Fonstad, N. O. (2020). How big old companies navigate digital transformation. In *Strategic information management* (pp. 133–150). Routledge. doi:10.4324/9780429286797-6

Sellar, C., & Battisti, G. (2023). Europe. The Many Reasons for an Epoch-Spanning Crisis. A Long-Term Geohistorical and Geoeconomics Analysis of the Obstacles to European Integration. In *Geopolitical Perspectives from the Italian Border: Introducing Gianfranco Battisti, Triestino Geographer* (pp. 89–102). Springer International Publishing. doi:10.1007/978-3-031-26044-5_9

Semjachkov, K. A. (2017). The digital economy and its role in the management of modern socioeconomic relations. *Modern Management Technology, 8*(80). https://sovman.ru/article/8001/

Sen, A. (1999). *Development as freedom.* Oxford University Press.

Senanayake, S. G. J. N. (2006). Indigenous knowledge as a key to sustainable development. *The Journal of Agricultural Sciences, 2*. Available at: https://www.researchgate.net/publication/265197993

Senghor, L. S. (1958). A Community of Free and Equal Peoples with the Mother Country. *Western World*, (18), 40.

Serneels, V. (2017). Metals. In A. S. Gilbert (Ed.), *Encyclopedia of geoarchaeology* (pp. 521–532). Springer. doi:10.1007/978-1-4020-4409-0_35

Serumaga-Zake, J. M., & Van der Poll, J. A. (2021). Addressing the Impact of Fourth Industrial Revolution on South African Manufacturing Small and Medium Enterprises (SMEs). *Sustainability (Basel), 13*(21), 11703. doi:10.3390/su132111703

Seth, S. (2013). Once Was Blind but Now Can See': Modernity and the Social Sciences. *International Political Sociology, 7*(2), 136–151. doi:10.1111/ips.12014

Settles, J. D. (1996). *The Impact of Colonialism on African Economic Development.* Chancellor's Honors Program Projects. https://trace.tennessee.edu/utk_chanhonoproj/182

Settles, J. D. (1996). *The Impact of Colonialism on African Economic Development*. Chancellor's Honors Program Projects.

Shaikh, A., & Karjaluoto, H. (2014). Mobile banking adoption: A literature review. *Telematics and Informatics*, 1–14. doi:10.1016/j.tele.2014.05.003

Shava, S. (2013). The representation of indigenous knowledges. In R. B. Stevenson, M. Brody, J. Dillon, & A. Wals (Eds.), *International Handbook of Research on Environmental Education*. Routledge.

Shava, S., & O'Donoghue, R. (2014). *Teaching Indigenous Knowledge and Technology. Fundisa for Change Programme*. Environmental Learning Research Centre, Rhodes University.

Sherif, K., & Xing, B. (2006). Adaptive processes for knowledge creation in complex systems: The case of a global IT consulting firm. *Information & Management*, *43*(4), 530–540. doi:10.1016/j.im.2005.12.003

Shillington, K. (2005). *History of Africa* (Rev. 2nd ed.). Palgrave Macmillan. Available at https://www.africaworld.net/afrel

Shimada, I., & Craig, A. (2013). The style, technology, and organization of Sicán mining and metallurgy, Northern Peru: Insights from holistic study. *Chungara (Arica)*, *45*(1), 3–31. doi:10.4067/S0717-73562013000100001

Shim, D. C., & Eom, T. H. (2008). E-Government and anti-corruption: Empirical analysis of international data. *International Journal of Public Administration*, *31*(3), 298–316. doi:10.1080/01900690701590553

Shishima, S. D. (2012). Challenges to religious freedom in Nigeria: The case of African religion. Department of Religion & Philosophy, Benue State University, Makurdi.

Shishima, S. D. (2014). African religions. Makurdi: Obeta Continental Press.

Shizha, E. (2016). African Indigenous Perspectives on Technology. In G. Emeagwali & E. Shizha (Eds.), *African Indigenous Knowledge and the Sciences*. Anti-colonial Educational Perspectives for Transformative Change. SensePublishers. doi:10.1007/978-94-6300-515-9_4

Shizha, E., & Emeagwali, G. (Eds.). (2016). *African indigenous knowledge and the sciences*. Sense Publishers.

Shoba, M. (2022). The Phenomenon of China Shops in South Africa: Development, Entrepreneurship and Social Cohesion. *African Renaissance, 19*(1).

Shongwe, M. M. (2016). An Analysis of Knowledge Management Lifecycle Frameworks: Towards a Unified Framework. *Electronic Journal of Knowledge Management*, *14*(3), 140–153.

Shore, D. (1983). Steel-Making in Ancient Africa. *Blacks in Science: Ancient and Modern*, 157-162.

Shore, D. (2013). Steel-making in ancient Africa. In *Great achievements in science and technology in ancient Africa.* ASMB. https://www.asbmb.org

Showers, K. B. (2014). Europe's long history of extracting African renewable energy: Contexts for African scientists, technologists, innovators and policy-makers. *African Journal of Science, Technology, Innovation and Development*, *6*(4), 301–313. doi:10.1080/20421338.2014.947200

Siekmann, F., Schlör, H., & Venghaus, S. (2023). Linking sustainability and the Fourth Industrial Revolution: A monitoring framework accounting for technological development. *Energy, Sustainability and Society*, *13*(1), 26. doi:10.1186/s13705-023-00405-4

Sifuna, D. N. (2008). The case of primary education in Africa from colonialism to globalisation. *African Traditional Herbal Research Clinic*, *10*(3), 20–33.

Signé, L. (2023, June 8). *What Does the Fourth Industrial Revolution Mean for Africa?* doi:10.1017/9781009200004.005

Sillitoe, P. (2006). Introduction: Indigenous Knowledge in Development. *Anthropology in Action : Newsletter of the British Association for Social Anthropology in Policy and Practice (BASAPP)*, *13*(3), 1–12. doi:10.3167/aia.2006.130302

Simelane, T., & Amutabi, M. N. (2019). Decolonizing Knowledge: The Use of Indigenous Knowledge in Sustainable Development. In Handbook of Research on Science Education and University Outreach as a Tool for Regional Development (pp. 1-18). IGI Global.

Simpson. (2004). Anticolonial Strategies for the Recovery and Maintenance of Indigenous Knowledge. *The American Indian Quarterly, 28*(3&4), 373-384. doi:10.1353/aiq.2004.0107

Sithole, J. (2017). *The challenges faced by African libraries and information centres in documenting and preserving indigenous knowledge*. Paper Presented at the World Library and Information Congress: 72nd IFLA General Conference and Council, 20-24 August 2006, Seoul, Korea. https://digitalcommons.unl.edu/libphilprac/1663

Sithole, J. (2007). The challenges faced by African libraries and information centres in documenting and preserving indigenous knowledge. *IFLA Journal*, *33*(2), 117–122. doi:10.1177/0340035207080304

Slavery Convention, 25 September 1926, 60 LNTS 253.

Smith, A. (2002). Power and hierarchy of knowledge. *Geoforum*, *40*(1), 230–248.

Smith, L. (1999). *Decolonizing methodologies: Research and indigenous peoples*. Zed Books.

Snively, G., & Corsiglia, J. (2000). Discovering indigenous science: Implications for science education. *Science Education, 85*(1), 6-34. . doi:10.1002/1098-237X(200101)85:1<6::AID-SCE3>3.0.CO;2-R

Sokolova, H. B. (2018). Some aspects of the development of the digital economy in Ukraine. *Economic Bulletin of Donbass*, *1*(51), 92–96.

South Africa's White Paper. (2016). *National Integrated ICT Policy White Paper*. Available at https://www.gov.za/sites/default/files/gcis_document/201610/40325gon1212.pdf

Southall, R. (1998). Moi's flawed mandate: The crisis continues in Kenya. *Review of African Political Economy*, *25*(75), 101–111. doi:10.1080/03056249808704295

Soyinka, W. (2012). *Slavery, Colonialism – Responsible for Africa's Underdevelopment*. ThisDay Journal.

Srai, J., & Lorentz, H. (2019). Developing design principles for the digitalisation of purchasing and supply management. *Journal of Purchasing and Supply Management*, *25*(1), 78–98. doi:10.1016/j.pursup.2018.07.001

Stan, E., & Kenneth, S. (2000). Institutions, factor endowments, and paths of development in the new world. *The Journal of Economic Perspectives*, *14*(3), 217–232. doi:10.1257/jep.14.3.217

Stasavage, D. (2020). *The decline and rise of democracy: A global history from antiquity to today*. Princeton University Press.

Statista. (2023). *M-PESA customer numbers from 2017 to 2022*. https://www.statista.com/statistics/1139190/M-PESA -customer-numbers/

Stelios, M., & Elias, P. (2016). The long-run effects of the scramble for Africa, 106 AM. *Economic Review (Kansas City, Mo.)*, *1802*, 1807.

Stevens, A. (2008). A Different Way of Knowing: Tools and Strategies for Managing Indigenous Knowledge. *Libri*, *58*(1), 25–33. doi:10.1515/libr.2008.003

Stilwell, S. (1999). *The Kano Mamluks: Royal Slavery in the Sokoto Caliphate, 1807-1909* [PhD Thesis]. York University.

Stilwell, S., Hamza, I., & Lovejoy, P. E. (2001). The Oral History of Royal Slavery in the Sokoto Caliphate: An Interview with Sallama Dako. *History in Africa*, *28*, 273–291. doi:10.2307/3172218 PMID:17657904

Suri, T., & Jack, W. (2011). *Mobile Money: The Economics of M-Pesa*. NBER Working Paper 16721.

Suri, T., & Jack, W. (2016). The Long-run Poverty and Gender Impacts of Mobile Money. *Science*, *354*(6317), 1288–1292. doi:10.1126/science.aah5309 PMID:27940873

Swan, M. (2015). *Blockchain: Blueprint for a new economy*. O'Reilly Media.

Sydella, B. (2013). *Great achievements in science and technology in ancient Africa*. ASBMB.

Tabellini, G. (2008). The Scope of Cooperation: Values and Incentives. *The Quarterly Journal of Economics*, *123*(3), 905–950. doi:10.1162/qjec.2008.123.3.905

Tauli-Corpuz, V. (2017). Cultural diversity: The right of indigenous people to remain different and diverse. In *African indigenous knowledge: An underutilized and neglected resource for development. Library Philosophy and Practice (e-journal)*. 1663. https://digitalcommons.unl.edu/libphilprac/1663

Tengeh, R. K. (2017). The role of indigenous knowledge and innovation in creating sustainable agri-food systems in Africa: Evidence from Cameroon. *Journal of Agriculture and Rural Development in the Tropics and Subtropics, 118*(2), 219–232.

Tharakan, J. (2017). Indigenous knowledge systems for appropriate technology development. *IntechOpen, 7*, 124–134. doi:10.5772/intechopen.69889

The Confucian Weekly Bulletin. (2020, February 17). *Technological Developments in Pre-Colonial Africa*. https://confucianweeklybulletin.wordpress.com/2020/02/17/technological-developments-in-pre-colonial

The Department of Trade, Industry and Competition (DTIC). (n.d.). *A Guide to the African Continental Free Trade Area*. Available at http://www.thedtic.gov.za/wp-content/uploads/Guide-to-the-ACFTA_2023.pdf

The World Bank. (2013). *Information and Communication Technologies: Results Profile*. Available at https://www.worldbank.org/en/results/2013/04/13/ict-results-profile#

Thomas, H. (1997). *The Slave Trade: The Story of the Atlantic Slave Trade: 1440-1870*. Simon & Schuster.

Thompson, M. (2023, May 31). *Impact of the Internet in Africa*. https://unconnected.org/blog/impact-of-the-internet-in-africa-2021#:~:text=Africa%20is%20the%20continent%20with,global%20average%20of%20nearly%2060%25

Thomson, A. (2010). *An Introduction to African Politics* (3rd ed.). Routledge. doi:10.4324/9780203857946

Thondlana, J., & Garwe, E. V. (2021). Repositioning of Africa in knowledge production: Shaking off historical stigmas — introduction. *Journal of the British Academy, 9*(s1), 1–17. doi:10.5871/jba/009s1.001

Thrupp, L. A. (1989). Legitimizing local knowledge: From displacement to empowerment for third world people. *Agriculture and Human Values, 6*(3), 13–24. doi:10.1007/BF02217665

Timothy, R. (1998). Views from the South: Intellectual hegemony and postmodernism in developing societies. *Revista de Antropologia, 20*(1), 61–78.

Tomei, L. (2009). *Lexicon of online and distance learning*. Rowman & Littlefield Education.

Tortora, D., Chierici, R., Farina Briamonte, M., & Tiscini, R. (2021). "I digitize so i exist:" Searching for critical capabilities affecting firms' digital innovation. *Journal of Business Research, 129*, 193–204. doi:10.1016/j.jbusres.2021.02.048

Touray, I. M., & Subramanian, V. (2019). Decolonizing renewable energy transitions: Perspectives from Africa. *Energy Research & Social Science, 47*, 125–134.

Triki, T., & Faye, I. (Eds.). (2013). *Financial inclusion in Africa.* Working papers of the African Development Bank (AfDB) https://www.afdb.org/fileadmin/uploads/afdb/Documents/Project-and- Operations/Financial_Inclusion_in_Africa.pdf

Tulinayo, F. P., Ssentume, P., & Najjuma, R. (2018). Digital technologies in resource constrained higher institutions of learning: A study on students' acceptance and usability. *International Journal of Educational Technology in Higher Education, 15*(1), 1–19. doi:10.1186/s41239-018-0117-y

Twizeyimana, J. D., & Andersson, A. (2019). The public value of E-Government–A literature review. *Government Information Quarterly, 36*(2), 167–178. doi:10.1016/j.giq.2019.01.001

Uleanya, C., & Ke, Y. (2019). Review of preparedness of rural African communities nexus formal education in the fourth industrial revolution. *South African Review of Sociology, 50*(3-4), 91–103. doi:10.1080/21528586.2019.1639074

UNCTAD. (2020, September 7). *The Role of Digitalization in the Decade of Action for Africa.* https://unctad.org/news/role-digitalization-decade-action-africa

UNDP. (2022, June 13). *Digital Technologies Key to Unlocking Transformation in Africa.* https://www.undp.org/africa/news/digital-technologies-key-unlocking-transformation-africa

UNESCO. (2019). *Indigenous knowledge for sustainable development.* UNESCO.

UNESCO. (2024, February 9). *Digital learning and transformation of education.* https://www.unesco.org/en/digital-education

United Nations Conference on Trade and Development (UNCTAD). (2019). *Digital Economy Report 2019: Value Creation and Capture: Implications for Developing Countries.* Geneva: United Nations Conference on Trade and Development.

United Nations Conference on Trade and Development (UNCTAD). (2020). *World Investment Report 2020: International Production Beyond the Pandemic.* Geneva: United Nations Conference on Trade and Development.

United Nations Department of Economic and Social Affairs. (2016). E-government survey 2016: E-government in support of sustainable development. New York: United Nations Department of Economic and Social Affairs (UNDESA).

United Nations Department of Economic and Social Affairs. (2018). E-government survey 2018: Gearing e-government to support transformation towards sustainable and resilient societies. New York: United Nations Department of Economic and Social Affairs (UNDESA).

United Nations Development Programme & Third World Network of Scientific Organisations. (1999), *Sharing Innovative Experiences: Examples of Successful Initiatives in Science and Technology in the South.* Author.

United Nations Development Programme. (2019). Digital governance in Zimbabwe: Analysis of strategies, actors and institutions. New York: United Nations Development Programme (UNDP).

United Nations Educational, Scientific and Cultural Organization. (2020). Artificial intelligence: A human rights perspective. Paris: United Nations Educational, Scientific and Cultural Organization (UNESCO).

United Nations Educational, Scientific and Cultural Organization. (2021). AI in education: Smart learning environments. Paris: United Nations Educational, Scientific and Cultural Organization (UNESCO).

United Nations E-Government Survey. (2020). *Digital government in the decade of action for the Sustainable Development Goals: E-government for building back better*. New York: United Nations Department of Economic and Social Affairs (UNDESA).

United Nations Environmental Programme (UNEP). (2004). *Capacity Building Through Partnership andInformation and Communication Technology for Using Indigenous Knowledge for Nature Conservation in Africa*. Available at: www.un.org/esa/devaccount/thematic%20 overview%20environment.html

Unwin, T. (2005). *Partnerships in development practice: Evidence from multi-stakeholder ICT4D Partnership Practice in Africa*. UNESCO Publications for the World Summit on the Information Society.

USAID. (2021). *Mali: Energy*. Retrieved August 3, 2021, from https://www.usaid.gov/mali/energy

Ushie, J. (2008). *Two Africans in One: Neo-colonialism and the African Writer*. http://www.africaresearch.org/ncol.htm

Utsua, T. P. (2015). The Evolution of African Indigenous Science and Technology. *Historical Research Letter, 16*. Available at: http://www.iis.te.org

Uwizeyimana, D. E. (2022). Black Africans' Contributions to Global Industrial Revolutions. *Think (London, England), 90*(1).

Van der Walle, N. (2003). Presidentialism and Clientelism in Africa's Emerging Party Systems. *The Journal of Modern African Studies, 42*(2), 297–321. doi:10.1017/S0022278X03004269

van Halm, I. (2023, January 19). *Zimbabwe joins the wave of resource nationalism*. Mining Technology. Retrieved from https://www.mining-technology.com/features/zimbabwe-critical-minerals-resource-nationalism/

Van Hove, L. & Dubus, A. (2019). *M-PESA and Financial Inclusion in Kenya: Of Paying Comes Saving?* doi:10.3390/su11030568

Van Sertima, I. (1983). The Lost Sciences of Africa: An Overview. *Blacks in Science: Ancient and Modern,* 7-26.

Van Sertima, I. (2013). The lost sciences of Africa: An overview. In *Great achievements in science and technology in ancient Africa*. ASMB. https://www.asbmb.org

van Stam, G. (2020). Power inequities. Observations on the development of information and communication technologies, from an African place. In *WebSci'20 Companion: 12th ACM Conference on Web Science Proceedings* (pp. 54–59). 10.1145/3394332.3402830

van Wyk, J.-A. (2007). *Political Leaders in Africa: Presidents*. Patrons or Profiteers.

Vassev, E., Hinchey, M., & Gaudin, B. (2012). Knowledge Representation for Self-Adaptive Behavior. *Proceeding of Conference on Computer Science & Software Engineering*.

Vázquez, R. (2011). Translation as Erasure: Thoughts on Modernity's Epistemic Violence. *Journal of Historical Sociology*, *24*(1), 27–44. doi:10.1111/j.1467-6443.2011.01387.x

Vodafone Foundation. (2023). *MPESA*. Available at: https://www.vodafone.com/about- vodafone/what-we- do/consumer-products-and-services/M-PESA

Vogelsang, K., Liere-Netheler, K., Packmohr, S., & Hoppe, U. (2019). *Barriers to digital transformation in manufacturing: development of a research agenda*. Academic Press.

Von Krogh, G., Nonaka, I., & Rechsteiner, L. (2012). Leadership in organizational knowledge creation: A review and framework. *Journal of Management Studies*, *49*(1), 240–277. doi:10.1111/j.1467-6486.2010.00978.x

Voronkova, V., Nikitenko, V., Oleksenko, R., Andriukaitiene, R., Kharchenko, J. & Kliuienko, E. (2023). Digital Technology Evolution of the Industrial Revolution From 4G to 5G in the Context of the Challenges of Digital Globalization. *TEM Journal, 12*(2), 732-742.

Wa Thiong'o Ngugi. (1986). *Decolonizing the mind: The politics of language in African literature*. James Currey.

Wa Thiong'o, N. (1993). *Moving the Centre: The Struggle for Cultural Freedoms*. James Currey.

wa Thiongó N. (1981). *Decolonizing the mind: The politics of language in African literature*. Portsmouth, NH: Heinemann.

Wallechinsky, D. (2006). *Tyrants: the World's 20 Worst Living Dictators*. Harper Collins Publishers.

Walter, D. (2002). Colonial and post-colonial discourses in social sciences: A cultural critique of colonialism. *Latin American Research Review*, *28*(3), 120–134.

Wandibba, S., Nangendo, S., & Mulemi, B. (2007). *Gender Empowerment and access to financial services in Machakos County, Eastern Kenya*. available at: https://www.imtfi.uci.edu/files/docs/2014/wandibba_imtfi_final_report.pdf

Wane, N. N. (2017). African Indigenous Knowledge: Claiming, writing, storing, and sharing the discourse. *Journal of Thought, 40*(2), 27-46. https://digitalcommons.unl.edu/libphilprac/1663

Wang, C., Medaglia, R., & Zheng, L. (2018). Towards a typology of adaptive governance in the digital government context: The role of decision-making and accountability. *Government Information Quarterly*, *35*(2), 306–322. doi:10.1016/j.giq.2017.08.003

Wang, H. H., & Du, M. (2021). Digital technology, employee participation, and enterprise innovation performance. *Research and Development Management*, (1), 138–148.

Wang, H. H., Li, Y., & Tan, Y. Y. (2022). The impact of digital transformation on enterprise performance based on meta-analysis. *Journal of System Management*, (1), 112–123.

Warren, D. M. (1991). *Using indigenous knowledge for agricultural development.* World Bank discussion paper 127.

Warren, D. M. (1991). *Using Indigenous Knowledge in Agricultural Development: World Bank Discussion Paper No. 127.* World Bank.

Warren, D. M. (1991). The role of indigenous knowledge in facilitating the agricultural extension process. Paper Presented at *International Workshop on Agricultural Knowledge Systems and the Role of Extension*, Bad Boll, Germany.

Warren, D. M., Slikerveer, L. J., & Brokensha, D. (Eds.). (1995). *The Cultural Dimensions of Development: Indigenous Knowledge Systems.* Intermediate Technology Publications. doi:10.3362/9781780444734

Waweru, N., & Kalani, V. (2009). Commercial banking crises in Kenya: Causes and remedies. *African Journal of Accounting, Economics, Finance and Banking Research, 4*(4). Available at: https://citeseerx.ist.psu.edu/document?repid=rep1&type=pdf&doi=ce4a3019b540f1c4e29741d19beb51e8653126d8

Waziyatawin, A. W. (2004). Introduction: Indigenous Knowledge Recovery Is Indigenous Empowerment. *American Indian Quarterly, 28*(3/4), 359–372. https://www.jstor.org/stable/4138922

Weitzman, M. L. (1998). Recombinant growth. *The Quarterly Journal of Economics, 113*(2), 331–360. doi:10.1162/003355398555595

Welbourne, M. (2001). *Knowledge.* Acumen. doi:10.1515/9780773585294

Weldegerima, B. (2017). Review on the importance of documenting ethno pharmacological information on medicinal plants. In *African indigenous knowledge: An underutilized and neglected resource for development. Library Philosophy and Practice (e-journal).* 1663. https://digitalcommons.unl.edu/libphilprac/1663

Wendorf, M. (2013). *Ancient Egyptian Technology and Inventions.* https://interestingengineering.com/lists/ancient-egyptian-technology-and-inventions

Westad, O. (2007). *The Global Cold War.* Cambridge University Press.

Whatley, W. (2012). *The transatlantic slave trade and the evolution of political authority in West Africa.* Academic Press.

Wikipedia. (2023). *M-PESA.* Available from: https://en.m.wikipedia.org/wiki/M-PESA

Wilk, R. (2004). *The Extractive Economy: An Early Phase of the Globalization of Diet* (Vol. 27). The Environment and World History.

Williams, A. (1996). An introduction to Technology Education. In J. Williams & A. Williams (Eds.), Technology education for teachers (pp. 1-26). Macmillan.

Williams, G. (1981). Nigeria: The Neo-colonial Political Economy. In Political Economy of Africa (Selected Readings). Longman.

Williams, P. J. (1996). *International developments in technology education: Lessons for South Africa*. Paper presented at the International Conference on Technology Education for Development in South Africa, Cape Town, South Africa.

Williams, H., & Torma, M. (2007). Trust and fidelity: From "under the mattress" to the mobile phone. In *The transformational potential of m-transactions* (Vol. 6, pp. 10–20). Vodafone.

Williams, J. (1980). *The educated and professional elite in the Gold Coast and Sierra Leone, 1885-1914*. University of California.

Williams, P. J. (2000). The only methodology of Technology? *Journal of Technology Education, 11*(2), 48–60. doi:10.21061/jte.v11i2.a.4

Wilson, V., & Harris, M. (2003). Designing the best: A review of effective teaching and learning of Design and Technology. *International Journal of Technology and Design Education, 13*(3), 223–241. doi:10.1023/A:1026133808945

Woods, G. (2013). Science in Ancient Egypt. In *Great achievements in science and technology in ancient Africa*. ASMB. https://www.asbmb.org

Woods, G. (1988). *Science in Ancient Egypt*. Franklin Watts.

World Bank. (1998). W*orld development report 1998/1999: Knowledge for development.* Retrieved 11th March, 2024, from https://www.worldbank.org/afr/ik/ikrept.pdf

World Bank. (2003). *Births attended by skilled health staff (% of total).* Available at: https://data.worldbank.org/indicator/SH.STA.BRTC.ZS

World Bank. (2010). *Kenya economic update: Kenya at the tipping point?* The World Bank.

World Bank. (2016). *World Development Report 2016: Digital Dividends*. World Bank Group.

World Bank. (2017). *World development report 2016: Digital dividends*. doi:10.1596/978-1-4648-0671-1

World Bank. (2020). *Digital Infrastructure Investment: Financing the Transformation of Emerging Economies*. World Bank Group.

World Bank. (2021). *Employment in agriculture, female (% of female employment) (modeled ILO estimate) – Kenya*. Available at: https://data.worldbank.org/indicator/SL.AGR.EMPL.FE.ZS?locations=KE

World Bank. (2023, April 25). *Accelerating the Use of Digital Technologies is Key to Creating Productive Jobs and Boosting Economic Growth in Africa.* https://www.worldbank.org/en/news/press-release/2023/03/13/accelerating-the-use-of-digital-technologies-is-key-to-creating-productive-jobs-and-boosting-economic-growth-in-africa

World Economic Forum. (2020, January). *Unlocking Technology for the Global Goals.* https://www.pwc.com/gx/en/services/sustainability/publications/accelerating-sustainable-development.html

World Health Organization. (2021). Artificial intelligence for health. Geneva: World Health Organization (WHO).

Wynter, S. (2003). Unsettling the coloniality of being/power/truth/freedom: Towards the human, after man, its overrepresentation: An argument. *CR (East Lansing, Mich.)*, *3*(3), 257–337. doi:10.1353/ncr.2004.0015

Xiao, X., & Qi, Y. D. (2019). Value dimension and theoretical logic of industrial digital transformation. *Reform*, (8), 61–70.

Xie, W., Zheng, D., Luo, J., Wang, Z., & Wang, Y. (2022). Research of cooperation strategy of government-enterprise digital transformation based on differential game. *Open Mathematics*, *20*(1), 791–808. doi:10.1515/math-2022-0480

Xie, X., Huo, J., & Zou, H. (2019). Green process innovation, green product innovation, and corporate financial performance: A content analysis method. *Journal of Business Research*, *101*, 697–706. doi:10.1016/j.jbusres.2019.01.010

Yadav, R. H. (2013). Medicinal plants in the folk medicine system of Ethiopia. *Journal of Poisonous and Medicinal Plants Research*, *1*(1), 7–11.

Yale, R., & Gestrin, P. (1998). *Into Africa: Intercultural insights.* Intercultural Press.

Yan, J. Z., Ji, W. Y., & Xiong, Z. (2021). Overview and prospect of digital innovation research. *Scientific Research Management*, (4), 11–20.

Yoo, Y., Boland, R. J. Jr, Lyytinen, K., & Majchrzak, A. (2012). Organizing for innovation in the digitized world. *Organization Science*, *23*(5), 1398–1408. doi:10.1287/orsc.1120.0771

Yoo, Y., Henfridsson, O., & Lyytinen, K. (2010). Research commentary-The new organizing logic of digital innovation: An agenda for information systems research. *Information Systems Research*, *21*(4), 724–735. doi:10.1287/isre.1100.0322

Young, C. (1982). *Ideology and Development in Africa.* Yale University Press.

Yu, V. F. (2012). E-government for good governance in developing countries: Empirical evidence from the eFez project in Morocco. *Government Information Quarterly*, 285-315.

Zaller, J. (1992). *The nature and origins of mass opinion.* Cambridge University Press. doi:10.1017/CBO9780511818691

Zaslavsky, C. (1983). The Yoruba Number System. *Blacks in Science: Ancient and Modern*, 110-127.

Zeleza, P. T. (2006). The disciplinary, interdisciplinary and global dimensions of African Studies. *International Journal of African Renaissance Studies*, *1*(2), 195–220. doi:10.1080/18186870608529717

Zhang, W., Li, G., & Guo, F. (2022). Does carbon emissions trading promote green technology innovation in China? *Applied Energy*, *315*, 119012. doi:10.1016/j.apenergy.2022.119012

Zhao, S., Peng, D., Wen, H., & Wu, Y. (2022). Nonlinear and spatial spillover effects of the digital economy on green total factor energy efficiency: Evidence from 281 cities in China. *Environmental Science and Pollution Research International*, *30*(34), 81896–81916. Advance online publication. doi:10.1007/s11356-022-22694-6 PMID:36029445

Zhou, K., Liu, T., & Zhou, L. (2015, August). Industry 4.0: Towards future industrial opportunities and challenges. In *Fuzzy Systems and Knowledge Discovery (FSKD), 2015 12th International Conference on* (pp. 2147-2152). IEEE

Zinyeka, G. (2013). Onwu and Mosimege on indigenous knowledge systems and science and technology education: A dialogue in some remaining issues. *Greener Journal of Educational Research*, *3*(9), 432-437.

Zinyeka, G., Onwu, G. O. M., & Braun, M. (2016). A truth-based epistemological framework for supporting teachers in integrating indigenous knowledge into science teaching. *African Journal of Research in Mathematics, Science and Technology Education*, *20*(3), 256-266. doi:10.1080/18117295.2016.1239963

Zinyeka, G. (2013). Onwu and Mosimege on Indigenous knowledge systems and science and technology education: A dialogue in some remaining issues. *Greener Journal of Educational Research*, *3*(9), 432–437.

Zissis, G. (2020). Power Distribution Systems: An Old Business With Modern Challenges [President's Message]. IEEE Industry Applications Magazine, 26(4), 5-6.

About the Contributors

Eric Blanco Niyitunga is a lecturer at the University of Johannesburg. He obtained his Doctorate from the University of Johannesburg in Political Studies, the title of his PhD research is "Exploring the Challenges of the African Union in Peace Processes: Cases of Burundi, Libya and Zimbabwe". His Masters research focused on testing the viability of the new preventive diplomacy in preventing African conflicts. Dr Niyitunga is a member of Initiatives of Change, including the Eastern Africa Youth Conference; the African Leadership Centre; the African Peacebuilding Network-Hub; and the American Political Science Association (APSA). Dr Niyitunga has played an active role in internationalization initiatives of the College of Business and Economics (CBE). Dr Niyitunga has presented papers at a number of local and international conferences and events. In April 2018, Dr Niyitunga facilitated the signing of the MOU between the University of Johannesburg (School of Public Management, Governance and Public Policy) and Daystar University. He continues to be actively involved in building collaborative networks with universities on the African continent. He acted as a non-assessing chair for Master's and doctoral students on a number of occasions at the University of Johannesburg. Dr Niyitunga attended a United Nations Humanitarian Peace Ambassador in 2018. His specialised research interests are in: Issues and cases in African politics and governance, politics of service delivery, law in International society, international mediation, theory and practice of diplomacy, Africa's international relations, foreign policy analysis, peace and security, politics of international development, and African studies.

* * *

Thivhavhudzi Muriel Badugela is a Lecturer in the Department of Educational Studies She has worked for over 4 years in the higher education sector. S/he specialises in Curriculum Studies and has published extensively in peer-reviewed and accredited journals. Her recent publications are: 1) Thesis: Exploring the viability of integrating Indigenous knowledge into Life Orientation Curriculum year:

2019. She has also written a book chapter. Re-Thinking the role of indigenous Systems in Life Skills Education Among the Youth of Local Communities, chapter 19, Year: 2022. She has supervised Honours students. Badugela is a Lecturer in the Faculty of Humanities, Social Sciences and Education at the University of Venda, South Africa.

Mkululi Gacula holds a BA degree in Public Management and Governance from the University of Johannesburg, he embarked on a career journey, initially as a Bulk Teller at Standard Bank South Africa before returning to the University of Johannesburg for a BA Honours degree in Public Management and Governance. Mkululi is currently pursuing a Masters degree in the same field, and also works as an Administrator for the School of Public Management, Governance and Public Policy, showcasing his dedication to academia and administration. His altruistic spirit led him to undertake a humanitarian mission in East Africa from 2012 to 2014, leaving an indelible mark on his character. Known for his honesty, patience, and proactive nature, Mkululi embodies the golden rule, believing in the inherent good of others and embracing lifelong learning as he aspires to attain a PhD in the near future.

Cleophas Gwakwara is a Post Graduate Research Fellow at the University of Johannesburg's prestigious School of Public Management, Governance and Public Policy. He also is a Principal External Relations Officer at the Parliament of Zimbabwe. He graduated from the University of Zimbabwe with a BA Honours in Religious Studies, Masters in Religious Studies and an Msc in International Relations from the same University. He also obtained a Graduate Certificate Education before graduating with a Doctor of Public Affairs from Tshwane University of Technology in 2022. He is part of the technical and administrative decision support systems to Presiding Officers, Members of Parliament, the Clerk of Parliament and Senior Management of Parliament on Protocol and Parliamentary Diplomacy issues. Dr. Gwakwara also serves as a SADC PF Desk Officer and was instrumental in writing the Strategic Lobby Document on the Transformation of the SADC PF into a Regional Parliament. The document became the most crucial strategic tool for driving the Transformation Agenda. Currently, he is a Researcher with the SADC PF, Sexual and Reproductive Health Rights (SRHR) Project.

Dimakatso Manthosi is an advocate for social, political, and legal justice. With a Bachelor of Arts in Political Sciences from the University of Pretoria, Dimakatso has dedicated his academic pursuits to understanding and addressing the needs of marginalized communities. Currently he is pursuing his BAHons in International Relations at the University of Pretoria, his research interest focuses on the intersection of digital technology, asset utilization, and international communication in the context of foreign policy. His commitment to advancing global understanding and cooperation has led him to participate as a delegate in several international diplomatic conventions held at various embassies in Pretoria. Notably, Dimakatso's contributions extend beyond academia. He served as a delegate to the 2023 UN ECOSOC in New York, engaging in critical discussions on global challenges and sustainable development. Additionally, he participated in a model UN simulation organized by UN Best Delegates in Turkey, further honing his diplomatic skills and fostering cross-cultural dialogue. Dimakatso's passion for positive change and his dedication to promoting equitable policies make him a rising force in the field of international relations and domestic politics. His commitment to bridging gaps and advocating for the rights of marginalized populations sets him apart as a promising leader in the global and domestic arena.

Mandisi Matyana is a lecturer in the Department of Public Administration at the University of University of Zululand.

Joshua Mawere is currently a lecturer at the University of Venda, South Africa. His research interest focuses on research on the efficacy of customary tenure as a tool for rural development and food security in South Africa. Skills and expertise are in African Indigenous Knowledge Systems and indigenous customary laws.

Lethiwe Mbongwa is a self-motivated and energetic educator. Her research interests include; education leadership and management, teenage pregnancy, and bullying.

Victor H Mlambo is a Senior Lecturer at the School of Public Management, Governance and Public Policy in the College of Business and Economics at the University of Johannesburg.

Logic Modibedi is a highly experienced and results-driven professional interested in driving positive change and transformation within organizations. Currently serving as the Senior Manager: Shared Services at the Johannesburg Business School. Logic has amassed a wealth of experience in the higher education sector. His academic background includes a B.A. Honours Degree in African Studies, a B. Com Degree in Industrial and Organizational Psychology, and a National Diploma in Information Management. Logic also holds an Advanced Project Management certificate with distinction and an MA in Politics and International Relations. He is a PhD candidate in the School of Public Governance, Management, and Public Policy at the University of Johannesburg. He possesses expertise spanning various areas such as Human Development, Policy Development, Financial Management, Community Engagement, Stakeholder Relations, Project and Programme Management, and Facilities Management. In addition to his professional accomplishments, Logic has held several senior leadership positions in the trade union movement, where he has made significant contributions to the transformational agenda. This experience has provided him with unique skills that have proven beneficial in his current role.

Sandile Moloi holds BA and BA Honours degrees in Politics and International Relations and an MA in Politics from the University of Johannesburg. He is pursuing his PhD studies with a topic focussing on Pan-Africanism at the University of Johannesburg. He teaches African studies courses in the Department of Politics and International Relations at the University of Johannesburg. His research interests are the African Union, Political Islam, Pan-Africanism, and Peace and Security in the Middle East & Africa.

Patrick Musembi is an accomplished educationist with over 20 years' teaching experience at university level. He founded department of the Peace and International Studies at Daystar in 2010 and developed its curriculum for BA in Peace and Conflict Transformation of Daystar University in 2008 making Daystar the first University in Kenya to offer a BA in Peace and Conflict Transformation. He is the Chair of the National Board of Trustees (BOT) of the Christian Chapels Kenya where I mentor young pastors, help them in the running of their local churches, as well as developing governance and accountability structures for the ministry in Kenya and beyond. He is an associate member of the Chartered Institute of Arbitrators of Kenya. Musembi holds PhD-Peace and Conflict Studies and a Master in Divinity and another MA in Peace Studies and International Relations.

Muzi Shoba is a lecturer in the Department of Development Studies at the Nelson Mandela University.

Xolani Thusi is a Senior Lecturer in the Department of Public Administration at the University of Limpopo.

Index

A

Africa 1-5, 8, 13, 20-22, 24-30, 32-33, 37-40, 42-44, 46-60, 62-75, 77-113, 115-118, 120-123, 125-191, 193-198, 200-203, 206-207, 212-217, 219-225, 227-228, 230, 232-250, 252-261, 264-265, 270-271, 282-283, 285-288

Archaeology 35, 64, 70, 98-99

C

Cold War 125-128, 134, 148-153, 155, 157

Colonialism 3-4, 13, 21, 25-26, 28-29, 32, 35, 44, 48, 50, 53, 56-57, 64, 67, 69-72, 78, 85-97, 99-101, 103-106, 108, 112-115, 118, 121-123, 126-128, 130-132, 134, 136-137, 140-141, 147, 152, 156, 161-162, 164, 169, 174-175, 178, 181, 195, 197-202, 213-216, 218-220, 222, 227-228, 234, 237, 255, 261

Coloniality 2-3, 13, 21, 26-29, 32, 74-75, 126-127, 129-133, 137, 140-142, 144, 146-152, 154-158, 194-195, 198-200, 202, 212-219, 221, 224, 226, 261

Contemporary Challenges 104, 112, 114, 169

Culture 1-2, 4, 6-7, 9, 13, 15, 20, 22-24, 28, 30, 32, 35, 41-42, 50, 53, 60, 63, 70, 78, 82-83, 90, 92, 94-96, 126, 129, 132, 135, 142, 144-145, 147-150, 153, 155, 201, 214, 216-217, 222, 236, 248, 252, 262

D

Decolonial Perspective 159-165

Digital Technology 1-3, 5-7, 11, 16-17, 19, 21-22, 24-25, 27-30, 32, 47-51, 54-55, 58-62, 66, 69-72, 75, 77, 79-87, 89-92, 94, 125-130, 133, 135-137, 139-142, 144, 146-149, 151-152, 159-160, 168, 181-186, 188-189, 193-197, 201, 203-206, 208-212, 214-215, 217-218, 220, 222, 226, 228, 230, 232-239, 244-245, 247-250, 254-256, 260-261, 264, 267, 279, 282

Digitalization 86, 126, 147, 151-152, 178, 185, 195-196, 208, 211, 218, 221-224, 226-227, 233, 240-241, 258

Documentary Analysis 103, 115

E

Economic Growth 13, 24, 72, 96, 136-137, 139, 146, 154, 165-166, 168, 170, 182, 193-194, 197, 203, 206-207, 209, 212-219, 221, 224, 233, 236, 238-239, 244-245, 252, 259, 273, 279-281

Electronic Technology 47

Environmental Degradation 26, 104, 109, 111-112, 168, 170, 199, 213, 221

Erosion of Traditional Practices 112, 114

G

Green Energy 159-160, 162-163, 168-173, 176

I

Indigenous Knowledge 1-11, 13-52, 55, 61-72, 77, 79-80, 82, 86, 102-105, 107-109, 111-123, 127-128, 130, 151-152, 159-160, 163-164, 167, 169, 173-177, 193-195, 197-200, 202, 214-218, 257, 261
Indigenous Knowledge Systems 4-6, 8, 16, 21, 26, 28-30, 34, 36, 38-40, 42, 45-46, 48, 51-52, 61, 65, 104, 107-108, 114, 119, 121, 123, 160, 173-174, 177
Industrialization 33, 61, 63, 86-87, 91-92, 96, 146-147, 178, 214-216, 220-221, 232-233, 240-241, 245

K

Knowledge Systems 1-2, 4-6, 8, 16, 21, 26-27, 29-30, 34-36, 38-40, 42, 45-46, 48, 50-53, 55, 57, 61, 65-66, 104, 106-108, 112-114, 119-121, 123, 130, 160, 162, 165, 170, 173-175, 177, 221

M

Marginalisation 51, 103-105, 108, 110-112, 123, 161, 242
Minerals 13, 59, 80-81, 85, 106, 111, 159-160, 168, 178, 180, 182-185, 188-189, 197, 202, 219, 232-233, 238

N

Neocolonialism 125-127, 129, 131-142, 144, 146-150, 152, 156-157, 201, 219

P

Post-Colonial Theory 105
Preserving Indigenous Knowledge 36, 42, 45, 68, 103, 113

R

Raw Minerals 13, 80, 178, 188

S

Sustainability 17, 24, 33-34, 36-37, 109, 111-112, 121-122, 147, 154, 161-162, 165-166, 168, 171, 175-176, 192-193, 195, 221, 223-224, 256, 258-259, 283

Recommended Reference Books

IGI Global's reference books are available in three unique pricing formats: Print Only, E-Book Only, or Print + E-Book.

Order direct through IGI Global's Online Bookstore at **www.igi-global.com** or through your preferred provider.

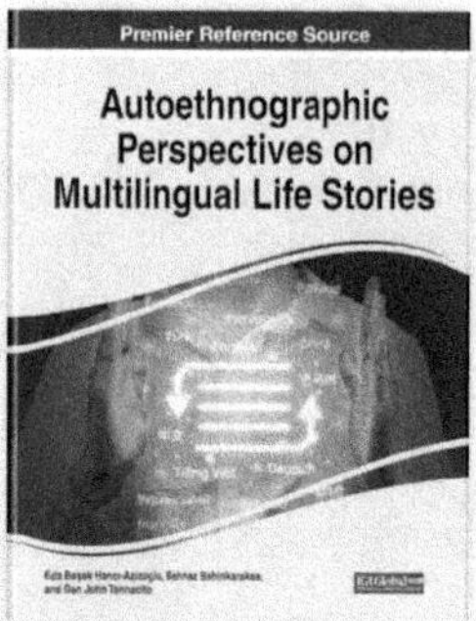

ISBN: 9781668437384
EISBN: 9781668437407
© 2022; 343 pp.
List Price: US$ 215

ISBN: 9781668433645
EISBN: 9781668433669
© 2022; 317 pp.
List Price: US$ 215

ISBN: 9781799884798
EISBN: 9781799884804
© 2022; 397 pp.
List Price: US$ 215

ISBN: 9781799885535
EISBN: 9781799885559
© 2022; 277 pp.
List Price: US$ 215

ISBN: 9781668459645
EISBN: 9781668459652
© 2023; 376 pp.
List Price: US$ 240

ISBN: 9781668438817
EISBN: 9781668438824
© 2022; 663 pp.
List Price: US$ 415

Do you want to stay current on the latest research trends, product announcements, news, and special offers?
Join IGI Global's mailing list to receive customized recommendations, exclusive discounts, and more.
Sign up at: **www.igi-global.com/newsletters.**

Publisher of Timely, Peer-Reviewed Inclusive Research Since 1988

IGI Global
PUBLISHER of TIMELY KNOWLEDGE

www.igi-global.com Sign up at www.igi-global.com/newsletters facebook.com/igiglobal twitter.com/igiglobal

Ensure Quality Research is Introduced to the Academic Community

Become an Reviewer for IGI Global Authored Book Projects

The overall success of an authored book project is dependent on quality and timely manuscript evaluations.

Applications and Inquiries may be sent to:
development@igi-global.com

Applicants must have a doctorate (or equivalent degree) as well as publishing, research, and reviewing experience. Authored Book Evaluators are appointed for one-year terms and are expected to complete at least three evaluations per term. Upon successful completion of this term, evaluators can be considered for an additional term.

If you have a colleague that may be interested in this opportunity, we encourage you to share this information with them.

Submit an Open Access Book Proposal

Have Your Work Fully & Freely Available Worldwide After Publication

Seeking the Following Book Classification Types:

Authored & Edited Monographs • Casebooks • Encyclopedias • Handbooks of Research

Gold, Platinum, & Retrospective OA Opportunities to Choose From

Easily Track Your Work in Our Advanced Manuscript Submission System With **Rapid Turnaround Times**

Double-Blind Peer Review by Notable Editorial Boards (*Committee on Publication Ethics* (COPE) Certified

Publications Adhere to All **Current OA Mandates & Compliances**

Affordable APCs ***(Often 50% Lower Than the Industry Average)*** Including Robust Editorial Service Provisions

Direct Connections with **Prominent Research Funders** & OA Regulatory Groups

Institution Level OA Agreements Available (Recommend or Contact Your Librarian for Details)

Join a **Diverse Community of 150,000+ Researchers Worldwide** Publishing With IGI Global

Content Spread Widely to Leading Repositories (AGOSR, ResearchGate, CORE, & More)

DID YOU KNOW?

Retrospective Open Access Publishing

You Can Unlock Your Recently Published Work, Including Full Book & Individual Chapter Content to Enjoy All the Benefits of Open Access Publishing

Learn More

Publishing Tomorrow's Research Today

IGI Global

e-Book Collection

Including Essential Reference Books Within Three Fundamental Academic Areas

Business & Management
Scientific, Technical, & Medical (STM)
Education

- Acquisition options include Perpetual, Subscription, and Read & Publish
- No Additional Charge for Multi-User Licensing
- No Maintenance, Hosting, or Archiving Fees
- Continually Enhanced Accessibility Compliance Features (WCAG)

Over **150,000+** Chapters	Contributions From **200,000+** Scholars Worldwide	More Than **1,000,000+** Citations	Majority of e-Books Indexed in Web of Science & Scopus	Consists of Tomorrow's Research Available Today!

Recommended Titles from our e-Book Collection

Innovation Capabilities and Entrepreneurial Opportunities of Smart Working
ISBN: 9781799887973

Advanced Applications of Generative AI and Natural Language Processing Models
ISBN: 9798369305027

Using Influencer Marketing as a Digital Business Strategy
ISBN: 9798369305515

Human-Centered Approaches in Industry 5.0
ISBN: 9798369326473

Modeling and Monitoring Extreme Hydrometeorological Events
ISBN: 9781668487716

Data-Driven Intelligent Business Sustainability
ISBN: 9798369300497

Information Logistics for Organizational Empowerment and Effective Supply Chain Management
ISBN: 9798369301593

Data Envelopment Analysis (DEA) Methods for Maximizing Efficiency
ISBN: 9798369302552

Request More Information, or Recommend the IGI Global e-Book Collection to Your Institution's Librarian

For More Information or to Request a Free Trial, Contact IGI Global's e-Collections Team: eresources@igi-global.com | 1-866-342-6657 ext. 100 | 717-533-8845 ext. 100

www.ingramcontent.com/pod-product-compliance
Lightning Source LLC
LaVergne TN
LVHW080340011026
849161LV00009B/2135